Curse Tablets and Binding Spells from the Ancient World

Contributors

Catherine F. Cooper
David Frankfurter
Derek Krueger
Richard Lim

Curse Tablets and Binding Spells from the Ancient World

Edited By

JOHN G. GAGER

New York   Oxford

OXFORD UNIVERSITY PRESS

1992

Oxford University Press

Oxford   New York   Toronto
Delhi   Bombay   Calcutta   Madras   Karachi
Kuala Lumpur   Singapore   Hong Kong   Tokyo
Nairobi   Dar es Salaam   Cape Town
Melbourne   Auckland

and associated companies in
Berlin   Ibadan

Copyright © 1992 by John G. Gager

Published by Oxford University Press, Inc.
198 Madison Avenue, New York, New York 10016-4314

Oxford is a registered trademark of Oxford University Press

Library of Congress Cataloging-in-Publication Data
Curse tablets and binding spells from the ancient world
edited by John G. Gager.
p.   cm.   Includes index.
ISBN 0-19-506226-4
1. Incantations.   2. Blessing and cursing.
BF1558.C87   1992
133.4′4—dc20   91-33236   CIP

2 4 6 8 9 7 5 3
Printed in the United States of America
on acid-free paper

Preface

This project arose initially from a desire to define a body of primary materials that might serve to illustrate the long and difficult debate about "magic" and "religion" in Western culture. It seemed clear to us that ancient *defixiones*—curse tablets and binding spells inscribed normally on thin metal sheets—offered a unique body of data. They are largely unknown, as much to general readers as to scholars; unlike the much more familiar spells written on papyrus and preserved in large collections of recipes for use by professional *magoi, defixiones* survived because they were actually put to use by individual clients; like ancient amulets on stone, they come to us largely unmediated by external filters; unlike ancient literary texts, they are devoid of the distortions introduced by factors such as education, social class or status, and literary genres and traditions. Most of all, they are intensely personal and direct.

Of course, we are not so naive as to believe that the *defixiones* were uninfluenced by cultural forces: their language is highly formulaic, and clients were often limited by the recipes that the local *magos* had available in his or her collection of recipes. One final advantage is that *defixiones,* for the most part, have been uncovered by modern archaeologists precisely where they were deposited by the ancient clients or their agents: in cemeteries, wells, or other appropriate sites.

For several reasons, we made the decision not to include the texts in their original languages. First, our intended audience includes not only—not even especially—scholars of Mediterranean antiquity but a broader range of students and general readers. Second, it soon became apparent that for many of these *defixiones* the published texts are not reliable. Indeed, in many cases, the tablets themselves are no longer available for inspection (for example, most of those published by Wünsch in DTA). The work of reexamining and reediting tablets published at the end of the nineteenth century (such as DTA) and early in

this century (such as *DT,* and Ziebarth, 1934) promises to reach far into
the next century. Third, for those who wish or need to consult the texts
in their primary languages, editions are readily available, along with
more recent catalogues and inventories (for example, SGD for Greek
tablets and Solin for the Latin ones). We have tried to provide exhaus-
tive bibliographical references for each of the texts included in this
collection; in addition, where individual words or phrases are crucial to
interpretation, we have provided them in transliteration.

The principle of organization within each chapter is basically geographical: we begin with ancient Greece (an arbitrary starting point) and
circle around the Mediterranean in clockwise fashion. The fact that we
have included a few objects from beyond the Mediterranean, such as the
bowls from Mesopotamia, indicates that we have not felt tightly bound
by any of our categories.

A brief word about the treatment of foreign and especially Greek
names is in order. In the texts themselves, we have not Latinized the
letters of personal names: Greek *kappa* remains *k, upsilon* is rendered
as *u* (rather than the more traditional *y*), *omega* becomes *ô,* and so on.
But in the introductions, annotations, and discussions, we have used the
Latinized conventions, especially with common and familiar names.
Thus a name will not uncommonly appear in two forms: Sôkratês (in the
text) and Socrates (in the discussion). *Voces mysticae*—those "words" or
"terms" in a spell that do not represent ordinary language—we have
rendered in upper-case letters. On some tablets it is difficult to decide
where one of these *voces* ends and the next one begins. On others,
separations are indicated by various scribal devices, for example, boxes
drawn around the *vox,* suprascript horizontal lines and colons. With a
number of tablets, we have attempted to format the translation so as to
reflect the unusual ways in which the text was inscribed; we have also
included a number of photographic reproductions that illustrate these
techniques.

One of the features that distinguishes this book from other collections
of ancients texts and documents—a feature that derives from our broad
conception of how the world of the ancient Mediterranean must be
studied and understood—is the extent to which its contents cross traditionally impervious barriers of language and culture. Thus we include
material written not just in Greek or Latin but also in Hebrew, Aramaic,
Coptic, and Demotic. Jewish, Egyptian, Greek, Roman, British, and
Christian tablets appear side by side, often employing the same formulas, mysterious names, and drawings. On a geographical scale, our tablets range from Britain to North Africa, from Mesopotamia to Spain.

Behind this mix lies a conscious intention on our part to undermine the confidence with which cultural, geographical, and chronological labels are applied to ancient texts and traditions, as if they represented clear, distinct, and nonoverlapping categories.

This book has been a collective effort from start to finish. In addition to the primary contributors, I express my gratitude to those whose generous assistance has proven invaluable and in many cases decisive. First among these come Christopher A. Faraone and David R. Jordan, whose impact has been immeasurable. Next we wish to mention the following: Gideon Bohak, Nancy Bookidis, Edward J. Champlin, Valerie Flint, Elizabeth R. Gebhard, Martha Himmelfarb, John J. Keaney, Israel Knohl, Robert Lamberton, Evasio de Marcellis, Joshua Marshall, Stephen G. Miller, Susan Rotroff, Michel Strickmann, and Emmanuel Voutiras. We also acknowledge generous support from Dimitri Gondicas, the Committee on Hellenic Studies and the Dean of the Faculty, all at Princeton University.

In the initial stages of this project, each of us translated and annotated a discrete set of texts. But in subsequent stages, translations and annotations relied on the collective wisdom of all. Thus we have decided not to indicate the initial translator of each text. As the volume editor, I assume full responsibility for the results.

Princeton   J. G.
May 1992

Contents

Abbreviations
and Conventions

AIBL	Académie des Inscriptions et Belles-Lettres
AJA	*American Journal of Archaeology*
AM	*Mitteilungen des deutschen archäologischen Instituts,* Athenische Abteilung
ANRW	*Aufstieg und Niedergang der römischen Welt,* ed. Hildegard Temporini and W. Haase
ARW	*Archiv für Religionswissenschaft*
B.C.E.	before the common era
BCH	*Bulletin de correspondance hellénique*
BE	*Bulletin épigraphique*
Bibl.	Bibliography
Bonner, *Amulets*	C. Bonner, *Studies in Magical Amulets, Chiefly Graeco-Egyptian* (Ann Arbor, 1950)
CAF	*Comicorum Atticorum Fragmenta,* ed. T. Kock
C.E.	of the common era
CIL	*Corpus Inscriptionum Latinarum*
CMRDM	*Corpus Monumentorum Religionis Dei Menis*
Davies, *Families*	J. K. Davies, *Athenian Propertied Families. 600–300 B.C.* (Oxford, 1971)
Delatte and Derchain	A. Delatte and Ph. Derchain, *Les intailles magiques gréco-égyptiennes* (Paris, 1964)

DTA

IG, vol. 3, pt. 3, Appendix: "Defixionum Tabellae" (Berlin, 1897)

DT

Defixionum Tabellae, ed. A. Audollent (Paris, 1904)

Dubois

L. Dubois, *Inscriptions grecques dialectales de Sicile* (Rome, 1989)

Faraone, "Context"

C. A. Faraone, "The Agonistic Context of Early Greek Binding-Spells," in *Magika,* pp. 3–32

GMP

The Greek Magical Papyri in Translation Including the Demotic Spells, ed. H. D. Betz (Chicago, 1986); translation and new edition of *PGM*

HTR

Harvard Theological Review

IG

Inscriptiones Graecae

JEA

Journal of Egyptian Archaeology

Jeffery

L. H. Jeffery, "Further Comments on Archaic Greek Inscriptions," *The Annual of the British School at Athens* 50 (1955): 69–76

JOAI

Jahreshefte des österreichischen archäologischen Instituts

Jordan, "Agora"

D. R. Jordan, "Defixiones from a Well near the Southwest Corner of the Athenian Agora," *Hesperia* 54 (1985): 198–252

Jordan, TILT

D. R. Jordan, "Two Inscribed Lead Tablets from a Well in the Athenian Kerameikos," *AM* 95 (1980): 225–39

JRS

Journal of Roman Studies

Kropp

A. Kropp, *Ausgewählte koptische Zaubertexte,* vols. 1–3 (Brussels, 1930–1931)

LSJ

A Greek-English Lexicon, 9th ed., ed. H. G. Liddell, R. Scott, H. S. Jones (Oxford, 1968)

MAMA

Monumenta Asiae Minoris Antiqua

Magika

Magika Hiera. Ancient Greek Magic and Religion, ed. C. A. Faraone and D. Obbink (New York, 1990)

Martinez	D. G. Martinez, *P. Michigan XVI. A Greek Love Charm from Egypt* (*P. Mich. 757*) (Atlanta, 1991)
Naveh and Shaked	J. Naveh and S. Shaked, *Amulets and Magic Bowls. Aramaic Incantations of Late Antiquity* (Jerusalem, 1985)
PA	*Prosopographia Attica*, ed. J. Kirchner (Berlin, 1901/1903)
PDM	spells in Demotic, often with Greek; part of the collection in GMP
Peek	W. Peek, *Kerameikos, Ergebnisse der Ausgrabungen*, vol. 3 (Berlin, 1941), pp. 89–100
PG	J. P. Migne, *Patrologia Graeca* (Paris, 1857–89)
PGM	*Papyri Graecae Magicae*, vol. 1, ed. K. Preisendanz (Stuttgart, 1928); rev. ed., A. Henrichs (1973); vol. 2, ed. K. Preisendanz (1931)
PL	J. P. Migne, *Patrologia Latina* (Paris, 1844–64)
Preisendanz (1930) and (1933)	"Die griechischen und lateinischen Zaubertafeln," *Archiv für Papyrusforschung* 9 (1930): 119–54, and 11 (1933): 153–64
Preisendanz (1972)	"Fluchtafel (Defixion)," *RAC* 8 (1972), cols. 1–29
RAC	*Reallexicon für Antike und Christentum*
RB	*Revue biblique*
RE	*Paulys Realencyclopädie der klassischen Altertumswissenschaft*
RM	*Rheinisches Museum*
Robert, *Froehner*	L. Robert, *Collection Froehner*, I: *Inscriptions grecques* (Paris, 1936)
Sepher ha-Razim	*Sepher ha-Razim. The Book of Mysteries*, ed. M. A. Morgan (Chico, 1983)

SEG	*Supplementum Epigraphicum Graecum.* References are by volume number and number of inscription within the volume, such as *SEG* 14.3.
SGD	D. R. Jordan, "A Survey of Greek Defixiones Not Included in the Special Corpora," *Greek, Roman and Byzantine Studies* 26 (1985): 151–97.
Solin	H. Solin, *Eine neue Fluchtafel aus Ostia* (Helsinki, 1968), esp. pp. 23–31: "Eine Übersicht über lateinische Fluchtafeln, die sich nicht bei Audollent und Besnier finden"
s.v.	*sub voce,* under the heading/word
SuppMag	*Supplementum Magicum,* vol. 1, ed. R. W. Daniel and F. Maltomini (Cologne, 1990)
TAPA	*Transactions of the American Philological Association*
Tomlin	R. S. O. Tomlin, "The Curse Tablets," in *The Temple of Sulis Minerva at Bath,* vol. 2: *The Finds from the Sacred Spring,* ed. B. Cunliffe (Oxford, 1988), pp. 59–277
Wilhelm	A. Wilhelm, "Über die Zeit einiger attischer Fluchtafeln," *JOAI* 7 (1904): 105–26
Wortmann	D. Wortmann, "Neue magische Texte," *Bonner Jahrbücher* 168 (1968): 56–111
Wünsch (1900)	R. Wünsch, "Neue Fluchtafeln," *RM* 55 (1900): 62–85 and 232–71
Wünsch, *Antike Fluchtafeln*	R. Wünsch, *Antike Fluchtafeln* (Bonn, 1912)
ZAW	*Zeitschrift für die alttestamentliche Wissenschaft*
Ziebarth (1899)	E. Ziebarth, "Neue attische Fluchtafeln," *Nachrichten von der Gesellschaft der Wissenschaften zu Göttingen, Phil.-hist. Klasse* (1899)

Ziebarth (1934)   E. Ziebarth, "Neue Verfluchungstafeln aus
                       Attika, Boiotien und Euboia,"
                       *Sitzungsberichte der preussischen Akademie
                       der Wissenschaften, Phil.-hist. Klasse* (Berlin,
                       1935): 1022–50

ZPE   *Zeitschrift für Papyrologie und Epigraphik*

*   An asterisk indicates *voces mysticae* discussed in the
                       glossary.

IAO   All non-standard words or terms, that is, the *voces
                       mysticae,* have been printed in upper case.

. . .   Spaced dots indicate a series of unreadable letters in the
                       original text; in general we have not specified the exact
                       number of missing letters.

( )   Parentheses indicate interpretive expansions or
                       clarifications of the original text.

[ ]   Square brackets are used to enclose letters not legible
                       but believed to have been in the original text.

{ }   Braces are used occasionally to indicate words or letters
                       mistakenly repeated in the original text.

< >   Angle brackets indicate letters or words mistakenly
                       omitted by the ancient author.

7 × 8 cm.   All measurements of *defixiones* give width first, followed
                       by height, both in centimeters.

Curse Tablets and Binding Spells from the Ancient World

Introduction

Defixiones, or *katadesmoi* as they are called in Greek, reveal a dark little secret of ancient Mediterranean culture.[1] At present the total number of surviving examples exceeds fifteen hundred.[2] Everyone, it seems, used or knew of them, yet only sporadically have they received serious attention from modern students.[3] One reason for this persistent neglect stems surely from the potentially harmful character of these small metal tablets—not so much the real harm suffered by their ancient targets but the potential harm to the entrenched reputation of classical Greece and Rome, not to mention Judaism and Christianity, as bastions of pure philosophy and true religion.

The Materials

David R. Jordan describes these curious objects as "inscribed pieces of lead, usually in the form of thin sheets, intended to bring supernatural power to bear against persons and animals."[4] Other materials could also be used—ostraca or broken sherds of pottery,[5] limestone,[6] gemstones,[7] papyrus,[8] wax[9] and even ceramic bowls[10]—but lead, lead alloys, and other metals remained the primary media for expressing a desire to enlist supernatural aid in bringing other persons (and animals, in the case of racehorses) under the control of the person who commissioned or personally inscribed the tablet. In fact, the vast majority of surviving tablets is made of lead or lead alloys.[11]

The preference for lead over other metals presents a complicated and revealing set of problems. First, analysis of the remarkable tablets from the spring of the goddess Sulis Minerva at Bath (England) has revealed that only one-fifth of the tablets contained as much as two-thirds lead.[12] The rest consist of alloys of lead and tin, sometimes fused with copper.[13] Perhaps this alloy was peculiar to England, a by-product of local pewter

3

industries, but these results should raise doubts about earlier claims that most tablets were made of pure lead. Second, the preference for lead seems due largely to its low cost and ready availability, whether as a by-product of silver mining in Greece or of pewter industries in England. Also, recipes for *defixiones* recommend "borrowing" (stealing?) lead from water pipes, presumably in the public domain.[14] Third, as Tomlin notes, "it was quite easy to make a tablet" by pouring hot lead into a mold and then rolling, hammering, or scraping the sheet to obtain a smooth surface.[15] Thereafter the sheet could be cut into smaller pieces to make individual tablets. Fourth, lead was a common medium, perhaps one of the very earliest, for writing of any kind, including private correspondence.[16] Fifth, certain obvious features of lead (it was cold, heavy, and ordinary) came to be seen, at a later time, as particularly suitable for the specific task of conveying curses and spells to the underworld. A tablet from the Athenian Agora pleads that "just as these names are cold, so may the name of Alkidamos be cold"[17]; others seek to render one's personal enemies as heavy as the lead[18]; and several early Greek tablets make use of what seems already to have become a formula: "Just as this lead is cold and useless, so let them (my enemies) be cold and useless."[19] But these formulas do not appear on the earliest tablets and probably represent a later stage of reflection.

The Inscribed Messages

Contrary to what one might expect, the process of inscribing metal tablets posed no great difficulty. The preferred instrument was a bronze stylus (*PGM* VII, lines 396ff.). In some cases, the letters are lightly scratched on the surface, but in others they are more deeply incised, with a clear buildup of metal visible at the end of the stroke. Tomlin notes that "a practiced scribe could write on the soft metal surface as easily as on wax."[20] Of course, the real issue here is to know who actually inscribed the letters, a professional scribe or the private individual seeking to enact the spell. We may begin with three observations. First, the range of skill exhibited on tablets is quite broad, with large, awkward letters on one extreme and fluent scripts on the other.[21] Second, professionals may have played a more important role in the Roman period (first to sixth centuries C.E.) than in classical and Hellenistic times, although Plato already indicates the presence of professionals in the fourth century B.C.E. who prepared *katadesmoi* for a fee.[22] Third, in most cultures the business of making spells has been an activity entrusted to specialists.

The tablets themselves suggest that it was not uncommon for individuals to turn to professionals, whether *magoi* or scribes. The work of professionals is evident in a tablet like DTA 55, whose hand is described as a *scriptura elegantissima* reminiscent of public monuments,[23] or in the "skillful, elegant, fluent semicursive" texts of the third century c.e., excavated from wells in the Athenian Agora.[24] The "clerical" characteristics of numerous tablets from the find at Bath suggest the presence of scribes, even though no two tablets appear to be by the same hand.[25] Also, highly formulaic texts, which cannot have been invented on the spot, must have been copied from formularies of the sort preserved in *PGM* and similar recipe collections used by professionals. Finally, several large caches of tablets found in one place clearly reflect the work of a local "cottage" industry. Common sense seems to point us toward Tomlin's cautious conclusion that we should expect to find "a mixture of professional and amateur scribes,"[26] but on balance the scales would appear to favor professionals, at least in the Roman period, both for inscribing the tablets and for providing the formulas.

Next we must ask what these scribes wrote on their tablets. In large part, the translations that follow in this collection will answer that question, but some general observations may serve to create a sense of broad patterns and of changes through time. The general rule is that the earliest examples are also the simplest: most of the early tablets from Sicily and Attica (fifth to fourth centuries b.c.e.) give only the name of the target, with no verb of binding and no mention of deities or spirits; some do include both a verb (usually a form of *katadein*) and the name of a deity (in Attica usually Hermes or Persephone).[27] Special forms of writing include either scrambling the names of the targets[28] or writing them, and sometimes the full text of the spell, backwards—that is, from left to right but with the individual letters facing in the proper direction. Such techniques clearly express a symbolic meaning like that attributed to the lead of the tablets themselves, that the fate of the targets should turn backward or be scrambled, just like their written names. We also find here yet another example of the way in which quite ordinary habits became "mystified" in time, gathering a significance and power quite unthinkable at earlier times.

With few exceptions, "mystical" words or formulas do not appear in Greek tablets of the classical and Hellenistic periods, in contrast to the richly variegated language of tablets in the Roman period (first century c.e. onward). But these exceptions are interesting and important. The first involves a set of six terms, called *ephesia grammata,* first attested in a fragment of Anaxilas, a comic poet of the fourth century b.c.e.: "(an unnamed person) . . . carries around marvelous Ephesian letters in sewn pouches."[29] These terms (*askion, kataskion, lix, tetrax, damnameneus,* and *aision/aisia*) were believed to possess the ability to endow those who wore them (especially boxers, so it seems) with great power, both defensive and aggressive.[30] They remained well known for centuries, appearing in several later spells and charms. The Christian writer, Clement of Alexandria (ca. 200 C.E.) not only lists the terms but calls them "famous among many people."[31] More significantly, they appear in at least one lead amulet from the Hellenistic period and thus clearly establish the use of "mystical" terms (*voces mysticae*) long before their widespread circulation from the first century C.E. onward.[32] This folded tablet from Crete, dating to the fourth century B.C.E., must have been carried as a protective amulet (line 20 speaks of protecting the wearer against hostile spells); it contains several of the *ephesia grammata: aski* and *kataski* (lines 9–10), *lix* (lines 5 and 10), forms of *tetrax* (lines 5, 11–12), and *damnameneus* (line 16).[33] A second exception is a curious stone monument from Greece, dating to the late fifth century B.C.E., which Jeffery takes to be a thank offering by an unnamed person for the successful punishment of a personal enemy.[34] She offers the following tentative reconstruction of the text: "The Ephesian vengeance was sent down (?); first Hecate injures (??) the possessions (??) of Megara in all things; then Persephone already is reporting all the (prayers?) to the gods."[35]

The significance of this evidence for the development of *defixiones* from those of the classical and Hellenistic periods to later Roman types cannot be exaggerated. As even a brief comparison of any early and late tablet from the following collection will reveal, the differences are real and many:

1. in some of the late Roman examples (esp. fourth to fifth centuries C.E.), *voces mysticae* and other forms of "unintelligible" writing can take up as much as 80 to 90 percent of the tablet, whereas in the amulet from Crete discussed previously, the *ephesia grammata* occupy no more than a line or two;

2. the names and invocations of the gods and spirits are notably longer, more complex, and aggressively international in the later examples;

3. drawings of human and animal figures, along with the probably astrological *charaktêres,* become omnipresent;

4. a general increase in Egyptian elements occurs, reflecting the fact that most of the surviving formularies were produced and copied in Egypt and thus reveal the fusion of Greek, Egyptian, and other cultures

typical of Egypt from the first century C.E. onward; among these Egyptian elements, one of the most notable is the use of threats against the gods.[36]

And yet we must not overemphasize these differences, as earlier interpreters have done, out of a desire to protect ancient Greek culture, even in its "lower" forms (such as the *defixiones*) from comparison with the "degenerate syncretism" of late Roman "magic" and "superstition." For underneath these differences, we can also detect clear lines of continuity.

For example, the earlier *ephesia grammata* continue to appear in the later texts[37] and may now be regarded as forerunners of the more elaborate *voces mysticae* so characteristic of them. Indeed, already on the early Hellenistic amulet from Crete, the originally impersonal *ephesia grammata* are addressed as powers in their own right: they have become the names of supernatural entities, just as the later *voces mysticae* come to function as the secret and powerful names of the gods invoked in the spells.[38]

Also, the relatively simple forms of the earlier tablets may be explained by the strong likelihood that the commissioning and depositing of tablets with simple written formulas were accompanied by *oral* prayers, invocations, and incantations. Gradually, with the growth of written language in Greek culture, these oral accompaniments were written down and took their place on the tablet alongside the traditional elements (the names of the targets and the deities and the verbs of binding).[39]

While the evidence for the use of *charaktêres* and engraved figures on earlier tablets is virtually nonexistent, the stone monument discussed previously does incorporate, in the midst of its text, the head of a ram. In this regard, it may also be worth noting that the use of dolls or figurines with early *defixiones* may provide another instance in which an originally separate item (the figurine) eventually moved onto the tablet itself (as a drawing of a human figure).

Still, there can be no mistaking the more elaborate forms of speech in tablets of the Roman period; in general, they are the most certain indicators of a late date. A partial catalogue of these nonstandard forms of speech would include the following:

1. palindromes;
2. *charaktêres* (see Figure 1);
3. vowel-series[40];
4. triangles, squares, "wings," and other geometric shapes made up of letters;

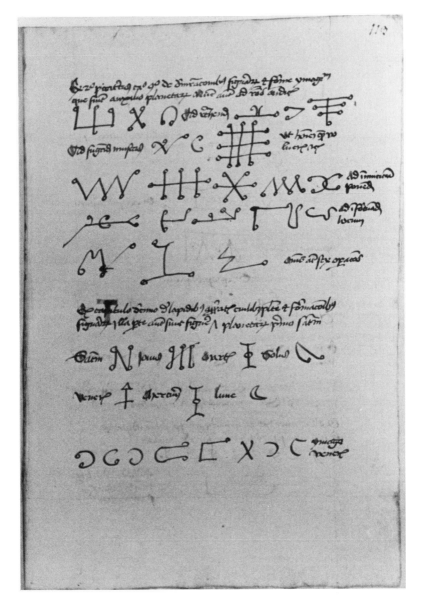

FIGURE 1. *Charaktêres* from a medieval Arabic manuscript of the *Ghayat al-Hakim* or *The Aim of the Sage*. Translated into Latin, it was known as the *Picatrix*. This elaborate treatise on celestial powers and their practical application develops a theory of correspondences between the celestial forces, especially the signs of the zodiac, and a set of written signs or symbols, that is, these *charaktêres*. (Österreichische Nationalbibliothek: Cod. 3317, fol. 113v. By permission.)

5. names ending in -*êl* and -*ôth,* clearly built on Jewish and Hebrew models;

6. *voces mysticae*—words not immediately recognizable as Greek, Hebrew, or any other language in common use at the time;

7. recurrent formulas (called *logoi* in *PGM*) consisting of several *voces mysticae;* these are often abbreviated in recipes, for example, "the abc-formula."

Traditionally, these "unintelligible" forms of speech have been treated as meaningless gibberish or nonsense. To be fair, such interpretations are not modern inventions but reach back to ancient critics. In his satire on such practices (*Philopseudes/Lover of Lies,* chap. 10), Lucian of Samosata has his protagonist protest: "Unless you can show me how it's physically possible for a fever to be frightened or a swelling to be scared away from the groin by a holy name or a word in some foreign language, the cases you quote are still only old wives tales." On the other hand, the Neoplatonic philosopher Iamblichus (ca. 300 c.e.), in his passionate defense of the same practices (he calls them "theurgy"), argues that *charaktêres* and foreign names,[41] when used properly, convey to the theurgist the powers of the gods. He adds that names lose their theurgical power when translated into Greek.[42] In the long run, however, the view of Lucian has prevailed and impressed itself on Sir James Frazer and his successors. Only recently have efforts been made to reverse these effects, efforts directed at understanding the foundations of such beliefs, without slipping into Iamblichus's posture of defending them as true.[43]

At the center of these efforts lies the work of the anthropologist Stanley J. Tambiah. His 1968 essay, "The Magical Power of Words," though written with no apparent knowledge of ancient Mediterranean *defixiones,* presents a definitive repudiation of the "gibberish theory" regarding *voces mysticae.*[44] For our purposes, we may concentrate on one aspect of Tambiah's argument, his analysis of special languages in Sinhalese spells employed to invoke demons of illness. In this case, where the healer is conscious of addressing a "foreign" audience of supernatural spirits, it would be entirely inappropriate to use one's native, human, ordinary language. Thus the mysterious language of spells does not violate the basic rule of speech communication, that the parties involved must understand one another, because the *voces mysticae* represent "the language the demons can understand."[45] Along similar lines, P. C. Miller and R. T. Wallis have recently shown in separate essays that in the culture of late antiquity it was precisely the use of unintelligible forms of speech that signaled the passage from the lower mundane realms into the sphere of true spiritual conversation with higher orders of being.[46] The words of the Christian philosopher Clement of Alexandria (ca. 200 C.E.) represent a wide consensus that normal human language was not appropriate in addressing gods or any higher beings: "Plato assigns a special discourse (*dialektos*) to the gods and he reaches this conclusion from the experience of dreams and oracles but most of all from those possesseed by *daimones,* for they do not speak their own language or discourse but rather the language of the *daimones* who possess them."[47]

Of course, this is not to say that either those who commissioned *defixiones* or those who prepared them understood the *voces mysticae.* That was not the point. What mattered was the belief that these invocations and secret names were understood by the spirits themselves. Just as in ordinary human encounters, the key to success was to address these superior beings by their proper names and titles. In this sense, although it is interesting to note that many of the *voces* have turned out to be "real" words borrowed and frequently distorted from other "real" languages (Hebrew, Aramaic, Persian, and various forms of ancient Egyptian and Coptic), this finding is quite irrelevant to understanding the attitudes of those who purchased the tablets. For the anxious client, what mattered was the belief that the *magos* possessed the special knowledge to get these names and titles right. But as the many variants reveal, we can see that even they did not always copy the *voces* with total accuracy. We should also consider the likelihood that there was an element of status enhancement for professionals in maintaining a core of "unintelligible" discourse, for this left the client with little choice but to assume that the specialist alone, through superior wisdom, understood the meaning and significance of this higher language. There is much more to Tambiah's essay, but on the single issue of "unintelligibility" he has pulled the rug from under smug interpreters who have, it turns out, vainly contrasted the benighted irrationality of superstitious and ignorant primitives with their own modern rationality.

Charaktêres occupy a special place in the symbol system of ancient spells, for their omnipresence—though not earlier than the second century C.E.—as well as for the scant attention they have received. They appear on amulets,[48] *defixiones,*[49] a private divination apparatus from Pergamum,[50] in recipes (including Greek, Hebrew, Coptic, and Arabic collections) for *defixiones* and other spells,[51] and in treatises of ancient Gnostics.[52] In addition, they appear in a public inscription on the wall of the theater at Miletus, where each of seven *charaktêres* is associated with two sets of vowels: under each set of *charaktêres* and vowels appears the following request, "Holy One, protect the city of Miletus and all its inhabitants," and under the full set of seven columns follows a one-line prayer, "Archangels, protect the city of Miletus and all its inhabitants."[53] Clearly, these *charaktêres* were seen as signs and sources of great power. They embody the classic definition of a religious symbol as embodying and transmitting power from the divine realm to the human. But what precisely did they represent—traditional gods, archangels, planets, or something else? No doubt, like other special forms of writing on *defixiones*, they were taken to be mysterious and powerful, which means that their "real" origins were not understood at all. Among competing interpretations regarding their origin, the most promising would appear to be astrological—that they symbolized various planetary powers, powers that were in turn commonly identified with angels and archangels by late Roman astrologers.[54] But whatever their origins, their presence as the sole powers invoked on tablets from Apamea ("most holy" and "lords"), Beth Shean ("fearsome"), and Hebron tells us that they had taken on a life of their own and were seen as personifying, representing, and embodying great power.[55]

A good number of the *defixiones,* most gem amulets, and many of the recipes in formularies also include drawings of human beings, animals, or mixed creatures (for example, the famous "Anguipede" figure of a human torso with head of a rooster and snakes for legs). In general terms, the meaning or function of these figures is obvious: like the *voces mysticae* which they represent, they embody and make present the reality of the various actors mentioned in the spell (the human target and the supernatural beings, rarely the client). Here again, and in contrast to early views, we see that the function of figures on the tablets is by no means unique or distinctive to them. For, as André Grabar has noted in his study of early Christian iconography, images in late antiquity "seem to have been used more frequently than at other historical periods and that an extraordinary importance was attributed to them. . . . [T]he portrait of the sovereign replaced the sovereign. . . . [P]ortraits of persons of such rank (i.e., Roman magistrates and Christian bishops) have the value of judicial testimony or of a signature."[56] On the role of images in Jewish and Christian settings, he notes that "images were intended to do more than recall events of the past: they were intended in some sense to perpetuate the intervention of God . . . just as the sacraments did."[57] In line with these observations, we may conclude that the drawings of mummies, dismembered bodies, and figures wrapped about with straps or snakes were intended to anticipate and enact the desired outcome of the spell itself, to bind or in some other way harm the target. But like many other features, these drawings have been little studied. A. D. Nock's plea of 1929 for a study of these drawings and their iconographic bearings remains largely unheeded.[58]

Gods, *Daimones,* and Spirits of the Dead

The role of images and figures as mediators of power brings us finally to the names of deities and other spiritual entities on *defixiones.* In discussing these names, it is essential to keep in mind three fundamental characteristics of the "spiritual universe" of ancient Mediterranean culture: first, the cosmos literally teemed, at every level and in every location, with supernatural beings; second, although ancient theoreticians sometimes tried to sort these beings into clear and distinct categories, most people were less certain about where to draw the lines between gods, *daimones,* planets, stars, angels, cherubim,, and the like; and third, the spirit or soul of dead persons, especially of those who had died prematurely or by violence, roamed about in a restless and vengeful mood near their buried body.

It has long been customary to distinguish ancient *defixiones* from other areas of ancient culture—that is, to separate magic from religion—by pointing out, as does H. Versnel, that the gods named in them "invariably either belong to the domain of death, the underworld, the chthonic or are reputed to have connections with magic."[59] But such observations tell us precious little, for the supernatural beings named in *defixiones* appear also in what we otherwise call ancient religion, where virtually every god or spirit reveals some connection with death and the underworld. In short, when Jewish (and later Christian) elements (angels, archangels, and the figure of IAO, the god of Israel) are taken into account, they will be seen to have almost no chthonic ties. In short, the presence or absence of chthonic deities offers no hope for a satisfactory differentiation between "religion" and "magic."

Once it became customary to write down, rather than recite, the names of the gods to whom the spells were addressed, a clear order of preference became apparent: Hermes is by far the most common; he is followed by Hekate, Kore and Persephone, Hades (also known as Pluto), Gê/Gaia, "the holy goddess" (at Selinus in Sicily), and finally Demeter (often cited together with "the gods with her"). Others addressed include Zeus,[60] "all the gods and goddesses," Kronos, the Mother of the gods, and the Furies (Erinyes).[61] On Latin tablets, the most common names are the Manes (spirits of deceased ancestors), Jupiter, Pluto (the Greek Hades), Nemesis, Mercury (the Greek Hermes), and various water nymphs. Now the 130 or so tablets from Bath, dedicated to the goddess Sulis (also called Minerva) must be added at the top of the list. Finally come those highly syncretized spells, primarily from North Africa and Egypt in the third to sixth centuries C.E., where gods, *daimones* with secret names, personified words (for example, EULAMON), *voces mysticae* containing elements of foreign deities (IAO, ERESCHIGAL), and especially a variety of Egyptian deities come together to form the rich international blend that is so characteristic of late antique culture in all of its dimensions. Among the Egyptian contributions, the most prominent are Thoth (commonly identified with Hermes), Seth, and Osiris. In addition to contributions from Egypt, one finds significant elements from Jewish sources, from Persia, and at a later date from Christianity. In general, two factors seem to have governed the selection of gods and spirits and their names: first, local customs and beliefs; and second, the recipes available through the formularies owned and used by local experts. In this sense, we may use what we read on *defixiones* as a reasonably accurate measure of prevailing beliefs at particular times and places.

Like other forms of human speech with which they show close similarities (legal,[62] cultic,[63] epistolary[64]), the language of *defixiones* is highly formulaic.[65] Various schemes have been proposed for organizing these formulas, most recently by C. Faraone. He proposes a simple yet comprehensive threefold division of styles or types, although he emphasizes that all three could be used at one and the same time, even on a single tablet[66]:

1. the direct binding formula ("I bind X!"): Faraone calls this a performative utterance, designed to operate automatically, through the effective force of the words themselves and without intervention from any supernatural source; here it should be recalled, however, that gods may have been invoked *orally*, when the tablet was either commissioned or deposited;

2. prayer formulas that appeal directly or indirectly for supernatural assistance ("Restrain X!");

3. persuasive analogies in which the client expresses the wish that the target should take on the characteristics of something mentioned in the spell ("As this lead is cold and useless, so may X be cold and useless!"); this, too, must have been regularly coupled, even if orally, with an appeal for divine assistance.[67]

To these basic types we may add a partial list of recurrent features in the language or discourse of the *defixiones:* repetition, pleonasm, metaphor and simile, personification, rhythmic phrases, exaggeration, threats, promises, prayers, and formal appeals.[68] Once again, it must be emphasized that these features are not distinctive of *defixiones* but instead mark them as part of the general culture of their time.

The last piece of information engraved on any tablet would be the name of the target and, in cases where it was appropriate, the name of the client. Not all tablets included a personal name, but it is clear, especially in the Roman period, that tablets were sometimes prepared in advance, with space left for inserting the names provided by paying customers. In cases where the names were too long for the space, the scribe was forced to squeeze the name or to write the final letters on a slant or even between lines.[69] We have seen that names were often written in symbolically significant ways on early Greek tablets—the letters could be scrambled or written backward—so that names clearly operated as more than labels. The name embodied the person or the animal and gave some measure of control over them. The same rule applied, as we have seen, to gods and spirits.

Another unusual characteristic in the treatment of names was the practice, from the second century c.e. onward, of identifying personal names by matrilineal descent ("I curse X, whose mother is Y").[70] Even in Jewish spells this was the custom, so much so that a noted rabbi, Abaye (ca. 325 c.e.), once reported the following: "Mother told me, 'All incantations which are repeated several times must contain the name of the patient's mother.' "[71] Various explanations have been advanced for this unusual custom: because precise identification was necessary, only the mother could be known for certain; influence may have come from Babylonia or Egypt, where matrilineal lineage appears in early spells; the practice was taken over from the world of slaves, who were regularly identified by matrilineal descent; and in Egypt, Jewish and Christian funerary monuments sometimes identify the deceased by descent from the mother.[72] Although several signs point in the direction of Egyptian influence, we must suppose that other forces were also at work, arising from social and psychological dynamics peculiar to the ancient Mediterranean. In the end, however, the practice must also be related to the countercultural and subversive character of the *defixiones* themselves.

Dolls, Hair, and Nails

As complex as the inscribed tablet sometimes appears, especially from the second century c.e. onward, it might require still more preparations before it was ready to leave the shop. Frequently these took the form of separate items attached to the tablet itself.

Figurines[73]

One of the most striking features of ancient *defixiones* emerges from the use of dolls or figurines as part of the binding process. A significant number of such figurines have survived, thus confirming and illustrating the literary texts[74] (see Chapter 8, esp. Plato) and formularies (see Chapter 2 and *PGM*), which speak of them as if they were a commonplace. The earliest discussion occurs in an Egyptian text of the Middle Kingdom (ca. 2133–1786 B.C.E.), which prescribes the making of a wax image of a personal enemy, to be buried in a grave for harsh treatment by Osiris.[75] Similar dolls were used by Greek cities in public binding ceremonies as early as the fourth century B.C.E.[76]

Three types of figurines were used in the private sphere of *defixiones*.[77] Almost all, whether made of lead, mud, or wax, had their hands tied ("bound") behind their backs; others appear to have been deliberately mutilated[78] (Figure 2). Frequently the name of the target was inscribed on the figurine.[79] The first type of figurine, which appears in love/sex spells, represents the desired object of one's affections. For example, a small human shape was found in Egypt, attached to a love spell written on papyrus[80]; another (see pp. 97–100) was produced in accordance with an elaborate recipe in *PGM*, which called for thirteen needles to be inserted at symbolically appropriate spots. Whatever else one might say about such puzzling "sex objects" found with *defixiones*, it is important to emphasize that they were not curses and were not always intended to harm the target.[81] The second type was used to assault personal enemies; a special subset of this type was discovered in graves from the Kerameikos (Athens) and dates from around 400 B.C.E. (Figure 3). The group includes four lead figurines (see pp. 127–29), each encased in a miniature coffin made of two lead sheets, with names of the targets inscribed on all of the sheets and on three of the figurines.[82] Finally, we note a unique instance of nine little horse figurines from Antioch. Six are inscribed with a single name: Huperochos, Pouklês, Pouandros, Dithurambos, Sundikos, and Pemphreôs; the other three have two names: Euaspis and Damastês, Aristenetos and Pontouchos, and Anchitheos and Thetis. In all probability the horses were intended to represent the teams of competitors in chariot races, for there are numerous examples of curse tablets from the world of ancient racing that name both drivers and steeds.[83]

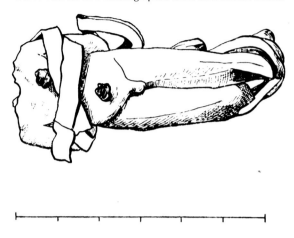

FIGURE 2. A deliberately decapitated male figurine made of lead, pierced by two iron nails; hands and feet are tied up by lead bands. The figurine was discovered in a grave from Attica (Greece) and dates to pre-Christian times. It might well have accompanied a *defixio* (DTA 86) of the same time and place: "I bind Dromôn . . . feet and hands . . ."

Hair and Clothing

When the female *magos* in Lucian of Samosata's *Dialogue of Courtesans* (4.4; see pp. 255–56) advises a client on techniques for winning back the affections of her alienated lover, she tells her to bring some of his hair or a piece of his clothing. This connection between hair or clothing (*ousia* or "stuff" as translated in this collection) is borne out by several surviving love spells on papyrus and lead[84]: *PGM* XVI (first to third

FIGURE 3. Three lead figurines in miniature coffins, recovered from graves in the Kerameikos (Athens). Two of the figurines have their hands tied behind their backs; all show exaggerated genitals. Each of the three boxes/coffins and two of the figurines are inscribed with a spell. Such figurines were a common feature of curse tablets in classical Athens. (Barbara Schlörb-Vierneisel, "Eridanos— Nekropole," *Athenische Mitteilungen* 81 [Berlin: Verlag Gebr. Mann, 1966], Beilage 51.1. By permission.)

century C.E.) was discovered folded around some hair; *PGM* XIXa, another love spell (fourth to fifth century C.E.), also contained hair; a lead *defixio* calls for hair from the head of the love target; an unpublished lead *defixio* from Egypt contained hair and other fibers (from clothing?); and a tablet from the Athenian Agora (third century C.E.) reveals both imprints left by hair as well as several of the strands themselves, presumably from the head of Tyche, the target of the spell. The symbolism of the hair is not difficult to comprehend. Like a modern wallet photograph, it makes present the absent person, and like the use of synecdoche in poetic and other forms of figurative speech, with which spells show numerous similarities, a part of something can always stand for the whole.

Sealing by Nails, Rolling, or Folding

Once all of the writing had been completed and the accompanying materials inserted or attached, almost all *defixiones* were rolled or folded; they might also be pierced by one or more nails.[85] Despite their corrodibility, a large number of these sealed and "fixed" tablets have survived intact.[86] What symbolic meaning was attached to these forms of sealing? In some cases, the rolling of the tablets, especially in cases where a "supernatural address" was added to the exposed outer sheet, may have simply followed the normal way of addressing a letter.[87] Yet the folding, rolling, and nailing was not designed primarily to prevent human eyes from reading the tablet's contents, for in virtually all cases the tablet would have been deposited where no human could have found it.[88] The use of nails, too, must have some special meaning. Kagarow, who provides a full discussion of nails as symbolic instruments in ancient as well as modern cultures, insists that their function here is to add pain and death to the spell. But as we have seen, this fails to do justice to their use in love spells. In fact, the root meaning probably derives from the ordinary function of nails, which is to fasten, to fix, to tie down, and thus to bind.[89] In any case, their universal application tells us that whatever their "original" purpose, they soon became a prescribed part of the process for preparing a *defixio*. In just one instance was a nail actually used for its ordinary purpose, to fix a tablet in place. Recent excavations in the circus at Carthage have recovered *in situ* a folded tablet, deliberately fixed to the compacted floor of the circus by a long bronze nail. Both are well preserved (Figure 4).

Depositing the Tablets

Defixiones differed from amulets in one fundamental respect; once they were inscribed, they were deposited in special locations where their powers took effect. *PGM* VII, lines 451–52, gives a typical list of such places: "have (the tablet) buried or [put in] a river or land or sea or stream or coffin or in a well."[90] Other places included, for love spells, the home of the desired target; for racing, the stadium floor[91]; and sanctuaries associated with chthonic deities.[92]

Many graves, which appear to have been the most common place of

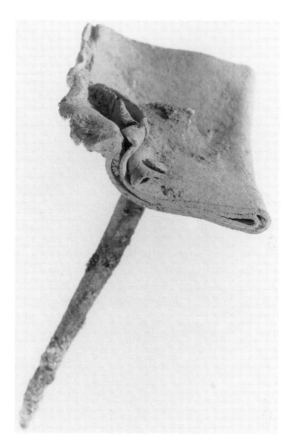

FIGURE 4. A unique example of an unpublished *defixio* from the circus at Carthage where the nail was used to fix the tablet to the floor of the racetrack itself, near the starting gates. (Used with the permission of Naomi Norman; photograph by Bill Wood.)

deposit, have yielded not only the remains of the bodies originally buried in them but *defixiones* inserted at a later time. Burial sites of those who had died young or by violent means were the preferred choices, because it was believed that their souls remained in a restless condition near the graves until their normal life-span had been reached.[93] In some cases, these souls were promised respite from their unhappy fate once they carried out their appointed task, as outlined in the tablet. The precise function of these "dead souls" (*nekudaimones* in Greek) is not always clear. Did they simply transmit the petition of the tablet to the gods, in their capacity as beings caught midway between the living and the dead? Or did they enact the spell themselves, under the watchful eye of the gods? Perhaps the answer is that their role changed over time and according to local beliefs. In any case, the gods were always involved in one way or the other. The standard technique consisted of using legal terminology, whereby the case of the target was handed over or transferred to a divine jurisdiction ("I hand over X before/to the presence of Hermes").[94]

In thinking about how tablets were deposited, we need also to ask whether it was the client or the scribe or *magos* who carried out the physical act of deposition. Like the related question of who made the inscriptions on the tablets, this one concerns and shapes our understanding of the total process. Unfortunately, there is little direct evidence to suggest an answer one way or another. In some cases, the depositing surely required no special skill—as in dropping the tablet into a well or spring—and we can well imagine the client carrying out the task alone. But in other cases, involving *defixiones* placed in graves or buried at the starting gates of race courses, the likelihood is much greater that scribes or *magoi,* or their assistants, undertook this task as part of their service to the client. It is unlikely that a private citizen would enter the Kerameikos at night, dig up the grave of a young person, and place the tablet in the corpse's right hand.[95]

Before leaving the question of where the tablets were placed, and who placed them, we must reemphasize one aspect of the process that has led to serious misunderstanding in the past. The business of commissioning a tablet entailed much more than buying a piece of folded metal and throwing it into a well. As the recipes from *PGM* clearly demonstrate, the total process could be quite complex. It required invocations, purifications, fumigations, prayers, instruments, rituals, and more. In an earlier literary source, the second *Idyll* of Theocritus, the story of Simaetha's efforts to regain the affections of Delphis is no less complex. Earlier still, when the sons of Autolycus heal Odysseus's wounds in the *Odyssey* (19.457–58), they apply bandages but also pronounce a spell (*epaoidê*) to stop the flow of blood. In short, the client's attempt to prepare a *defixio,* from the initial decision to its actual commission, must be located in a series of actions, a total flow of events rather than a single isolated act. It is a serious mistake to focus attention solely on the innocuous piece of corroded lead and not to perceive other actions that accompanied its commission.[96]

A related and equally serious issue concerns the balance between private and public aspects of commissioning and depositing a *defixio*. It has generally been assumed, with important consequences for interpretation ("Those foolish people believed that by simply tossing a hunk of useless metal into a well and mumbling a few senseless words they could affect the lives of others!"), that the process was entirely private, secret, and hidden from the public eye. Yet there are numerous indications that such was not the case, not least among them the undeniable indicators that *defixiones* worked, or were believed to work, which amounts to much the same thing. First, almost everyone, from aristocrats and philosophers to slaves and jockeys, seems to have known the *defixio* as a social fact. Second, the public monuments that speak of curses and binding spells belong not to the private realm but to the public (see pp. 246–48). Third, the fact that in the famous case involving the unexpected illness of Germanicus (see pp. 254–55) someone thought to look for tablets or spells—and found them—suggests that people knew both who their enemies were and what actions they might take against them. Fourth, to return briefly to the jilted Simaetha, she does not limit her countermeasures to rituals and spells but goes down to the gymnasium to confront her faithless lover face-to-face. Love spells in particular present a strong case for supposing that the use of a *defixio* was but one strategy among several for achieving one's goal. Fifth, one suspects that the perpetrators (if not the client, at least the scribe or *magos*) were less than totally discreet about their business, perhaps intentionally so, and let it be known that a "fix" had been put on so-and-so.[97] If this more public aspect of commissioning and depositing *defixiones* is at all true to reality, we will need to revise our thinking about whether and how they worked.

What Effects Did Tablets Seek?

We know that all *defixiones* express a formalized wish to bring other persons or animals under the client's power, against their will and customarily without their knowledge. In some cases, the wish is expressed as an intention to inflict personal harm or death. Kagarow lists the following examples: death, illness (fever, consumption, blindness, dumbness, lameness, broken limbs), loss of memory, various forms of mental suffering, sleeplessness, involuntary celibacy, loss of family and house, public humiliation, defeat in war and athletic competition, failure in business, conviction in public courts, denial of an afterlife, and general lack of success.[98] In racing spells, horses are to fall down, break their limbs, and rear up at the start. In the relatively few cases of tablets directed against inanimate targets, public baths are to be restrained and hindered (see pp. 173–74), and the gates of Rome are to be struck (see pp. 171–72).

But how are we to take these "wishes" and who is the real audience of the invocations? Once again, the tendency among interpreters has been to read them literally. Here we might begin with our own forms of cursing. What do we mean when we blurt out, "Screw you!"? Is this an expression of our desire for sexual intercourse? When we hear team-mates or sports fans shout, "Kill the bum!", do we load our rifles? Two considerations seem in order here. First, the audience may not be exclusively, or even primarily, "out there." As Tambiah, among others, has argued, spells are directed primarily to the *human* participants in all ritual action.[99] Second, the function of verbal speech-forms represents a unique feature of human language, namely, its ability to communicate and give form to the expressive and metaphorical aspects of human experience. We take this treatment of language for granted when we read poetry or novels. Perhaps we should apply it to the language of spells and curses. One is reminded of an anecdote told by Mary Douglas: "Once when a band of !Kung Bushmen had performed their rain rituals, a small cloud appeared on the horizon, grew and darkened. Then rain fell. But the anthropologists who asked if the Bushmen reckoned the rite had produced the rain, were laughed out of court. . . . How naïve can we get about the beliefs of others?"[100]

Did They Work?

Until recently, the very idea of asking such a question would have seemed absurd. Of course this stuff doesn't work! Indeed, from the time of Sir James Frazer to the present, the ruling assumption has been that spells, charms, and amulets cannot work—by definition. Once again, the initial assumption sets the agenda for the ensuing discussion and interpretation: because the beliefs are assumed to be false and because the practices are taken to be ineffective, how are we to explain the persistent irrationality of those who pursued them through so many centuries.

What would happen, however, if we changed our initial assumption and began with the idea that these beliefs and practices must have worked *in some sense;* if we indicated that we can no longer accept the notion that those who hold to them are irrational; and if we recovered our sense of poetic language and expressive ritual as fundamental con-stituents of all human experience?

Roger Tomlin's introduction to the Bath tablets may be taken as a measure of a new climate regarding the treatment of the effectiveness of ancient spells and curses. "Did the Bath tablets work?" he asks. And answers, "the practice of inscribing them for two centuries . . . implies that they did work. Or rather that they were believed to work; and perhaps, that this belief was justified."[101] Of course, we need not assume that they worked in the same way that the participants themselves believed. But neither are we justified in imposing simplistic or literalistic preconceptions on these participants and their beliefs. Tomlin offers several useful insights as to how the tablets might have worked. The process "removed intolerable tensions."[102] It "allowed a transfer of emotion."[103] More pragmatically, he continues, "to inscribe a curse tablet and throw it into the sacred pool relieved the injured party's feelings; something at least had been done."[104]

Much more has been written on this subject, including Lévi-Strauss's remarkable essay on "The Sorcerer and His Magic,"[105] and Geza Roheim's treatment of what he calls "the *magical principle,*" described variously as "the counterphobic attitude,"[106] and "the ever-present matrix of our actions."[107] In line with these neo-Freudian views, we may also mention the work of E. E. Evans-Pritchard, the British anthropologist whose work on African witchcraft has shown that curses and spells play an important role in explaining personal misfortune and in excusing public failure.[108]

Defixiones and the Law

It has often been assumed that the use and preparation of *defixiones* were against the law, despite the fact that classical Greek law codes never mention *katadesmoi* as such.[109] Perhaps, some have suggested, there was no need for such legislation, specific cases being covered by laws against impiety (*asebeia*).[110] Or perhaps only those that fell under the rubric of homicide, those that called for the death of the target, were treated as illegal. On the other hand, in a law code of 479 B.C.E., the community of Ionian Greeks at Teos decreed as a capital offense the preparation and use of *pharmaka*.[111] It is worth noting here that while *pharmaka* sometimes referred to poisonous potions, it was also used frequently in an extended way to designate spells and curses. Plato's proposed legislation against curse tablets, outlined in the *Laws* (see pp. 249–50), yields uncertain results on this point. Does his law mirror existing but unknown laws of his time or is he seeking to fill a void in current jurisprudence? At a later time, under the influence of Roman laws, the *defixiones* were clearly illegal. Yet, like astrology, which was also generally illegal, they flourished, both early and late.[112]

But it is not enough to say that *defixiones* were illegal because they sought to do harm to, even to kill, other persons, for that is not true of either love spells or prophylactic amulets.[113] Two important questions thus remain unanswered: first, why did they flourish and, second, why were they declared to be illegal or, at the very least, dangerous and threatening? We will find a single answer to both questions.

We have already seen that *defixiones* must be treated as a familiar feature of ancient Mediterranean cultures. What is more, they cut across all social categories; on this point there is virtual unanimity. The reason for their pervasive presence lies in the observation that they worked, or that they were believed to work, which comes to the same thing. Their success and effectiveness also explains why they were treated as illegal or dangerous. Dangerous not because they always intended harm but because they worked. Better yet, they worked in ways that could not be controlled by the legal, social, and political centers of ancient society. Indeed, at times they stood outside, perhaps in direct opposition to those centers. The idea that *magoi* could dispense power on matters of central importance to human life; the idea that any private person, for nothing but a small fee, could put that power to use in a wide variety of circumstances; and the idea that all of these transactions were available to individuals who stood outside and sometimes against the "legitimate" corporate structures of society—all of these ideas presented a serious threat to those who saw themselves as jealous guardians of power emanating from the center of that society, whether Greek, Roman, Antiochene, or Rabbinic. Here was power beyond their control, power in the hands of freely negotiating individuals.

"Magic" and "Religion"

We have avoided the use of the term "magic" in this volume. Customarily, *defixiones* have been treated as an expression of ancient "magic" and thus contrasted with ancient "religion." But it is our conviction that magic, as a definable and consistent category of human experience, simply does not exist. For as one critic has put it,

> the scientific debate over the relation between "magic" and "religion" is a *discussion of an artificial problem created by defining religion on the ideal pattern of Christianity*. The elements of man's beliefs and ceremonies . . . which did not coincide with the ideal type of religion was—and is—called

"magic." . . . "Magic" became—and still becomes—a refuse-heap for the elements which are not sufficiently "valuable" to get a place within "religion."[114]

By substituting the generic phrase "sanctioned or official beliefs" for "Christianity" in Petersson's lament, we may extend it to non- or pre-Christian societies as well—the beliefs and practices of "the other" will always be dubbed as "magic," "superstition" and the like. Even those definitions that speak of an overlap between magic and religion must presuppose them somehow to be distinct and definable entities.[115] By implication, we also reject as useless all evolutionary schemes, whether the classic theory according to which magic precedes first religion and then science[116] or the converse view that magic represents religion in a state of decay[117] or the related stance according to which religion is the work of the literate and cultured classes whereas magic and superstition—often used interchangeably—represent the stunted circumstances of the ignorant and the unlettered.[118] Thus the use of the term "magic" tells us little or nothing about the substance of what is under description. The sentence, "X is/was a magician!" tells us nothing about the beliefs and practices of X; the only solid information that can be derived from it concerns the *speaker's* attitude toward X and their relative social relationship—that X is viewed by the speaker as powerful, peripheral, and dangerous.[119] Thus the only justifiable (answerable) historical question about magic is not "What are the characteristics of, for example, Greek magic?" but rather "Under what conditions, by whom, and of whom does the term 'magic' come to be used?" Thus our treatment of ancient *defixiones* does not locate them in the category of magic, for in our view no such category exists.[120] Instead, our approach has been more descriptive—of what they say and claim to do, of who made and used them, and of what various people thought of them. Of course, this is not to argue that some of the material in ancient spells, *defixiones*, and formularies—the sorts of things customarily labeled as magical—do not derive from "religious" (cultic) sources. There was a great deal of borrowing in every corner of late antiquity and the *magoi* were certainly no exception.[121]

The Pre-history and Post-history of Cursing

Defixiones represent but a small sample of the universal human practice of swearing during moments of anger, frustration, uncertainty, and intense competition. Like all forms of cursing, our own curious and distinctive swearwords and oaths represent highly formalized, sometimes even quite ancient, verbal formulas. Ancient Mediterranean cultures no doubt had their own forms of private, spur-of-the-moment swearing,[122] but they also developed more highly formalized, even ritualized expressions of their desire for protection, aid, and revenge against enemies real or imagined. The *defixiones* in this volume present these formalized expressions as they were deployed by private individuals against personal and sometimes public enemies. As such, they are to be distinguished from public forms of cursing, even though the techniques used are often the same. In the first century c.e., for instance, the citizens of Syedra (Asia Minor) consulted the oracle of Apollo at Claros when plagued by pirates. The god advised them to set up a statue of Ares, the traditional god of war, bound with chains of Hermes and supplicating the figure of Dike/Justice.[123]

Much earlier, the civilizations of the ancient Near East and Egypt display a wide variety of techniques for cursing and binding others and for protecting individuals against hostile spells. Like those in the period covered by the present collection, these spells were used both in public settings and by private parties. In some cases, the parallels with later Mediterranean *defixiones* seem more than coincidental. A Hittite ritual, presumably used by a professional, for counteracting spells uses the familiar formula, "Just as . . . so also," four times in its prescriptions ("Just as I have burned these threads and they will not come back, even so let also these words of the sorcerer be burned up!").[124] Various Babylonian rituals prescribe the use of figurines and amulets in the dialectic of spell and counterspell.[125] Equally striking are fragments of Egyptian pots and figurines from the nineteenth and eighteenth centuries b.c.e.; inscribed with the names of hated foreigners (among them the ruler of Jerusalem and his retainers), the pots were then smashed in order "to break the power of their enemies."[126] The practice of formalized cursing was also well established in ancient Israel, most notably in the blessings and curses which conclude the giving of the Mosaic covenant in Deuteronomy 28.[127] Curse formulas, with clear parallels in other Near Eastern cultures, appear in a number of biblical prophecies where they are directed against enemies of Israel, Israelites who violate the covenant, and criminals of various sorts.[128] An early student of Greek and Latin *defixiones* concludes from his examination of six biblical curses (Judges 17:1–2; Malachi 3:8–9; Zechariah 5:2–4; Jeremiah 51:60–64; Ezekiel 4:1–3; 2 Kings 13:17–19) that these texts "contain all the principal elements of *defixiones* . . . (that) the Hebrews were familiar with the written curse, of near kin to *defixio,* as early as the first part of the sixth century B.C."[129] All in all, there seems little reason to doubt the word of S. A. B. Mercer, given already in 1915, that

> the malediction in Babylonian and Assyrian times was a highly developed legal and religious ceremony, universally practiced and respected. It not only figured in ceremonies of great occasions, but also penetrated into the everyday life of the people. It seemed to have served almost the same purpose as Common Law does among modern people, for it acted as a restraint, corrective, and stimulant to better deeds.[130]

If ancient Near Eastern and Egyptian cultures provide the background for many aspects of later Mediterranean curse tablets and binding spells, we must not neglect the distinctive developments of this later period. The use of lead tablets; the appeal to Greek, Roman, Egyptian, and Jewish *daimones*, spirits, and deities; the proliferation of occasions deemed appropriate for the deployment of *defixiones;* their increasing popularity in the private as opposed to the public sphere; and the related emergence of official disapproval toward their use—these features separate later Mediterranean *defixiones* from their Near Eastern and Egyptian forerunners.

There can be no doubt that the use of *defixiones* in the private realm survived the rise of Christianity. Of this there is ample proof in surviving texts on papyrus[131]; in a late sermon that specifically mentions their use by Christians and argues that "whoever, during the time of the moon's increase, thinks that it is possible to avert (harm) through the use of inscribed lead tablets . . . they are not Christians but pagans"[132]; and in other literary texts.[133] But from the eighth century on, there is a curious absence of physical evidence for *defixiones*. Still, they did not disappear without a trace, not even in the Latin West. First, a little-known set of Latin prayers, known as *loricae* (from the Latin for "breastplate") and especially characteristic of Celtic Christianity, shows clear traces of being both used against and influenced by earlier *defixiones*.[134] Intended for the protection of individuals against harm of various kinds, these *loricae* sometimes enumerate at great length the parts of the body to be protected, in ways reminiscent of similar lists in curse tablets. Two passages from one of the *loricae* (the *Lorica of Laidcenn,* dating from the seventh century C.E.) may even be seen as containing explicit references to the use of *defixiones:* "Then be a most protective breastplate for my limbs and for my inwards, so that you drive back from me the invisible nails of the shafts that the foul fiends fashion" (lines 51–54); and "Protect my bladder, fat and all the innumerable rows of connecting parts; protect my hairs and the remaining members whose names I have perchance passed over" (lines 79–82). The mention of "nails" in the first passage reminds us of the use of nails in many *defixiones,* while the use of the "safety clause" in the second ("whose names I have perchance passed over") may derive from similar formulas in spells and charms used to cover names or items not specifically mentioned.

In addition, in a late Greek manuscript (Codex Parisinus Gr. 2419), edited by G. Midiates in the fifteenth century and no doubt copied from earlier sources, there appears the following recipe for an amulet against fevers: "take the stone and inscribe on it an erect headless spirit with its hands and feet bound behind; and around it write these signs and on the back the preceding names . . . wear it under a lead plate and you will be far (removed) from any such harm. . . ."[135] To be sure this is not a *defixio,* but it does contain unmistakable elements (signs, the headless spirit, mysterious names, bound hands and feet, lead plate) from much earlier curse tablets, binding spells, and amulets, which might suggest that such practices had remained alive after all.

Finally, there is a small body of *defixiones* from early modern England, written on lead as well as paper, which are striking in their similarity to our ancient tablets[136] (Figure 5). May we take these few examples as evidence for a continous tradition of *defixiones,* otherwise unattested from the eighth to the seventeenth centuries, or do they represent a "revival," inspired perhaps by the chance discovery of ancient tablets? Perhaps not, given the absence of physical evidence for the continued use of metal tablets in the Western and Eastern Middle Ages. Still, the practice of formalized cursing remained alive throughout this period.

To some extent, we can trace this continuity by noting a shift of media, from metal tablets to hand-written books, while, at the same time, curses on gravestones remained in constant use. Book curses, directed by scribes against those who might modify, deface, or destroy their precious writings may reach back as far as 3800 B.C.E.[137] The earliest clear example dates to the seventh century B.C.E. And the practice of attaching such curses to the colophons of medieval manuscripts of all sorts was widespread among both Jewish and Christian copyists. The New Testament Book of Revelation concludes with the following admonition: "I warn everyone who hears the words of the prophecy of this book: if any one adds to them, God will add to him the plagues described in this book, and if any one takes away from the words of the book of this prophecy, God will take away his share in the tree of life and in the holy city which are described in this book."[138]

In the end, it would be a mistake either to overemphasize the similarities between curses in books and our *defixiones* or to ignore their differ-

FIGURE 5. Seventeenth-century *defixio* with *charaktêres* from Wilton Place (England). This early modern tablet may well represent a conscious imitation of ancient *defixiones*. (City of Gloucester, City Museum & Art Gallery. By permission.)

ences. The fact remains that after a continuous history of more than twelve hundred years, physical evidence for their use disappears from the record. An explanation for this break may lie in a Christian sermon from the eighth century C.E. (no. 167): "Whoever ties around the neck of humans . . . any characters . . . on metal tablets made from bronze, iron, lead or any other material, such a person is not a Christian but a pagan." Here we may be confronted with a rare instance where ecclesiastical authorities actually succeeded in suppressing a "pagan" practice. For unlike many other practices, this one neither slipped through the antipagan filters nor found an acceptably Christian guise. It was simply *too* pagan.

Notes

1. In Greek, the standard term in modern discussion for an inscribed metal strip with spells directed against others was *katadesmos,* derived from the verb *katadein,* meaning "to bind up," or "to tie down." In Latin, *defixio* is quite rare, although forms of the verb *defigere* ("to fasten" or "to nail down") do appear. Latin tablets used other terms, e.g., *devotio, donatio,* or *commonitorium;* see the discussion in Tomlin, p. 59. Jordan notes in a private communication that *katadesmos* appears to derive from the idea of binding by rolling the metal sheet; when used of other media such as stones, the term must be understood in a broader sense. By contrast, *defixio,* when used in the narrow sense, seems to imply the act of "fixing" by driving a nail through the tablet. Of course, there are rolled tablets pierced by nails, although Jordan observes that no Athenian tablets of the Roman period make use of nails. "*Defixio,*" he concludes, "may well refer to a different kind of operation . . . from what *katadesmos* refers to." In this volume, we use *defixio* and *katadesmos*—sometimes "tablets"—in the generic sense to designate spells and curses inscribed on a variety of media.

2. See the most recent discussion by Tomlin, p. 59.

3. Preisendanz (1930), pp. 119–20, dates the first scholarly notice of *defixiones* to 1796, when Nicolo Ignarra published a few lines on tablets discovered in Italy in 1755 (*DT* 212). Jordan, in his review of Tomlin (*Journal of Roman Archaeology* 3 [1990]: 440) contends that *DT* 212 is not a *defixio* and that the first scholarly treatment dates to 1813, when J. D. Åkerblad published an Attic tablet, *Iscrizione greca sopra una lamina piombo trovata in un sepolcro nelle vicinanze de Atene* (Rome, 1913). C. T. Newton's publication of the tablets from the temple of Demeter on Cnidos (1862; see no. 89) and L. Macdonald's treatment of sixteen tablets from Cyprus (1890; see pp. 132–37) mark important turning points in the study of these neglected documents. R. Wünsch's collection of the Attic tablets in 1897 (DTA) and A. Audollent's subsequent collection of many more *defixiones* from the Mediterranean region in 1904 (*DT*) made it possible for the first time to examine all of the known evidence. The following general treatments should be noted: M. Jeanneret, "La langue des tablettes d'exécration latines," *Revue de philologie* 40 (1916): 225–58, and 41 (1917): 5–99; E. Kagarow, *Griechische Fluchtafeln* (Leopoli, 1929); M. Besnier, "Récents travaux sur les *defixionum tabellae* latines," *Revue de philologie* 44 (1920): 5–30; Solin, who produced a catalogue of Latin tablets published between 1920 and 1968; Preisendanz's articles of 1933 and 1972 (see abbreviations); D. R. Jordan, SGD, published in 1985; and Faraone, "Context," pp. 3–32.

4. Jordan, "Agora," p. 206.

5. A piece of Attic black-ware, dating from the late fifth or the early fourth century B.C.E., and published by M. Nilsson, *Geschichte der griechischen Religion*, vol. 1 (Munich, 1967), p. 801 (with photograph), bears the message, "I lay upon Aristiôn a deadly (*es aida* or until death) quartan fever." From a much later period ("late Roman") stem two binding spells on ostraca from Egypt, one (*PGM*, vol. 2, Ostraka 1) urging the god Kronos to restrain (*katechein*) a certain Hori from speaking against Hatros (see no. 111), while the other apparently seeks to bind (no verb is used) the tongue and power of Sitturas and Epikratês (*PGM*, vol. 2, Ostraka 5). A photograph of the spell against Hori appears in A. Deissmann, *Light from the Ancient East* (London, 1911), opposite p. 309. Finally, see pp. 92–94, for a love spell inscribed on a piece of pottery.

6. The spells from Tell Sandahannah (see pp. 203–5).

7. See Bonner, *Amulets,* pp. 103–22 ("Aggressive Magic"), where several engraved gemstones bear spells normally encountered on metal tablets.

8. *PGM* VIII, lines 1–63, bears the title of an erotic binding spell (*philtrokatadesmos*), which it attributes to Astrapsoukos, a known *magos* of Persia; such spells are quite common on lead tablets. In fact, the text appeals for worldly success of various sorts. See also *PGM* CIX, a love spell directed at a certain Kalemera, and a plea for justice and revenge, dating from the sixth century C.E., by a Christian man, directed against his daughter, Severina, and another man named Didymos (published with an extensive commentary by G. Björck, *Der Fluch des Christen Sabinus—Papyrus Upsaliensis 8* [Uppsala, 1938]).

9. DTA 55a (see pp. 158–59) states "I bind all of these people in lead and in wax," while Ovid (*Loves* 3.7.29) wonders whether he himself has been victimized (*defixit*) by a specialist who has written his name in red wax and pierced it with a needle (see no. 142). See the discussion in Faraone, "Context," p. 7.

10. See pp. 205–7; such bowls were normally used for protective, not aggressive purposes.

11. The recipes used in the later Roman empire in producing *defixiones* do not always specify lead. *PGM* X, lines 24–35, a spell for restraining anger, calls for a gold or silver strip; *PGM* IV, lines 2145ff., a multipurpose spell for obtaining oracles, wrecking chariots during races, and undoing other people's binding spells (!), calls for iron, whereas an eighth century Christian sermon against pagan practices condemns the use of metal tablets made from bronze, iron, or lead. By and large, gold and silver were reserved for protective amulets and for medical spells to cure various diseases and infirmities, although other materials could be used for the same purposes. On the use of amulets for healing and other purposes, see R. Kotansky, "Incantations and Prayers for Salvation on Inscribed Greek Amulets," in *Magika*, pp. 107–37. On gold and silver tablets in particular, see Kotansky's Ph.D. dissertation, "Texts and Studies in the Graeco-Egyptian Magic Lamellae: An Introduction, Corpus and Commentary on the Phylacteries and Amulets, Principally Engraved onto Gold and Silver Tablets" (University of Chicago, 1988).

12. See the full discussion in Tomlin, p. 824.

13. An earlier study by R. H. Brill and J. M. Wampler, "Isotope Studies of Ancient Lead," *AJA* 71 (1967): 63–77, analyzed an unpublished lead tablet from the Princeton excavations at Antioch (pp. 69 and 76, no. 135). The tablet itself contained .02 percent silver and the lead probably came from the famous mines at Laurion, south of Athens.

14. See *PGM* VII, line 397, and *Sepher ha-Razim,* p. 49, second firmament, line 68.

15. Tomlin, p. 83. In his review of Tomlin (pp. 439–40), Jordan doubts whether the tablets would have been heated before being inscribed. He notes that several recipes in *PGM* demand the use of cold lead, e.g., *PGM* XXXVI, line 2.

16. See Jordan, TILT, pp. 226–28, and Faraone, "Context," p. 4, for a discussion of lead as a medium for writing documents other than curses and spells.

17. Jordan, "Agora," p. 207.

18. See Tomlin, p. 82, n. 3.

19. DTA, nos. 105, 106, 107.

20. Tomlin, p. 81.

21. Such a range can appear even within a single find, as with the tablets from Bath, which include a few described as "calligraphic," a majority as "clerical," and a number as "clumsy." See Tomlin, p. 100.

22. See pp. 249–50.

23. See pp. 158–59.

24. Jordan, "Agora," p. 210. More recently, in a paper presented to the American Philological Association (1989, Baltimore), Jordan has argued that these texts were inscribed by the very same professional, civic scribes·(moonlighting after hours?) who worked in buildings close to the site (a well in front of the Middle Stoa) where the *defixiones* were discovered. For Jewish amulets and spells, Michael Swartz has reached similar conclusions. He argues that the so-called magical materials from the Cairo Geniza were copied out by the same hands that produced other, more "secular" or public documents for the Jewish community; see "Scribal Magic and its Rhetoric: Formal Patterns in Medieval Hebrew and Aramaic Incantation Texts from the Cairo Geniza," *HTR* 83 (1990):163–80.

25. So Tomlin, pp. 98–99.

26. Tomlin, p. 100.

27. Jeffery, pp. 72–75, gives a list of 25 *defixiones* which can be safely dated to the fifth century B.C.E.; Jordan, SGD, extends the list to 1985.

28. SGD, p. 162, mentions an early *defixio* (fifth or fourth century B.C.E.) from the Athenian Agora that has deliberately scrambled the names of the targets.

29. From his lost play, *The Harp Maker,* as cited by Athenaeus, *Learned Banquet* 548c (*CAF,* frag. 18); slightly later, the comic poet Menander also mentions them, adding that they were used with newlyweds to ward off spells (*CAF,* Menander frag. 371 = *Suda,* s.v. *alexipharmaka*). For general discussions see C. C. McCown, "The Ephesia Grammata in Popular Belief," *TAPA* 54

(1923): 128–40; Jeffery, pp. 75–76; K. Preisendanz, "Ephesia Grammata," *RAC* V (1962), cols. 515–20; and Kotansky, "Incantations," pp. 111–12.

30. A similar phenomenon, dating from approximately the same time, appears in Latin texts, first mentioned by Cato in his *On Agriculture* 160 (a spell for dislocated bones): "Take a green reed four or five feet long and split it down the middle, and let two men hold it to your hips. Begin to chant: 'MOTAS UAETA DARIES DARDARES ASTATARIES DISSUNAPITER' and continue until they meet"; see the discussion in A.-M. Tupet, "Rites magiques dans l'antiquité romaine," *ANRW* II.16.3 (1986), pp. 2596–98.

31. Clement, *Stromata* 8.45.2.

32. The *voces mysticae* appear not just in spells but also in the literature of certain Christian Gnostic groups, notably the so-called Sethians; see H. M. Jackson, "The Origin in Ancient Incantatory *Voces Magicae* of Some Names in the Sethian Gnostic System," *Vigiliae Christianae* 43 (1989): 69–79.

33. See discussions in Wünsch (1900), pp. 73–85; McCown, "Ephesia Grammata," pp. 132–36 (with full translation and much reconstruction); and Kotansky, "Incantations," pp. 111–12. For full publication, see M. Guarducci, ed., *Inscriptiones Creticae,* vol. 2 (Rome, 1939), p. 19, no. 7. Some of the same hexameters and *ephesia grammata* of the Cretan tablet occur on an unpublished lead tablet from Selinus (fourth century B.C.E.), now in the Getty Museum; see D. R. Jordan, "A Love Charm with Verses," *ZPE* 72 (1988): 256–58, and Kotansky, "Incantations," p. 127, n. 27 (with the enticing suggestion that "the text cites words from the hitherto 'meaningless' *Ephesia grammata,* which can now be understood as perfectly good Greek hexameters").

34. "Perhaps in this case, the suppliant, whose *defixio* had been entirely successful, offered the stone in gratitude; on it he had a representation made of the leaden scroll bearing the original curse, and below came the details of the fulfillment" (Jeffery, p. 75).

35. The monument is *IG* 4.496, found near modern Phychtia, a town near the site of ancient Mycenae; see the discussion in Jeffery, pp. 69–76.

36. On Egyptian "magic," a category that seems indistinguishable from what might elsewhere be called "religion," see J. F. Borghouts, *Ancient Egyptian Magical Texts* (Leiden, 1978), and C. Jacq, *Egyptian Magic* (Chicago, 1985). There are very few aspects of Greco-Roman "magic" in Egypt that are not anticipated in earlier Egyptian tradition. It is also true that we are much better informed about early Egyptian beliefs and practices; if early Greek or Etruscan handbooks had survived, they might have yielded a similar picture.

37. Some examples: *PGM* VII, lines 451–52 (in a recipe for an all-purpose *defixio*): "Write the Orphic formula, saying '*askei kai taskei*' . . ."; *PGM* LXX, line 12 (a spell to protect against punishments in the underworld): ASKEI KATASKEI ERÔN . . . ; the term *damnameneus* appears frequently (*PGM* II, lines 163–64; III, line 511, etc.).

38. See the intelligent discussion of F. Graf, "Prayer in Magic and Religious Ritual," in *Magika,* pp. 188–97.

39. The same point is made persuasively by Kotansky, "Incantations," pp. 109–10, and Faraone, "Context," pp. 5–6. The early separation of oral and written aspects of curses might explain the unusual cache of forty blank tablets (*DT* 109), each rolled up and pierced with a nail, found at Rom in France (Deux-Sèvres). If these represent real *defixiones,* we can imagine that they were deposited with no written inscription and only a series of oral spells, which in this case would have included the name to the target as well; see the cautious remarks of Faraone, "Context," p. 24, n. 19.

40. On the use of letters, especially vowels, in spells see the still essential work of F. Dornseiff, *Das Alphabet in Mystik und Magie* (Leipzig, 1925). A text from the rhetorical writer, Demetrius (*On Style,* chap. 71; ca. first century C.E.), should be noted in connection with the frequent appearance of vowel series: "In Egypt the priests, when singing hymns in praise of the gods, employ the seven vowels, which they utter in due succession (i.e., in order); and the sound of these letters is so euphonious that men listen to it in place of flute and lyre." The liturgical use of vowels was also practiced by the Valentinian Christian, Marcos, whose congregants lined up and recited the vowels in praise and evocation of the father of all (Irenaeus, *Adversus Haereses* I.14.1ff.). The seven Greek vowels were also associated with planets, angels, and sounds and were eventually invoked as powers in their own right. Their number (seven) also carried mythological and cosmological significance. In Egypt, the use of vowels, whose invention was attributed to Thoth (cf. Plato, *Philebus* 18B–C), allowed the precise pronunciation of ritual texts. See C.-É. Ruelle, "Le chant des sept voyelles grecques," *Revue des études grecques* 2 (1889): 38–44, 393–95; Dornseiff, *Das Alphabet,* pp. 35–60; Patricia Cox Miller, "In Praise of Nonsense," in *Classical Mediterranean Spirituality,* ed. A. H. Armstrong (New York, 1986), pp. 481–505.

41. The fact that Lucian, in his *Dialogue of Courtesans* 4.6, uses the same term probably indicates that the phrase must have had a technical meaning.

42. *On the Mysteries of Egypt* 3.14 (on *charaktêres*) and 7.5 (on foreign names). That Iamblichus's theory of the power of foreign names was not limited to theurgical circles is indicated by a comment of the Christian philosopher Origen (ca. 250 C.E.), certainly no friend of "magic," who observes in his *Against Celsus* that "a man who pronounces a spell in its native language can bring about the effect that the spell is claimed to do. But if the same spell is translated into any other language whatever, it can be seen to be weak and ineffective" (1.25).

43. See in particular the work of J. L. Austin, *How To Do Things with Words* (Cambridge, 1962), and Susan Stewart, *Nonsense: Aspects of Intertextuality in Folklore and Literature* (Baltimore, 1979).

44. Stanley J. Tambiah, "The Magical Power of Words," *Man* 3 (1968): 177–206.

45. "Magical Power," p. 179.

46. Miller, "In Praise of Nonsense," pp. 481–505; and R. T. Wallis, "The Spiritual Importance of Not Knowing," in *Classical Mediterranean Spirituality,* ed. A. H. Armstrong (New York, 1986), pp. 460–80.

47. *Stromata* 1.143.1.

48. See examples in Delatte and Derchain, pp. 360–61.

49. See nos. 84 and 115.

50. See R. Wünsch, *Antikes Zaubergerät aus Pergamon* (Berlin, 1905). A similar apparatus has been discovered at Apamea in Syria; see G. Donnay, "Instrument divinatoire d'époque romaine," in *Fouilles d'Apamée de Syrie,* ed. Janine Balty (Brussels, 1984), pp. 203–210, and plates.

51. E.g., *PGM* VII, lines 396–404; VII, lines 795–845; the Hebrew *Sepher ha-Razim* used *charaktêres* freely; for Arabic examples, see D. Pingree, ed., *Picatrix: The Latin Version of the Ghayat al-Hakim* (London, 1986), pls. 1 and 2.

52. See C. Schmidt, ed., *The First Book of Jeu* (Berlin, 1954), chaps. 33–52 (pp. 290–329).

53. See the publication by H. Grégoire, *Receuil des inscriptions grecques chrétiennes d'Asie Mineure* (Paris, 1922), no. 221. A photograph and discussion are given in Deissmann, *Light from the Ancient East,* pp. 448–55.

54. See the discussion by W. van Rengen, "Deux défixions contre les bleus à Apamée (VIᵉ siècle apr. J.-C.)," *Apamée de Syrie* (1984), pp. 213–34.

55. See pp. 168f.–203.

56. A. Grabar, *Christian Iconography. A Study of Its Origins* (Princeton, 1968), p. 64.

57. Grabar, *Iconography,* p. 21.

58. One notable exception to this rule is C. Bonner's study of figures on gem-amulets in *Amulets.* Two proposed studies now promise to fill this void. P. Corby Finney has undertaken the task of producing an iconographic catalogue of all ancient magical documents, and Richard L. Gordon is completing a study that will shed new light on all aspects of Greco-Roman magic, including images and designs.

59. H. Versnel, "Beyond Cursing," in *Magika,* p. 64.

60. *DT* 7a, line 12; the reading is uncertain.

61. For lists of gods addressed, see Kagarow, *Fluchtafeln,* pp. 67–75, and the indices to DTA, p. 47, and Audollent, *DT,* pp. 461–64.

62. See the excellent discussion of legal terms in the Bath tablets; Tomlin, pp. 70–71.

63. Again, see Tomlin, p. 70.

64. See Faraone, "Context," p. 4.

65. In general, see Kagarow, *Fluchtafeln,* pp. 28–49; Tomlin, pp. 63–74; and Faraone, "Context," pp. 4–10.

66. Kagarow, *Fluchtafeln,* pp. 28–34, shows the many possible combinations of the basic forms.

67. The notion of performative utterances and persuasive analogies has been developed most recently by Stanley J. Tambiah in his essay, "Form and Meaning of Magical Acts: A Point of View," in *Modes of Thought: Essays on Thinking in Western and Non-Western Societies,* ed. R. Horton and Ruth Finnegan (London, 1973), pp. 199–229.

68. See the extensive treatment in Kagarow, *Fluchtafeln*, pp. 34–44.

69. So, for example, an unpublished tablet discussed by Jordan, "Agora," p. 251, where the letters have been squeezed together.

70. For a full discussion see D. R. Jordan, "CIL VIII 19525 (B).2QPVULVA = Q(UEM) P(EPERIT) VULVA," *Philologus* 120 (1976): 127–32.

71. Babylonian Talmud, Shabbat 66b.

72. On this issue, see the brief discussions in M. Guarducci, *Epigrafia greca IV: Epigrafi sacre pagane e cristiane* (Rome, 1978) p. 245, n.1, and Alan Cameron, *Porphyrius The Charioteer* (Oxford, 1973), pp. 157–58.

73. See the forthcoming essay of C. Faraone, "Binding and Burying the Forces of Evil: The Defensive Use of 'Voodoo Dolls' in Ancient Greece," *Classical Antiquity* 10 (1991); in an appendix to his essay, Faraone includes a survey of the Greek, Etruscan, and Roman dolls.

74. Theocritus (ca. 275 B.C.E.), in *Idyll* 2.28, records the feverish efforts of Simaetha to win back her lover, Delphis. Her efforts amount to an elaborate ritual, consisting of many separate actions and invocations. At one point she exclaims, "As I melt, with the goddess's help, this wax (probably a figurine representing the faithless Delphis), so may Delphis of Myndus waste away at once from love." Among other things, this text warns us not to read such actions literally, for if we were to find just the melted wax we might well suppose that its intent was to destroy Delphis!

75. See the discussion in Kropp, vol. 3, pp. 114–15.

76. See the discussion in Faraone, "Context," pp. 7–9.

77. In his survey published in 1915, C. Dugas ("Figurines d'envoûtement trouvées à Délos," *BCH* 39 [1915]: 412–23) was able to list thirty-five examples. Preisendanz (1933), pp. 163–64, and Jordan, SGD, give more recent surveys.

78. R. Wünsch, "Eine antike Rachpuppe," *Philologus* 71 (1902): 26–31.

79. See B. Nogara, "Due statuette etrusche di piombo trovate recentemente a Sovana," *Ausonia* 4 (1909): 31–39; and L. Mariani, "Osservazioni intorno alle statuette plumbee sovanesi," ibid., pp. 39–47 (with several additional examples).

80. A. S. Hunt, "An Incantation in the Ashmolean Museum," *Journal of Egyptian Archaeology* 15 (1929): 155–57 and plate XXXI.1; the two pieces were probably found in the cemetery at Hawara and date from the second or third century C.E.

81. Here it may be useful to mention an object from Zaire (no. 347) in the Royal Ontario Museum in Toronto. The figure represents a human body and is penetrated by numerous nails and needles. The accompanying description states that the purpose of the needles was "to arouse the figure to action or to mark events such as concluding a treaty." In short, we need to avoid reading such objects with an overly literalistic eye. Much the same point has now been made by Faraone, "Binding and Burying."

82. See the discussion in D. R. Jordan, "New Archaeological Evidence for the Practice of Magic in Classical Athens," in *Praktika tou XII diethnous synedriou klasikês archaiologias* (Athens, 1988), pp. 273–77.

83. See H. Seyrig, "Notes archéologiques," *Berytus* 2 (1935): 48.

84. See the discussion in Jordan, "Agora," pp. 251–52.

85. Jordan, SGD, p. 182, reports a tablet from Rome with six nails. Curiously, the many recipes in *PGM* for producing *defixiones* fail to mention the practice of piercing the rolled tablet with a nail.

86. See the discussion in Kagarow, *Fluchtafeln*, pp. 10–16, and Preisendanz (1933), pp. 162–63.

87. DTA 107 and 109; SGD 62; see the discussion in Faraone, "Context," p. 4.

88. Faraone, "Context," p. 17, may be correct in suggesting that the placement of tablets in deep wells and graves fulfilled the additional purpose of making it impossible for the target to uncover them and take appropriate counteraction to undo the spell.

89. See the discussion in Faraone, "Binding and Burying," n. 101.

90. On wells, see Jordan, TILT, p. 232, n. 24, and "Agora," pp. 207–10.

91. Modern excavations have found several *defixiones* within ancient stadiums, near the starting gates, where they must have been deposited for maximum effect; see Faraone, "Context," p. 23, n. 9.

92. See Jordan, TILT, pp. 231–32, n. 23. A set of fourteen tablets, directed against women, has been excavated in the temple of Demeter on the lower slopes of Akrocorinth (SGD, p. 166). The translation of one has been published by N. Bookidis and R. S. Stroud, *Demeter and Persephone in Ancient Corinth* (Princeton, 1987), pp. 30–31 (plate 32). It reads as follows: "I consign and entrust Karpime (corrected by Stroud) Babia, the weaver of garlands, to the Fates who exact justice (*praxidikai*) so that they may expose her acts of insolence (*hubreis*), and to Hermes of the Underworld, to Earth, to the children of Earth, so that they may overcome and completely destroy her . . . and her heart and her mind and the wits of Karpime Babia, the weaver of garlands. I adjure (*enarômai*) and I implore you and I beg of you, Hermes of the Underworld, (to grant) heavy curses." We are indebted to Bookidis and Stroud for providing us with a Greek transcription of this tablet.

93. Jordan, "New Archaeological Evidence," p. 273, remarks that "in every period of antiquity when we have been able to estimate the ages of the dead who have curse tablets placed in their graves . . . , those ages have been proved to be young."

94. See the full discussion in Versnel, "Beyond Cursing."

95. As in two instances described by Jordan, "New Archaeological Evidence," pp. 273–74.

96. Much the same point is made by Kotansky, "Incantations," pp. 108–9.

97. A similar point is made by Tomlin, p. 102. Speaking of the "climate of belief" regarding connections between guilt, psychosomatic illness, and divine punishment, he concludes that "[s]omeone who stole a cloak from the baths, if he had any imagination, might suspect that he had been cursed by his victim."

98. Kagarow, *Fluchtafeln*, pp. 55–58. Faraone, "Context," p. 8, n. 38, has

argued that spells that call for the death of the target are relatively rare, though not altogether lacking, in tablets before the fourth century B.C.E.

99. "Magical Power," p. 202.

100. *Purity and Danger* (New York, 1966), p. 58.

101. Tomlin, p. 101.

102. Here quoting M. Henig, *Religion in Roman Britain* (London, 1984), p. 145. The full statement reads as follows: "Thus the temples of Britain helped to resolve conflicts, to punish transgressions which would sometimes be hard to bring to court without risking a breakdown in social order."

103. Here quoting G. Webster, *The British Celts and Their Gods under Rome* (London, 1986), p. 136.

104. Ibid., pp. 101–2. A more elaborate and sophisticated version of the idea that "doing something" was better than doing nothing was developed by the late John Winkler in connection with love charms.

105. In his collected essays, C. Lévi-Strauss, *Structural Anthropology* (New York, 1976), pp. 167–85.

106. Geza Roheim, "The Origins and Function of Magic," in *Magic and Schizophrenia* (Bloomington, 1955), p. 3.

107. Ibid., p. 85.

108. *Witchcraft, Oracles, and Magic among the Azande* (Oxford, 1976), pp. 18–55 ("The Notion of Witchcraft Explains Unfortunate Events" and "Sufferers from Misfortune Seek for Witches among Their Enemies"). As C. Faraone puts it, appeals to curses commissioned by enemies and rivals served to provide "professional performers with an easy opportunity for face-saving in the event of a radically poor performance"; see his "An Accusation of Magic in Classical Athens (Ar. *Wasps* 946–48)," *TAPA* 119 (1989): 154. Much the same point is made by Peter Brown in his essay, "Sorcery, Demons, and the Rise of Christianity," in *Religion and Society in the Age of Saint Augustine* (London, 1972), esp. pp. 24–33. Both make use of Evans-Pritchard.

109. Faraone, "Context," p. 20, argues that the use of *defixiones* in classical Greek society was in fact not illegal; instead "they fit easily into the popular competitive strategy of survival and dominance . . . within the rules of the game for intramural competition." A similar view is proposed, for *mageia* in general, by T. Hopfner, "Mageia," *RE* 14 (1928), cols. 384–85.

110. So C. R. Phillips III, "*Nullum Crimen sine Lege:* Socioreligious Sanctions on Magic," in *Magika*, p. 264 and n. 10.

111. W. Dittenberger, ed., *Sylloge Inscriptionum Graecorum*, 3d ed., vol. 1 (Leipzig, 1915), no. 37; the law speaks of *pharmaka dêlêteria.*

112. On the general subject, see A. A. Barb, "The Survival of Magic Arts," in *The Conflict between Paganism and Christianity in the Fourth Century,* ed. A. Momigliano (Oxford, 1963), pp. 100–25, and Phillips, "*Nullum Crimen,*" pp. 262–78.

113. Later Roman emperors wavered on the issue of whether all forms of *magia* were illegal and punishable by death or whether the law applied only to so-called harmful types; see Barb, "Survival," pp. 102–3.

114. O. Petersson, "Magic—Religion. Some Marginal Notes to an Old Problem," *Ethnos* 3–4 (1957): 119. To Petersson's observations it might be added that while the pattern described by him (i.e., the beliefs and practices of "others" are labeled "magical") certainly precedes the birth of the Christian movement, it attained its canonical status in and around the first centuries of the movement. Recent years have seen a spate of useful literature on the topic: Robert-Léon Wagner, "*Sorcier*" *et* "*magicien*." *Contribution à l'histoire du vocabulaire de la magie* (Paris, 1939); M. Smith, *Jesus the Magician* (New York, 1978), esp. p. 80: "the lines between *goes,* magus, and divine man shifted according to the sympathies of the speaker"; J. Z. Smith, "Towards Interpreting Demonic Powers in Hellenistic and Roman Antiquity," *ANRW* 16.1 (1978), pp. 423–39; idem, "Good News Is No News," in *Map Is Not Territory* (Leiden, 1978), pp. 190–207; D. Harmening, *Superstitio. Ueberlieferungs- und theoriegeschichtliche Untersuchungen zur kirchlich-theologischen Aberglaubensliteratur des Mittelalters* (Berlin, 1979); D. Aune, "Magic in Early Christianity," *ANRW* 22.2 (1980), pp. 1507–57; Anitra Kolenkow, "Relationships between Miracle and Prophecy in the Greco-Roman World and Early Christianity," ibid., pp. 1482–91; G. Poupon, "L'accusation de magie dans les actes apocryphes," in *Les actes apocryphes des apôtres,* ed. F. Bovon (Geneva, 1981), pp. 71–85; E. Gallagher, *Divine Man or Magician? Celsus and Origen on Jesus* (Chico, Calif., 1982); H. Remus, *Pagan-Christian Conflict over Miracle in the Second Century* (Philadelphia, 1983); and A. Segal, "Hellenistic Magic: Some Questions of Definition," in *Other Judaisms of Late Antiquity* (Atlanta, 1987), pp. 79–108.

115. See Dorothy Hammond, "Magic: A Problem in Semantics," *American Anthropologist* 72 (1970): 9.

116. *The Golden Bough,* 1-vol. ed. (1922; New York, 1967); see, e.g., p. 824: "the movement of higher thought, so far as we can trace it, has on the whole been from magic through religion to science."

117. This widely held position is perhaps best illustrated by Barb in his essay "The Survival of Magic Arts," p. 101: "on the contrary magic derives from religion, which, as it becomes tainted by human frailty, deteriorates into so-called white magic."

118. Perhaps best illustrated by K. Thomas, *Religion and the Decline of Magic* (New York, 1971); see the critical review of Thomas by Hildred Geertz, "An Anthropology of Religion and Magic," *Journal of Interdisciplinary History* 6 (1975): 71–89.

119. It may be noted here that there are cases in the period where the terms *mageia/magia* and *magos/magus* are used positively or, at the very least, neutrally; e.g., *PGM* IV, line 2453: "Pachrates, the prophet of Heliopolis, revealed it (a ritual) to the emperor Hadrian, revealing the power of his own divine magic"; and Apuleius, in his *Apology* (chap. 26), who defends himself against the criminal charge of practicing magic by replying that for the ancient Persians magic "is an art acceptable to the gods, full of all knowledge of worship and of prayer, full of piety and wisdom in things divine, full of honour and glory since the day of Zoroaster." These and other passages deserve further attention.

120. In line with this procedure are the comments on *defixiones* and magic by Faraone, "Context," pp. 17–20; Kotansky, "Incantations," pp. 119–122; and Phillips, "*Nullum Crimen.*"

121. See the useful discussion by Betz, in *GMP*, pp. xliv–xlviii. We may simply call attention to the remarkable autobiographical tale of Thessalus, a young aristocrat, whose spiritual and geographical migrations culminate in a visionary confrontation with the god Asclepius, arranged for Thessalus by a priest in the Egyptian city of Thebes. Whether or not the story is accurate in all its details, it offers a possible channel though which spells and rituals could have moved from traditional cult sites in Greco-Roman Egypt into formularies like those published in *PGM/GMP*. On the Thessalus tale, see J. Z. Smith, "The Temple and the Magician," in *Map Is Not Territory* (Leiden, 1978), pp. 172–89; on the role of Greco-Egyptian priests in general, see P. W. van der Horst, ed., *Chaeremon: Egyptian Priest and Stoic Philosopher* (Leiden, 1984).

122. Faraone, "Context," pp. 17–18, cites several examples of purely oral curses from early Greek literature; we may suppose that these admittedly literary texts reflect common usage.

123. See the discussion in Faraone, "Context," p. 9, along with the literature cited there.

124. *Ancient Near Eastern Texts,* ed. J. B. Pritchard (Princeton, 1955), p. 347; the translation is by A. Goetze.

125. See especially Erica Reiner, "Magic Figurines, Amulets, and Talismans," in *Monsters and Demons in the Ancient and Medieval Worlds,* ed. A. E. Farkus, P. O. Harper, and E. B. Harrison (Mainz, 1988), pp. 27–36. Beyond mere parallels between Babylonian and later Mediterranean materials, Reiner is interested in the possibilities of direct influence. See also J. S. Cooper, *The Curse of Agade* (Baltimore, 1983); J. van Dijk, A. Goetze, and M. I. Hussey, *Early Mesopotamian Incantations and Rituals* (New Haven, 1985) and several works of T. Abusch, most notably his *Babylonian Witchcraft: Case Studies* (Atlanta, 1987).

126. *Ancient Near Eastern Texts,* p. 328; the translation is that of John A. Wilson.

127. Among the copious literature on the subject, see F. C. Fensham, "Malediction and Benediction in Ancient Near Eastern Vassal-Treaties and the Old Testament," *ZAW* 74 (1962): 1–9.

128. See L. Fensham, "Common Trends in Curses of the Near Eastern Treaties and *kudurru*-Inscriptions Compared with Maledictions of Amos and Isaiah," *ZAW* 75 (1963): 155–75.

129. W. Sherwood Fox, "Old Testament Parallels to *Tabellae Defixionum,*" *American Journal of Semitic Languages* 30 (1913–14): 111–24.

130. "The Malediction in Cuneiform Inscriptions," *Journal of the American Oriental Society* 34 (1915): 309.

131. See the discussion by Versnel, "Beyond Cursing."

132. See pp. 263f.

133. On the persistence of pre-Christian practices, including the language of binding and loosing, in Christian Europe, see now Valerie I. J. Flint, *The Rise of Magic in Early Medieval Europe* (Princeton, 1991).

134. The view was first advanced by W. M. Lindsay, *Early Welsh Script* (Oxford, 1912), p. 23; more recently it has been proposed by M. W. Herren's study of the *loricae, The Hesperica Famina: II. Related Poems* (Toronto, 1987), pp. 26–31. I am most grateful to my friend and colleague, Michel Strickmann, for drawing my attention to this material.

135. The text has been published by A. Delatte, *Anecdota Atheniensia* (Paris, 1927), pp. 489–90.

136. See the discussion in R. Merrifield, *The Archaeology of Ritual and Magic* (London, 1987), pp. 147–58.

137. On the subject of book curses, see M. Drogin, *Anathema: Medieval Scribes and the History of Book Curses* (Totowa, N.J., 1983), who provides numerous examples.

138. Revelation 22:18–19. On similar curses in ancient Jewish and Christian writings, see R. H. Charles, *The Revelation of St. John,* vol. 2 (Edinburgh, 1920), pp. 222–24.

1

Competition in Theater and Circus

> The city's two strongest rivals are the Goose . . . and the Tower. Of dozens of stories illustrating the depth of the hatred between Goose and Tower, perhaps the best known is about an old priest from Goose who was compelled to miss a Palio in order to conduct a funeral service for one of his parishioners. . . . As the deceased was being interred, the priest walked over to the coffin, where the dead man's relatives were standing. The family assumed that he was about to whisper some words of comfort to them. Instead, as the coffin was lowered, the priest addressed the deceased: "Holy soul, who is certainly going to Heaven, tell the Lord not to let Tower win!"[1]

In the major cities of the ancient Mediterranean world, much of life unfolded in public settings—theaters, amphitheaters, hippodromes, odeums, stadiums, and circuses.[2] Whereas large installations like stadiums and circuses tended to be limited to cult centers (Greece) and large cities (Rome), theaters and odeums (small covered lecture halls) were much more common.[3] Depending on the size of the building, crowds could vary considerably: several hundred in small theaters; several thousand in larger theaters, such as the one at Pompeii; perhaps 50,000 in the Roman Colosseum and the stadium of Herodes Atticus at Athens; and as many as 250,000 (almost one-quarter of the city's population) for chariot races in the Circus Maximus at Rome.

In Greece, from classical to Roman times, games were celebrated not so much in the cities as at major cult sites (Olympia, Delphi, Isthmia/ Corinth, Nemea, Epidaurus), although the later Panathenaic games did take place in Athens. Such games were great festivals for all Greeks, and they occurred only once every two or four years—in short, they were relatively infrequent events and heavy with political, national, and religious trappings. Originally created for wealthy aristocrat-citizens who competed as amateurs, the games took on an increasingly professional character so that by the second century B.C.E. the athletic competitors were all professionals; the theatrical events were probably "professional" from the beginning.[4] A list of prizes from Athens in the fifth century B.C.E. reveals the comprehensive character of these games—rhapsodists reciting Homer, singers with harp or flute, instrumentalists on harp or flute, athletes in various events (foot races, pentathlon, wrestling, boxing, and the *pankration,* a free-for-all that combined elements of boxing and wrestling) for men and boys,[5] equestrian events such as four- and two-horse chariot racing, various team competitions, and a regatta for ships.[6] Complementing this cycle of biennial and quadrennial competitions were regular dramatic festivals, such as the annual festival of Dionysus at Athens at which dramatists competed for the right to stage their plays,[7] and other public events at which poetry was recited and songs chanted. All of these occasions involved competition of one kind or another, usually among choruses of singers, dancers, reciters, and their leaders. As we shall see, the competitive nature of these occasions, which placed both employment and status on the line, prompted the use of *defixiones* in order to hinder one's opponents and to enhance one's own chances of success. No less a figure than Augustine of Hippo, Christian bishop and former professor of rhetoric, illustrates the long history of the ties between such competitions and the use of curse spells. In his *Confessions* (4.2), written eight hundred years after the earliest Greek tablets, he relates that "once, when I had decided to enter a competition for reciting theatrical verse, a sorcerer (*haruspex*) sent to inquire of me how much I would pay him to guarantee a victory."[8]

In the period that concerns us here, the second century C.E. onward, Roman hegemony in the Mediterranean meant that distinctively Roman forms of entertainment, especially chariot racing and gladiatorial contests, could be found everywhere, west and east. Unlike their Athenian counterparts, Romans enjoyed a calendar literally bursting with various forms of competition—festivals (*feriae*), games (*ludi*), and shows (*munera*) of every kind and combination.[9] By the year 300 C.E. there were 177 days of games, including 66 days of racing in the circus. To complete this list of public celebrations, we must finally take account of the many lavish and lusty shows and gladiatorial contests, involving both men and wild animals, that occurred on a nonregular basis several times a year, sometimes as part of the games and sometimes separately.[10]

What Greek and Roman performances shared was a keen sense of competition, copious rewards, and enormous popularity. The number of actual competitors may have been limited, but the tally of those interested in the events was enormous, encompassing the full spectrum of the social order from emperor to slave. For the winning performers the tangible rewards were fame and fortune, while for the spectator-participant there was the suspense over the results, exultation at the competition, and, depending on the outcome, delight in victory or despair at defeat.

For all involved, much depended on the outcome, and competitors and fans sought advantages wherever they could find them.[11] Among these advantages was the use of *defixiones,* here understood quite literally as an effort to bind one's competitors—their limbs, their sinews, their courage—through spells addressed to gods, spirits, and *daimones.* Curse tablets of this sort played a regular and persistent role in the life of the circus. The number of surviving examples is considerable; their findspots stretch right across the Roman world, including several excavated in and around hippodromes where they were originally deposited in accordance with prescribed procedures.[12] Literary testimonies of various kinds support the picture sketched by the tablets themselves.[13]

As for *defixiones* related to athletic competition among humans, all of the surviving tablets are relatively late and concern a variety of events: five from Athens in the third century C.E. are directed against professional wrestlers in specific matches against named opponents (no. 3)[14]; one from Oxyrhynchus in Egypt a century later names a runner (no. 8).[15] But by far the greatest number of "athletic" tablets, whose total now exceeds eighty, concern chariot racing in circuses and hippodromes of the Roman world.[16]

It would be difficult to overestimate the cultural significance of chariot racing in the Greco-Roman world. The earliest literary account appears in Homer's *Iliad* (23.262ff.) in the lengthy account of the funeral games in honor of the warrior-friend of Achilles, where it was the first event. More than a millennium and a half later, Byzantine civilization could still be described, in the words of Norman Baynes, as honoring two heroes—the Christian holy man and the triumphant charioteer.[17] Certainly Roman authors, even when they professed dislike for this facet of Roman life, recognized and described it in telling detail. Late in the first century C.E., Tacitus speaks disparagingly of "the peculiar vices found in our city, which seem to be conceived already in the mother's womb—a partiality for the stage and a passion for gladiators and horse racing."[18] Some three hundred years later, Ammianus Marcellinus mockingly depicts race fans who argue that the state itself will fall unless their favored team is first from the starting gates and negotiates the turns in proper fashion. For such, he writes, "their temple, their dwelling, their assembly, and the height of all their hopes is the Circus Maximus."[19] For winning drivers, the gain was enormous fortune—Juvenal complains that the Red driver, Lacerta, earned one hundred times the fee of a lawyer.[20] Betting was omnipresent and riots not infrequent. Emperors regularly proclaimed their loyalty to favorite drivers, while powerful factions (Reds, Whites, Blues, Greens) organized the financial, technical, and professional side of the sport and spread eventually to every major city of the empire.[21] In the sixth century C.E., the historian Procopius could write that "in every city the population has been divided for a long time into the Blue and Green factions."[22] Vivid accounts of races survive in a youthful poem of Sidonius, later a Christian bishop in Gaul in the 470s, and in the epic poem entitled *Dionysiaca* by Nonnus, a writer of the fifth or sixth century C.E. Such accounts make it easy to understand the remarkable persistence of chariot racing into the Christian empire despite the resistance of figures like John of Ephesus (sixth century C.E.) who responded to the patriarch of Antioch's plan to build a hippodrome by labeling it "the church of Satan."[23] Indeed, it is a Christian writer, Cassiodorus, in the sixth century C.E., who provides an elaborate interpretation of the circus and its races as an astrological and astronomical symbol of the entire universe—the twenty-four races each day are the twenty-four hours; the seven laps of each race are the days of the week; the twelve portals at the entrance are the signs of the zodiac; the turning posts are the tropics and so on.[24]

For these reasons and more, curse tablets played a potent and abiding role in the world of chariot racing.[25] Whether commissioned and deposited by supporters[26] or by the drivers themselves,[27] they were a regular feature of competition, with factions, fans, and charioteers seeking advantage not just by tricks and skills on the course but by hampering the performance of man and beast by *defixiones*.

There is good reason to assume that everyone believed them to be effective, even when they disapproved. How else to understand a practice attested over such a stretch of time and geography? Certainly the legal evidence points in this direction. For in the fourth century C.E., a high point in the construction of Roman circuses,[28] Roman emperors began to issue legal decrees specifically aimed at the notorious and well-attested connection between charioteers and the use of curses.[29] A remarkable passage in Cassiodorus's edition of important documents written by him during his years as secretary to Theodoric illustrates several facets of this connection:

King Theodoric to Faustus, Praetorian Prefect (of Rome): Since constancy in actors is not a very common virtue, therefore with all the more pleasure do we record the faithful allegiance of Thomas the Charioteer, who came here long ago from the East, and who, having become champion charioteer, has chosen to attach himself to the seat of our Empire; we therefore decide that he shall be rewarded by a monthly allowance. He embraced what was then a losing side in the chariot races and carried it to victory— victory that he won so often that envious rivals declared that he conquered by means of witchcraft. For they were driven to attribute his victories to magic when they could not account for them by the strength of his horses.[30]

"The circus was indeed," in Alan Cameron's words, "a microcosm of the Roman state."[31] But it does not require "a mystical turn of mind" to see this, nor is the symbolism limited to the visible realm of social and political life displayed explicitly in the emperor, senators, and so on "down" to slaves and children. The inconspicuous lead tablet, inscribed, folded, and buried in the dust beneath the starting gates, symbolized the invisible world of Rome—a world of gods, spirits, and *daimones* on the one side, of aspirations, tensions, and implicit power on the other—in short, a world where emperors, senators, and bishops were not in command.[32]

Notes

1. L. Harris, "Annals of Intrigue: The Palio," *New Yorker,* June 5, 1989, p. 86.

2. On Greek and Roman public performance, see E. N. Gardiner, *Athletics of the Ancient World* (Chicago, 1955); J. P. V. D. Balsdon, *Life and Leisure in Ancient Rome* (New York, 1969), esp. pp. 244–339; H. A. Harris, *Sport in Greece and Rome* (Ithaca, 1972); H. W. Parke, *Festivals of the Athenians* (Ithaca, 1977); D. C. Young, *The Olympic Myth of Greek Amateur Athletics* (Chicago, 1984); W. Sweet, *Sport and Recreation in Ancient Greece: A Sourcebook with Translations* (New York, 1987); and D. Sansone, *Greek Athletics and the Genesis of Sport* (Berkeley, 1988). On the symbolic aspects of public games, especially at Rome, see Keith Hopkins, *Death and Renewal* (Cambridge, 1983), esp. chap. 1 ("Murderous Games").

3. For a brief review of the various types of structures used for games, racing, and theater, see Harris, *Sport,* pp. 161–72; Balsdon, *Leisure,* pp. 252–61; and J. H. Humphrey, *Roman Circuses: Arenas for Chariot Racing* (Berkeley, 1986), pp. 1–24.

4. On the difficulty of using the modern terms "amateur" and "professional" of ancient Greek games, see Young, *Olympic Myth.*

5. Harris, *Sport,* pp. 40–41, argues that before the Christian era, women had competed in separate games at Olympia; from the first century c.e. onward,

women's athletic events, including wrestling, took place alongside men's at the same games.

6. See Parke, *Festivals*, pp. 34–37.

7. See A. Pickard-Cambridge, *The Dramatic Festivals of Athens* (Oxford, 1968), pp. 40–42 and 74–83, for a discussion of the competitive aspects of dramatic festivals at Athens and other Greek cities.

8. Of course, Augustine declined the offer, disgusted by the horrible rites associated with such procedures, specifically the killing of small animals. On the use of animals in curses and spells, see Faraone, "Context," p. 21, n. 3.

9. See especially Balsdon, *Leisure*, pp. 244–52, 267–70.

10. On gladiatorial shows, see Balsdon, *Leisure*, pp. 288–313, and Hopkins, *Death and Renewal*, pp. 1–30.

11. Faraone, "Context," pp. 16–17, suggests that early Greek curses were used in order to even up contests where the opponent seemed to have an unfair advantage. The wide usage of curse tablets indicates that while "evening up the odds" may have been an occasional motive, perhaps even a rationalization, their real goal was to create an unfair advantage, to "fix" the outcome. At the same time, it must be recalled that those who resorted to the use of *defixiones* did so in the full knowledge that their opponents were up to the same tricks. Both also used defensive phylacteries and spells to guarantee victory.

12. See now the discussion in D. R. Jordan, "New Defixiones from Carthage," in *The Circus and a Byzantine Cemetery at Carthage*, ed. J. H. Humphrey, vol. 1 (Ann Arbor, 1988), pp. 117–20.

13. Faraone, "Context," p. 20, observes that the first literary instance of a curse in athletic competition occurs in Pindar (fifth century B.C.E.), in his first *Olympian Ode* (lines 76–78) where Pelops, the son of Tantalus, competes with Oenomaos for the hand of his daughter, Hippodameia. In preparation for this all-or-nothing battle with spears and chariots, Pelops prays for help to Poseidon: "Block the bronze spear, and grant me the swifter chariot . . . and surround me with power."

14. The tandem of curses and wrestlers shows up again in later Christian saints' lives. In one (Life of Saint George of Choziba), a professional wrestler is released from spells cast on him by his opponents only by becoming a monk himself. In the other (Life of Saint Theodore of Sykeon), a wrestler unable to compete because of pain in his body, is released from a demon, introduced no doubt by a curse tablet. For a discussion of both texts, see H. J. Magoulias, "The Lives of Byzantine Saints as Sources of Data for the History of Magic in the Sixth and Seventh Centuries: Sorcery, Relics and Icons," *Byzantion* 37 (1967): 245–46.

15. Jordan, "Agora," p. 214, notes the discovery of a *defixio* from Isthmia against a runner. In addition, one of the tablets from the find in the Athenian Agora, directed against a certain Alkidamos, probably concerns a foot race (pp. 221–22).

16. On chariot racing, see especially Alan Cameron, *Porphyrius the Chario-*

teer (Oxford, 1973) and *Circus Factions: Blues and Greens at Rome and Byzantion* (Oxford, 1976).

17. N. Baynes, *The Byzantine Empire* (London, 1925), p. 33.

18. Tacitus, *Dialogue on Oratory* 29.

19. Ammianus Marcellinus's account of racing and its popularity in Rome (37.4.28–31) appears in his catalogue of the vices of the Roman people (38.4). His history of the later Roman empire, of which only the portions covering the years 353–378 have survived, was written in Rome sometime after 378 C.E.

20. *Satires* 7.114; see the discussion in Cameron, *Porphyrius*, p. 244.

21. In their discussion of a lead tablet discovered in the circus at Lepcis Magna, J. H. Humphrey, F. B. Sear, and M. Vickers propose the attractive idea that the use of Greek in the *defixiones* from the Latin West may indicate that most of the professional charioteers came from the Greek-speaking East; see *Libya Antiqua*, 9–10 (1972–1973): 97.

22. Procopius, *Wars* 1.24.

23. On Christian attitudes to racing, see Magoulias, "Lives," pp. 242–45, and Harris, *Sport*, pp. 227–37. Tertullian's *On the Games/Spectacles* (written ca. 200 C.E.) illustrates the sharply antagonistic attitude, whereas Cassiodorus's history of chariot racing, written in the early sixth century C.E. by a Roman aristocrat who served as secretary to Theodoric and later founded a monastery in Italy, indicates just how central an institution the circus remained in Christian Rome.

24. Spelled out by Cassiodorus, *Variae Epistolae* 51, and discussed by Cameron, *Factions*, pp. 230–31.

25. Brief discussions of *defixiones* and chariot racing in Harris, *Sport*, pp. 234–37; Balsdon, *Leisure*, pp. 318–19; Cameron, *Porphyrius*, p. 173, n. 3 and p. 245, and Cameron, *Factions*, pp. 56, 61–62, 194, 200, and 345 note. Once again, Christian texts provide further evidence regarding the use of curse tablets in racing; see Cameron, *Porphyrius*, p. 245, and *Factions*, p. 345 note.

26. So Cameron, *Porphyrius*, p. 245.

27. So in Humphrey, Sear, and Vickers, *Libya Antiqua*, p. 97.

28. See Humphrey, *Roman Circuses*, chap. 11.

29. See the *Theodosian Code* 9.16.11, an imperial decree issued in 389 C.E. and renewed subsequently by later emperors. The decree requires anyone with knowledge of persons practicing magic to expose them publicly. It goes on to forbid charioteers from seeking to contravene the edict by carrying out the punishment on their own authority. In line with this, the historian Ammianus Marcellinus, writing in the 370s, records three instances in which charioteers were punished for involvement with illegal spells (26.3.3; 28.1.27; 29.3.5); see the discussions in Harris, *Sport*, pp. 234–35, and Cameron, *Porphyrius*, p. 245.

30. The translation is adapted from Thomas Hodgkin's condensed rendering of Cassiodorus's edition, known commonly as *Variae Epistolae* (3.51), written in the years 507–511 C.E.; see T. Hodgkin, *The Letters of Cassiodorus* (London, 1886), p. 226.

31. Cameron, *Factions,* p. 231; see also G. Dagron, *Naissance d'une capitale. Constantinople et ses institutions de 330 à 451* (Paris, 1974), pp. 330–47.

32. On charioteers and curses, see the perceptive comments of Peter Brown, "Sorcery, Demons, and the Rise of Christianity," in *Religion and Society in the Age of Saint Augustine* (New York, 1988), pp. 128–29. Dagron (*Naissance,* p. 347) remarks that the emperor Theodosius expressed horror at the spectacles in the hippodrome, in part due to their passionate outbursts but even more because "they contradicted both the Empire and the Church by introducing an alternative into a system of unity."

1. Greece, Attica; original location uncertain. Probably fourth or third century B.C.E. Lead tablet measuring 7 × 4 cm.; originally folded.[1] The first and last lines are written left to right; the middle two lines, right to left. There is no verb of binding; the cursed person is simply cited in the accusative case, as the direct object of an implied (spoken?) verb. Whether Theagenes was a director or a financial backer is not stated. Neither is it indicated what activity was involved, although competition between theatrical choruses seems certain. Formal, public contests between choruses and their sponsors and leaders took place regularly.[2] This tablet suggests rivalry between directors. *Bibl.:* DTA 34; Faraone, "Context," p. 12.

All the choral directors[3] and assistant choral directors[4] with Theagenês, both the directors and the assistant choral directors.

1. A similar tablet (DTA 33), probably by the same person and of the same size and form, curses (without verb) "all the teachers with Si . . . (?) and all the youths/sons (*paidas*)." Choruses of youths (boys) were a regular feature of dramatic productions.

2. On formal competitions between choruses and their leaders, see E. Reisch, "Chorikoi agones," *RE 3* (1899), cols. 2431–38.

3. The Greek term is *didaskalos,* which normally means "teacher." Here it refers to a director of a dramatic chorus, a task frequently carried out by the poets themselves.

4. The Greek term is *hupodidaskalos.* Together, the two terms generally designate directors and trainers of choruses in the Greek theater. The terms occur together in Plato, *Ion* 536a: "Thus there is a great linked chain of dancers and masters and undermasters." Theagenes may have been a *chorêgos,* a citizen who undertook to underwrite the expenses for training and maintaining the chorus or he may have been one of the directors. On the use of these terms to designate theatrical specialists in the world of Greek theater, see A. Pickard-Cambridge, *The Dramatic Festivals of Athens* (Oxford, 1968), pp. 91, 291, 303–4.

2. Greece, Athens (Patissia); original location not known. Lead tablet measuring 6 × 4 cm.; written on both sides; originally rolled up and pierced by a nail. The editor gives no date, but it can hardly be later than the second century B.C.E. No deity is invoked and the simple formula, "I bind!" is used. The occasion is not clear; the only hint is that the primary target, Euandros, is twice identified as an actor. This label raises the possibility that the tablet may have arisen out of rivalry between actors. The actor's son, Asteas, is also cited. *Bibl.:* DTA 45; Faraone, "Context," p. 12.

(*Side A*) I bind Euandros with a leaden bond[5] and . . . Euandros the actor

(*Side B*) and all the . . . of Euandros . . . Asteas, son of Euandros the actor.

3. Greece, Athens; discovered in excavations of Roman wells in the Athenian Agora. Altogether some one hundred lead curse tablets have been recovered from the Agora, rolled up as scrolls. Jordan dates this one to the mid-third century C.E. The tablet measures 13.9 × 11.5 cm. Jordan has described the writing as a "skillful, elegant, fluent semicursive,"[6] as if produced by a professional scribe. The tablet involves a curse against a professional wrestler; several others are directed against wrestlers and lovers. In a number of these, the same names and formulas reappear, indicating that they were copied, though not exactly, from the same formulary in the scribe's possession.[7] Another tablet in Jordan's collection (no. 6) is aimed at Alkidamos, a runner about to participate in the athletic games of the city. That spell, presumably commissioned by one of Alkidamos's competitors, expresses the desire that he may not pass the starting line and that if he does he may veer off the course and disgrace himself. The connection between binding spells and wrestlers appears also in several Christian saints' lives.[8] The deity invoked is the widespread Greco-Egyptian figure of Seth-Typhon. *Bibl.:* Jordan, "Agora," no. 1, pp. 214–15.

5. The language here is suggestive of cases in which *defixiones* were accompanied by figurines of the intended victim, frequently with bound hands and feet. Yet the phrase may simply refer to the binding action of the tablet itself.

6. Jordan, "Agora," p. 210.

7. Jordan, "Agora," pp. 236–40, has produced a synoptic version of ten different curses, which reveals their basic similarities, together with minor differences.

8. H. J. Magoulias relates two incidents involving professional wrestlers who found themselves suddenly unable to compete, presumably because they had come under spells like those illustrated here; they were cured only by resorting to saints who required them to give up their careers as athletes; see Magoulias, "The Lives of Byzantine Saints as Sources of Data for the History of Magic in the Sixth and Seventh Centuries A.D.: Sorcery, Relics and Icons," *Byzantion* 37 (1967): 245–46.

*BÔRPHORBABARBORBABARPHORBABORBORBAIÊ, powerful BETPUT[9] I deliver to you Eutuchianos, to whom Eutuchia gave birth, that you may chill him and his resolve,[10] and in your gloomy air also those who are with him. Bind him in the unlit realm of oblivion, chill[11] and destroy the wrestling which he is about to do in the De . . . ei[12] this coming Friday. And if he does wrestle, I hand over to you, MOZO[U]NÊ ALCHEINÊ PE[R]PERTHARÔNA IAIA,[13] Eutuchianos, to whom Eutuchia gave birth, in order that he may fall down and make a fool of himself. Powerful Typhon KOLCHOI[14] TONTONON Seth[15] SATH[AÔCH][16] EA Lord APOMX[17] *PHRIOURIGX, regarding the disappearing and chilling of Eutuchianos, to whom Eutuchia gave birth, KOLCHOICH[EILÔPS, let Eutuchianos grow cold and not be in condition this coming Friday, but let him be weak. As these names grow cold, so let Eutuchianos grow cold, to whom Eutuchia gave birth, whom Aithalês promotes.

4. Apheca (also Fiq), Syria. The lead tablet measures 23 × 29 cm. The writing on both the left and right margins has been eroded. Audollent dates it in the third century C.E. This fragmentary tablet, first interpreted by Audollent as dealing with circus factions, has since been identified by Ganszyniec and Louis Robert as referring to factions of pantomimes. *DT* 16 is very similar and targets Hyperechios the actor, binding various parts of his body necessary for successful performance. Another tablet relating to the rivalry of pantomime actors comes from Gaul (see no. 16). The immediacy of this spell is striking, as the event is referred to as taking place "tomorrow." The confused language suggests that the inscriber clumsily copied the curse from a formulary. The original publication included neither a photograph nor a drawing; located today in the Louvre, it is not available for inspection. There is thus every reason to

---

9. Not otherwise attested, but probably one of Seth-Typhon's secret names.

10. The Greek term is *gnomê;* its closest modern analogue might be "spirit."

11. Here there is an obvious symbolic connection between the cool and dark character of the well where the tablets were deposited, the subterranean deities invoked to carry out the curse, and the desired effects on Eutychian. Jordan notes that the use of the verb *katapsuchô* occurs regularly in curses associated with Seth-Typhon. Whether "the gloomy air" and "the unlit realm" refer to two different subterranean zones, or to this world (so Jordan) as opposed to the underworld, is not clear.

12. Jordan suggests a place name where the bouts would be staged.

13. These words occur on each of the ten tablets edited by Jordan, with slight variations from one version to the next.

14. Part of the secret name of Seth-Typhon.

15. Seth-Typhon, the great deity of magical texts; a highly syncretized figure, with accretions from many sources.

16. This word appears as SABAOTH in another tablet from the Agora and as BASAOTH in another.

17. Sometimes spelled *apomps*, this term is commonly used of Seth-Typhon; cf. P. Moraux, *Une défixion judiciaire au Musée d'Istanbul* (Brussels, 1960), p. 17, n. 3.

suppose that numerous readings should be treated as uncertain. *Bibl.: DT* 15; R. Ganszyniec, "Magica," *Byzantinische-Neugriechische Jahrbücher* 3 (1922): 164; L. Robert, *Études épigraphiques et philologiques* (Paris, 1938), pp. 99–102; R. A. Maricq, "Notes philologiques," *Byzantion* 22 (1952): 360–68.

> . . . their tongues . . . the voice . . . if someone introduced him before heaven or before all the earth[18] or he already made rituals of sending or turning away for his sake,[19] destroy, cancel[20] all assistance and . . . [for the sake of] Huperechios[21] the bewigged pantomime[22] of the Blue (team). If for the sake of the faction . . . it was done for his sake or if someone had introduced him either before heaven or before all the earth, or already made rituals of sending or turning away for his sake, destroy, cancel all the assistance for Huperechios the . . . for the sake of his faction, or anyone of the thirty-six Decans[23] . . . the *pekkrateritôr*[24] [of the tombs?] or some of the five planets or one of the two luminous bodies[25] are his assistance, destroy, cancel all his assistance . . . of the one gazed upon in the morning,[26] and if someone . . . you . . . for his sake . . . destroy . . . [of his own party] to become a physical force for the sake of Huperechios the bewigged pantomime[27] . . . he attempted [to give] the same assistance (?) . . . [leave] let other assistance [not?]

18. The Greek verb *sunhistêmi* probably refers to a ritual for introducing a person into the company of a deity; see S. Eitrem, "Die *sustasis* und der Lichtzauber in der Magie," *Symbolae Osloenses* 8 (1929): 49–51.

19. The Greek words *apopompai êdei apotropai* refer to rituals performed for the targeted victim's sake. The words might also refer to phylacteries and apotropaic amulets. In either case, our tablet seeks to neutralize any rituals that the target, Huperechios, may have undertaken on his own behalf.

20. The verbs are *luô* and *analuô;* an instance of the tendency to repeat a simple verb with its compound, which has a similar meaning, e.g., "bind" and "bind together" (*deô* and *sundeô*) and "turn" and "overturn" (*strephô* and *katastrephô*).

21. The name of the targeted victim. It is spelled with slight variations throughout this text.

22. Audollent reads the word as *remmachchos*, an otherwise unattested Greek word. David Jordan suggests that it may in fact be a corruption of *emmallos*, a term used of bewigged pantomimes in the theater. The two lambda's could easily be confused with chi's. The rho remains unexplained.

23. Mythological figures derived from the ten-degree sections of the astrological zodiac.

24. Perhaps from *pagkrateutês* or *pagkratiastês*, an athlete who participated in all-out wrestling or boxing matches.

25. That is, the sun and the moon.

26. Perhaps a reference to the Morning Star.

27. The Greek *emmollos* is taken by Audollent (in *DT*, p. 25) to represent a transliteration of the Latin word *aemulus*, either a partisan or adversary of the named faction. However, Maricq (pp. 364–68), following Robert, argues that it should be read *emmallos*, an adjective designating pantomime roles in the early Byzantine period; for example, a *chrusomallos* was a gold-wigged actor, a *karamallos* a black-wigged one. Thus *emmallos* = *en* + *mallos* would be a "bewigged" actor. See A. Alföldi, *Die Kontorniaten* (Budapest, 1943), p. 193, no. 585: a medallion with the exclamation, "Karamallos, conquer!"

be furnished for his sake, destroy, cancel . . . before the party of the Green to
the column . . . , tranquil, unmoved . . . [by?] the bewigged pantomime
daimon (?) lest his enemy and his faction be moved . . . they remain . . .
adoring him . . . Huperechios the bewigged pantomime[28] . . . the
tongue . . . affliction on account of . . . of his three hundred sixty-five members.[29] . . . Bind his neck, his hands, his feet, bind, bind together his . . . his
sinews, his . . . his pulse, his ankles, his steps, the bottom of his feet.[30] Let
him cry out "Enough!" . . . his [provision] . . . unite . . . to him . . . the end
of him for which let us be eager . . . his stomachs, his mind, his midriff . . .
[do not?] bestow his joy. But . . . upon him and his supporters and the chorus
with its leader . . . miss the end, strike Huperechios the dancer . . . the stomachs of all, their [spectators?] . . . [their] voice . . . lock up their tongues . . .
block the threshold[31] of their mouth, their jaws . . . their cursed . . .
their . . . the chorus, but neither this man nor his entire chorus nor his faction
but . . . (The rest is even more fragmentary).

5. Beirut, Syria; possibly discovered near the ancient race course. Lead
tablet measuring 9 × 15.8 cm. The date is the late second or early third
century C.E. The figures invoked are "holy angels," referred to by their
mysterious names. The spell is directed against a long list of horses and
drivers of the Blue faction. Directly under the title and between two
columns of mysterious names at the top there appears a bound human
figure seemingly under attack by a second, incomplete figure (probably
a snake), showing only a head and an open mouth (Figure 6). According
to Mouterde, the small circles on the body and the protrusions from the
head represent nails, symbolizing the binding process.[32] *Bibl.:* R.
Mouterde, "Le glaive de Dardanos," *Mélanges de l'Université Saint Joseph* 15 (1930–1931): 106–23; *SEG* 7.213; A. Maricq, "Tablette de
défixion de Beyrouth," *Byzantion* 22 (1952): 368–70; *SEG* 15.847; *BE*
(1954) no. 21, pp. 100–101; SGD 167.

28. Here Audollent in *DT* reads *ommadôn* (*ommatôn*, "of the eyes"); but this makes no
sense and is probably a corruption of *emmallos*.
29. Audollent's text in *DT* here reads NELIOUA . T. David Jordan suggests that the text here
should read *melê autou*, as in Wortmann, no. 12, lines 15–16 (see no. 8), which also binds the 365
members of the target's body.
30. Audollent reads *tôn aposiôn* ("unholy things"), which makes no sense; Jordan proposes
instead *tôn podôn* ("feet").
31. Reading the Greek word here as *oudon* for *ouron*.
32. The so-called Sethian tablets from Rome offer a number of close parallels; cf. esp. no. 11
in Wünsch, with a figure bound by straps or bands, three nails emerging from the head, and a
snake underneath. Nos. 17 and 20B in Wünsch also show interesting similarities. No. 15 in *DT*
offers further parallels.

FIGURE 6. Design of bound figure and snake head on *defixio* from Beirut. The protrusions from the head and the circles on the body represent nails, commonly used to "fix" the target, in this case horses and jockeys of a rival team. The total number of nails is twelve, thirteen if one can be assumed between the legs. The snake is reminiscent of snakes attacking targets on several tablets from Rome. The lines on the body also depict in graphic terms the binding process. (A. Maricq, "Tablette de défixion de Beyrouth," *Byzantion* [Brussels: Fondation Byzantine et Néo-grecque, 1952], facing p. 368. By permission.)

For Restraining Horses and Charioteers.[33]

(*Column 1*) PHRIX[34] PHÔX[35] BEIABOU[36] STÔKTA NEÔTER whether above the earth or below *DAMNÔ DAMNA LUKODAMNA MENIPPA[37] *PURIPIGANUX

(*Column 2*) *EULAMÔ
    EULAÔ
    EULA
      EULAMÔ
      ULAMÔ
        AMÔ
        MÔ
          Ô [38]

*OREOBARZAGRA*AKRAMMACHARI PHNOUKENTABAÔTH ÔBARABAU, you holy angels, ambush and restrain LULATAU AUDÔNISTA them. The spell—OIATITNOUNAMINTOU *MASKELLI MASKELLÔ PHNOUKENTABAÔTH OREOBARZA, now attack, bind, overturn, cut up, chop into pieces the horses and the charioteers of the Blue colors—Numphikos, Thalophoros, Aêtôtos, Mousotrophos, Kalimorphos, Philoparthenos, Pantomedôn, Hupatos, Philarmatos, Makaris, Omphalios, Hêgemôn, Ôkeianos, Turanos, Chôrikis, Kalimorphos, Aurios, Aktinobolos, Egdikos, Zabadês, Chôrikis, Nomothetês, Barbaros, Eieronikês, Xaes, Makaris, Dênatos, Antheretos, Phôsphoros, Lukotramos, Germanos, Obeliskos, Astrophoros, Anatolikos, Antiochos[39]—CHRAB,[40] bind and CHRAB,

33. The copyist clearly erred in writing out these words, which constitute the title from the recipe book from which the tablet was drawn. Properly speaking, the title is not part of the spell at all. Such errors of transcription are not at all uncommon; see Jordan, "Agora," p. 235, n. 20.

34. The word appears several times in *PGM*, e.g., I, line 230 (where it should be read as a separate word); III, line 413; and IV, line 1196. The combination of PHRIX PHÔX appears also on a lead tablet from Athens, dated by its editor to a relatively late period; cf. Ziebarth (1934), no. 24, l. 24. Some *voces mysticae* in our tablet, including PHRIX PHÔX, appear also in *PGM* XIX a, a love spell from Egypt dating to the fourth or fifth century C.E.

35. An exclusively Egyptian spirit or *daimon*, invoked in *PDM* xiv, lines 105ff., as "him who is seated in the fiery cloak on the serpentine head of the Agathos Daimon, the almighty, four-faced, highest *daimon*, dark and conjuring, PHÔX."

36. The letters IABOU may be taken as a form of the tetragrammaton or four-letter name of the god of ancient Israel, otherwise mentioned as IAO in the papyri; cf. *PGM* V, line 102 ("You are Iabas. You are Iapas . . ."). The first two letters of the word probably represent the Hebrew preposition *be*, "in/through." Thus the original phrase may have been an invocation, "in (the name of) Yahweh."

37. Possibly a reference to the moon-god, Selene, who was commonly addressed as Mên; here, however, the first syllable is spelled *men* rather than *mên*. In *PGM* IV, lines 2545ff., Selene is addressed both as Mene and as horse-faced (*hippoprosopos*). In the same hymn (line 2301), Selene is invoked simply as horse (*hippos*).

38. An imperfect attempt to create a wing-shaped figure based on the name EULAMO.

39. Mouterde assumes that the thirty-five names refer to horses; Jordan questions whether some may also designate charioteers; cf. Mouterde, pp. 118–21, for a full discussion.

40. The same four Greek letters precede the two verbs of binding, though they do not make Greek words. Mouterde (p. 116) suggests, appropriately in a region where Hebrew and Aramaic would have been familiar, that behind these letters lies the Hebrew or Aramaic verb ḥrb, "to destroy."

damage(?) the hands, feet, sinews of the horses and charioteers of the Blue colors.

6. Apamea, an important Greek city in Syria, on the Orontes River. Discovered in fill (material not in its original location), along with a second smaller tablet and other miscellaneous debris. Lead tablet measuring 11.8 × 5.2 cm., found rolled up (Figure 7). The smaller of the two has a hole in the center, probably made by a nail. Dated to the late fifth or early sixth century C.E. The figures invoked are especially interesting. The text appeals to the *charaktêres,* known from other tablets and texts. At the end of the text, two unusual names appear: *topos* ("the place"), a designation of the highest god in other texts; and Sablan/Zablan, an uncommon name with angelic and astrological associations in Jewish texts of the same period. The setting of the spell is chariot racing in the hippodrome of Apamea. Intense rivalry between the major teams or factions, the Blues and the Greens, is known from a passage from the sixth-century historian, Procopius (*Wars* 2.11.31–35). On a "visit" to Apamea, the Persian general Chosroes issued orders for a special series of races in the hippodrome of the city. Knowing that the Roman emperor Justinian favored the Blues, Chosroes decided to support the Greens. When the Blues took an early lead, Chosroes commanded his agents to slow down the Blue team in order to guarantee a Green victory. The targets of this spell are named—Porphyras, Hapsicrates, and Eugenius, all of the Blue team. The unnamed client must have

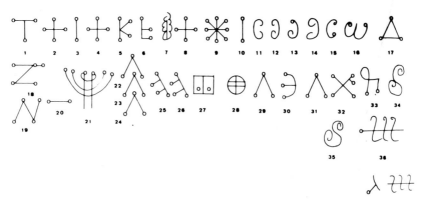

FIGURE 7. Drawing of 36 *charaktêres* on a *defixio* from Apamea (Syria). Similar, even identical, *charaktêres* are invoked on numerous tablets from scattered sites thoughout the Mediterranean region. (*Apamée de Syrie,* ed. Janine Balty [Brussels, 1984], p.216. By permission.)

represented or supported the Greens. Above the first line of text, which invokes the *charaktêres*, appear two lines of signs or, more precisely, the *charaktêres* themselves. These signs are identical on the two tablets and similar to representations of the *charaktêres* elsewhere. The text itself is virtually complete. *Bibl.:* W. van Rengen, "Deux défixions contre les bleus à Apamée (VIᵉ siècle apr. J.-C.)," *Apamée de Syrie* (Brussels, 1984), pp. 213–34; *SEG* 34.1437; SGD, p. 192.

Most holy Lord Charaktêres,[41] tie up, bind the feet, the hands, the sinews, the eyes, the knees, the courage, the leaps, the whip (?),[42] the victory and the crowning of Porphuras[43] and Hapsicratês, who are in the middle left, as well as his codrivers of the Blue colors in the stable of Eugenius.[44] From this very hour, from today, may they not eat or drink or sleep[45]; instead, from the (starting) gates may they see *daimones* (of those) who have died prematurely,

41. We must suppose a close connection between the signs on the first two lines of the tablet, above the text, and their use here. The signs embodied the higher powers invoked to carry out the spell. What exactly these powers were thought to be is not clear, although the papyri make it clear that all superior beings were believed to possess their own characters or signs as empowered signatures. As symbols of higher power, they are addressed here as "Lords, most holy." Elsewhere, they are called "divine" (*theioi;* see pp. 000–000) and "terrifying" (*phoberoi;* see pp. 000–000). The presence of exactly thirty-six signs on the two tablets from Apamea leads van Rengen to argue that the connection or association here may be with the thirty-six decans or divisions of the heavens common in Egyptian astrology. Elsewhere, the *charaktêres* are associated with the protective powers and functions of angels and archangels. In this tablet, there may well be a tie between archangels and decans, with the former understood as in charge of the latter.

42. The text reads *taura,* a word otherwise unattested. The editor suggests a miswriting for *tauria/taureia,* "whip." The whip was, of course, the most important instrument for charioteers and appears frequently in drawings of charioteers in ancient art.

43. Binding spells from hippodromes regularly give the names of horses and drivers, sometimes both. The question as to who these two names are meant to designate—horses and/or drivers—depends in part on the rare word "in the middle left" (*mesaristeros*) that follows. In the *Book of Ceremonies* of Constantine VII Porphyrogenitus (ca. 940 C.E.), the word appears to designate the favored starting position: "Once the jar has been turned, the ball that comes out (first) designates which one will occupy the position in the middle left (*mesaristeron*)." On the other hand, as van Rengen notes, the term might also have been used to specify the lead horse in the team of four. Porphyras is a late form of Porphyrius, a common name for horses and jockeys. The most famous jockey of this name was a professional who enjoyed phenomenal success in the mid-sixth century and received numerous public honors; cf. A. Cameron, *Porphyrius the Charioteer* (Oxford, 1973).

44. The Eugenius named here in connection with the *stablon* (a Latin word taken into Greek) is no doubt the *factionarius* or professional manager, later the leading jockey, of a team; on the term and its development, see Alan Cameron, *Circus Factions: Blues and Greens at Rome and Byzantion* (Oxford, 1976), pp. 5–13.

45. In this case, the client's wish is that the spell begin immediately, not just at the time of the race. Thus, the horses are marked for trouble not just during the race but for unfavorable conditions beforehand. Similar commands to prevent eating, drinking, and sleeping show up in love spells (see *PGM* IV, lines 354ff.).

spirits (of those) who have died violently,[46] and the fire of Hephaestus[47] . . .
in the hippodrome at the moment when they are about to compete[48] may they
not squeeze over,[49] may they not collide,[50] may they not extend,[51] may they
not force (us) out, may they not overtake, may they not break off (in a new
direction?)[52] for the entire day when they are about to race. May they be
broken, may they be dragged (on the ground), may they be destroyed; by
Topos[53] and by Zablas.[54] Now, now, quickly, quickly!

46. The start was the critical moment of the race and thus the subject of particular anxiety.
Other spells specify a bad start as their goal. In this case, the designated technique takes the form
of causing the horses to see the terrifying *daimones* of those who had died prematurely or by
violence.

47. Hephaestus, the Greek god of fire and the foundry. Here he appears in his familiar role
as the first Greek "magician," though his name appears in just one other surviving magical text
(*PGM* XII, lines 177–78). The point is clear: the horses are to see the fire of Hephaestus as they
leave the starting gates and thus be frightened into a bad race. On Hephaestus, see M. Delcourt,
Héphaistos ou la légende du magicien (Paris, 1957).

48. The following series of verbs is not merely repetitive. It lists the various techniques or
tricks used by charioteers to gain an advantage over their competitors. There is a close parallel in
DT 187 (Rome): "May they not get a good start, may they not pass."

49. The editor is not certain how to read the Greek at this point and offers two possibilities,
chiazein or *piazein/piezein*. The first would indicate a maneuver of crossing or zigzagging from
lane to lane; the second would signify much the same thing—pressing across into another lane.
The effect of either action would be the same, to cut off the path of one's competitors by forcing
them out of their lanes.

50. The verb *parabuzein* here means to force one's own chariot, from an outside passing
position, into the chariot and the horses of one's competitor so as to overturn them and put them
out of the race. The technique is described by the poet Nonnus (fifth or sixth century C.E.) in his
The Story of Dionysus 37.351ff.: Actaeon (one of the drivers) caught up and passed from the
outside; he then swerved to the inside, touching his rival's chariot and scraping its horse's legs
with his own wheel. The chariot overturned and three of the four horses, along with the driver,
stumbled and fell.

51. The Greek is *tathôsi(n)* from *teinô* ("to strain, exert, stretch"). The meaning here seems
to be "to stretch," "to strain," or "to exert oneself to the fullest."

52. The verb *periklan* might mean either to break away or to make the sharp turns around
the two end posts of the course. Other spells from the world of racing use various forms of the
same verb in a "positive" meaning—to wish that their competitor's chariots will break apart; cf.
DT 187 and in our tablet, line 12.

53. The phrase *kata topo(n)* and the following phrase *kata Zablan* are unusual and somewhat
puzzling. The sense seems to be that the figures are here invoked as the divine agents through
whose authority (*kata*) the *charaktêres* are to carry out the spell. *Topos* ("the place") as a divine
name occurs frequently in Jewish texts, translating Hebrew *maqom*, commonly used to designate
the deity (e.g., Philo, *Dreams* 1.63). The same term appears in Valentinian Gnostic texts where
it designates the demiurge or creator-god (cf. Clement of Alexandria, *Excerpts of Theodotus* 34,
37–39). And in the *Hermetic Tractate* 2.12 (from Egypt and dating to the same period as the
curse tablets), *topos* is described as "that in which everything moves . . . the incorporeal . . . a
mind that contains its own self entirely, free from any form of bodily nature, unerring, above
emotion, intangible, immutable in itself, containing all things and redeeming all things . . . from
which the good, the truth and fountainhead of what is spiritual emanate like rays." Thus, while
topos appears in no other binding spell, it certainly falls within the range of those powerful
spiritual beings that could be useful in circumstances like these.

54. The only other known occurrence of Zablas/Sablas is in a Coptic amulet from around 600
C.E. (Kropp, vol. 1, text F, line 37). It appears in a list of the seven angels who assisted God in the

7. Egypt, from the Jewish collection of spells and recipes, *Sepher ha-Razim* (see p. 106).[55] Properly speaking, the recipe is not for a curse tablet directed against other competitors but a spell to guarantee victory for oneself by forcing horses to run even when they are tired. A similar recipe for producing metal tablets guaranteed to "fix" horse races—but certified in addition as useful in producing dreams and generating hatred—appears in *PGM* III, lines 15–30.

> If you wish to race horses, (even) when they are exhausted, so that they will not stumble in their running, that they will be swift as the wind, and that the foot of no living thing will pass them, take a silver *lamella* and write upon it the names of the horses and the names of the angels and the name of the prince who is over them and say:
>
> > I adjure you angels of running, who run amid the stars, that you will gird with strength and courage the horses that N is racing and his charioteer who is racing them. Let them run and not become weary nor stumble. Let them run and be swift as an eagle. Let no animals stand before them, and let no other magic or witchcraft affect them.
>
> Take the *lamella* and conceal it in the racing lane (of the one) you wish to win.

8. Egypt, Oxyrhynchus; exact origin not known. Lead tablet measuring 8×13 cm. and originally rolled up. Dated to the fourth century C.E. The tablet is poorly preserved. The spirit invoked at the beginning is addressed by the familiar name EULAMÔ. The only other legible name is also known from similar tablets—CHUCH BAZACHUCH. The occasion is competition between athletes, in this case runners. In addition to the

creation of Adam; the theme of angelic cooperation in the act of creation is a common one in Jewish and Christian texts of this period. This helps us to understand the connection between *topos*, taken as referring to the biblical god, and Zablas, an angel. What about their connection to the *charaktêres*, depicted in the first two lines above the text and invoked in the first line of the text? As van Rengen observes, angels and archangels were thought to command and control not only spirits and demons generally, but the thirty-six decans specifically—that is, the thirty-six equally divided portions, also known as world rulers, of the heavenly sphere or zodiac according to Egyptian astrological tradition. And the thirty-six *charaktêres* of the first two lines of our text almost certainly represent the thirty-six astrological decans; see the convincing argument of van Rengen, pp. 216–19. Furthermore, in *The Testament of Solomon*, the thirty-six decans are summoned to appear before Solomon, where they announce their names, their powers, and the particular angel to whom they are subjugated: "I am the first decan of the zodiac and I am called Ruax. I cause the head of men to suffer pain and I cause their temples to throb. But if I even hear, 'Michael, imprison Ruax,' I retreat immediately" (18:5). Unfortunately, none of the thirty-six decans summoned before Solomon specialized in chariot racing and none is named Zablas. But it is evident that there was great variability in the names and functions of spirits from one text to another. In short, there is a close connection in our tablet between the *charaktêres* summoned to carry out the spell, and *topos* and Zablan, under whose power and authority they stand.

55. This passage is from the third firmament, lines 35ff.; *Sepher ha- Razim*, p. 64.

customary organs and faculties, this spell also binds "the 365 limbs and sinews of the body." The atmosphere of the tablet is markedly Greco-Egyptian. *Bibl.:* Wortmann, no. 12, pp. 108–9; SGD 157; D. R. Jordan, "Inscribed Lead Tablets from the Games on the Isthmus of Corinth," (forthcoming in *Hesperia*).

> EULAMÔ . . . I command. Accomplish (this) for me . . .
> ULAMÔE
> LAMÔEU
> AMÔEUL
> MÔEULA
> ÔEULAM[56]

> . . . *CHUCH BAZACHUCH . . . Bind and tie up the sinews, the mind, the thoughts, the thinking, the 365[57] limbs and sinews of those athlete runners under[58] (name missing), to whom Taeias gave birth, and under Ephous, to whom Taeias gave birth, so that they might have neither power nor strength. Keep them up all night long and keep them away from all nourishment, [so that they will have no strength] but fall behind . . . and restrain all the . . . to whom Taeias [gave birth] . . . [to whom] Taeias gave birth . . . hold them . . . so that they may have no strength . . .

9. North Africa, Carthage; found with *DT* 233; lead tablet measuring 7.7 × 7.7 cm. as folded up. Inscribed with seventy-five lines, decreasing in length, so that the text forms a long triangle flush with the top and the left margin. The setting is horse racing; steeds of the Reds and Blues are bound, together with their drivers. The names of the horses (and drivers) are mostly Latin forms transliterated into Greek. The person who deposited the tablet clearly did not know in whose grave he was placing it. *Bibl.:* DT 237; CIL 8.12508; R. P. Delattre, "Inscriptions imprécatoires trouvées à Carthage," *BCH* 12 (1888): 297–300.

> I invoke you, spirit of one untimely dead, whoever you are, by the mighty names SALBATHBAL AUTHGERÔTABAL BASUTHATEÔ ALEÔ SAMABÊTHÔR. Bind the horses whose names and

---

56. This configuration of letters is known in the papyri as a *plinthion* or square (see *PGM* IV, line 1305, with instructions for making a *plinthion* of the seven Greek vowels). Here, each letter of EULAMO, in succession, begins the next line. The identical figure appears in a lead tablet from Syria (see pp. 000–000) and in two of the so-called Sethian curse tablets (from Rome) published by Wünsch (DTA 33 and 49).

57. Egyptian texts commonly list various parts of the body along with their divine overseers. The most complete list, numbering exactly 365, appears in the Apocryphon of John from Nag Hammadi (Codex II.1, lines 15–19). Our text does not mention such overseers, but they are probably to be assumed in the background. The same number appears in the tablet from Apheca in Syria (no. 4), where parts of the body are also to be bound.

58. The targets appear to be the two runners and their supporters ("those around . . .").

images/likeness[59] on this implement I entrust to you; of
the Red (team): Silvanus, Servator, Lues, Zephyrus, Blandus,
Imbraius, Dives, Mariscus, Rapidus, Oriens, Arbustus; of the
Blues: Imminens, Dignus, Linon, Paezon, Chrysaspis, Argutus,
Diresor, Frugiferus, Euphrates, Sanctus, Aethiops,
Praeclarus. Bind their running, their power, their
soul, their onrush, their speed. Take away their victory,
entangle their feet, hinder them, hobble them, so that
tomorrow morning in the hippodrome they are not able to run
or walk about, or win, or go out of the starting gates,
or advance either on the racecourse or track,
but may they fall with their drivers, Euprepês, son of
Telesphoros, and Gentius and Felix and
Dionusios "the biter" and Lamuros.[60] For AMUÊKARPTIR
ERCHONSOI RAZAABUA DRUENEPHISI NOINISTHERGA
BÊPHURÔRBÊTH command you. Bind the horses whose
names and images I have entrusted to you on this
implement; of the Reds: Silvanus,
Servator, Lues, Zephyrus, Blandus, Imbraius,
Dives, Mariscus, Rapidus, Oriens, Arbustus;
and of the Blues: Imminens, Dignus,
Linon, Paezon, Chrysaspis, Argutus,
Derisor, Frugiferus, Euphrates, Sanctus,
Aethiops, Praeclarus. Bind their running,
their power, their soul, their onrush,
their speed. Take away their victory,
entangle their feet, hinder them,
hobble them, so that tomorrow
morning in the hippodrome they
are not able to run or walk
about, or win, or go out
of the starting gates, or
advance either on the racecourse,
or circle around the turning point;
but may they fall with their
drivers, Euprepes, son of
Telesphoros, and Gentius and
Felix, and Dionysius "the
biter" and Lamuros. Bind
their hands, take away
their victory, their exit,

59. The meaning of *eidaias* is unclear, but seems to refer to images of the horses, perhaps
drawn on some accompanying instrument.
60. The word here is Greek, *lamuros* ("glutton").

their sight, so that they
are unable to see their
rival charioteers, but
rather snatch them up
from their chariots
and twist them to
the ground so that
they alone fall,
dragged along
all over the
hippodrome,
especially
at the turning
points, with
damage to
their body,
with the
horses
whom
they
drive.
Now,
quickly.

10. North Africa, near Carthage; 12.5 × 15.1 cm. and originally rolled up. This lead tablet, inscribed in Greek, was probably produced in the third century c.e. Various deities are invoked, including several with Jewish connections. A number of *voces mysticae* in this tablet are known to us from the collections of *PGM*. The tablet was deposited to ensure that a rival circus faction, the Reds, would not triumph in races on a certain eighth of November. Both the charioteers and their horses, with particular reference to the parts of their bodies that would be needed to run a race, are cursed by the commissioner of the tablet. *Bibl.:* R. P. Molinier, "Imprécation gravée sur plomb trouvée à Carthage," *Mémoires de la Société des Antiquaires de France* 58 (1897): 212ff.; *DT* 242; Wünsch, *Antike Fluchtafeln*, no. 4; see also Wünsch (1900), pp. 248ff.

I invoke you, whoever you are, spirit of the dead, IÔNA,[61] the god who established earth and heaven. I bind you by oath, NEICHAROPLÊX, the god who holds the power of the places down beneath. I bind you by oath, . . . , the god . . . of the spirits. I bind you by oath, great *AROUROBAARZAGRAN,[62]

61. The Greek text reads *iôna*. Wünsch thinks that this does not refer to the prophet Jonah, but should be emended to IAON, a form of the familiar IAO.
62. A variation of OREOBAZAGRA.

the god of Necessity. I bind you by oath, BLABLEISPHTHEIBAL, the firstborn god of Earth "on which to lie(?)" I bind you, *LAILAM, the god of winds and spirits.[63] I bind you, . . . RAPÔKMÊPH (?)[64] the god who presides over all penalties of every living creature. I bind you, lord *ACHRAMACHAMAREI, the god of the heavenly firmaments. I bind you, SALBALACHAÔBRÊ, the god of the underworld who lords over every living creature. I bind you, ARCHPHÊSON (?) of the underworld, the god who leads departed souls, holy Hermes, the heavenly AÔNKREIPH, the terrestrial. . . . I bind you by oath, *IAÔ, the god appointed over the giving of soul to everyone, GEGEGEGEN. I bind you, *SEMESEILAM, the god who illuminates and darkens the world. I bind you, *SABAÔTH, the god who [brought] knowledge of all the magical arts. I bind you, SOUARMIMÔOUTH, the god of Solomon. I bind you, *MARMARAÔTH, the god of the second firmament who possesses power in himself. I bind you, THÔBARRABAU, the god of rebirth. I bind you, . . . , the god who . . . the whole wine-troughs. . . . I bind you, AÔABAÔTH,[65] the god of this day in which I bind you. I bind you, ISOS (Jesus?), the god who has the power of this hour in which I bind you. I bind you, IAÔ IBOÊA, the god who lords over the heavenly firmaments. I bind you, ITHUAÔ, the god of heaven. I bind you, NEGEMPSENPUENIPÊ, the god who gives thinking to each person as a favor. I bind you, CHÔOICHAREAMÔN, the god who fashioned every kind of human being. I bind you, ÊCHETARÔPSIEU, the god who granted vision to all men as a favor. I bind you, THESTHENOTHRIL . CHEAUNXIN, who granted as a favor to men movement by the joints of the body. I bind you, PHNOUPHOBOÊN, the Father-of-Father god. I bind you, NETHMOMAÔ, the god who has given you (the corpse) the gift of sleep and freed you from the chains of life. I bind you, NACHAR, the god who is the master of all tales. I bind you, STHOMBLOÊN, the god who is lord over slumber. I bind you, ÔÊ IAO EEÊAPH, the god of the air, the sea, the subterranean world, and the heavens, the god who has produced the beginning of the seas, the only-begotten one who appeared out of himself, the one who holds the power of fire, of water, of the earth and of the air. I further bind you, AKTI. . . PHI *ERESCHEICHAL NEBOUTOSOUANT,[66] throughout the earth (by?) names of triple-form Hekate, the tremor-bearing, scourge-bearing, torch-carrying, golden-slippered-blood-sucking-netherworldly and horse-riding (?) one. I utter to you the true name that shakes Tartarus, earth, the deeps and heaven, *PHORBABORPHORBABORPHOROR BA SUNETEIRÔ MOLTIÊAIÔ Protector NAPUPHERAIÔ Necessity *MASKELLI MASKELLÔ PHNOUKENTABAÔTH OREOBARZARGRA ÊSTHANCHOUCHÊNCHOUCHEÔCH, in order that you serve me in

63. Here the connection with the Greek word *lailaps* ("storm, hurricane") seems inescapable; whether our *vox mystica* was originally derived from it, or merely associated with it secondarily, is another matter.
64. The -KMEPH ending suggests that this deity is connected with Kmeph, the Egyptian snake-god; see *PGM* II, line 142; IV, lines 1705 and 2094; VII, col. 17.
65. A variation of the familiar IAO ABAOTH.
66. Probably meant to be the familiar *vox mystica* NEBOUTOSOUALÊTH; cf. *PGM* IV, lines 2603ff. and 2666ff.

the circus on the eighth of November and bind every limb, every sinew, the shoulders, the wrists, and the ankles of the charioteers of the Red Team: Olympos, Olympianos, Scorteus, and Iuvencus. Torture their thoughts, their minds, and their senses so that they do not know what they are doing. Pluck out their eyes so that they cannot see, neither they nor their horses which they are about to drive: the Egyptian steed Kallidromos and any other horse teamed with them; Valentinus and Lampadios . . . Maurus who belongs to Lampadius; Chrysaspis, Juba and Indos, Palmatus and Superbus . . . Boubalus who belong to Censorapus; and Ereina. If he should ride any other horse instead of them, or if some other horse is teamed with these, let them [not] outdistance [their foes] lest they ride to victory.

11. North Africa, Hadrumentum. Lead tablet measuring 5.8 × 8 cm. Bilingual text: spell in Latin; *voces mysticae* and names of most horses in Greek. Unknown date, but clearly late Roman. The spell addresses numerous chthonic powers. From the final lines it is apparent that the tablet must have been deposited in the grave of someone who had died prematurely or by violence. The goal is to incapacitate racehorses; the horses are named, but the client and the horses' owner(s) remain anonymous. The main binding verb is *obligate*. A primary interest of the spell is its bilingual nature: the author evidently knew Greek and considered its letters to be intrinsically powerful. *Bibl.: DT* 295.

*HUESSE[M]IGAD[Ô]N IA[Ô AÔ BAUBÔ EÊAÊIE . . . SOPESAN KANTHARA *ERÊSCHI-GAL SANKISTÊ *DÔDE[K]AKÊTÊ *AKROUROBORE KODÊRE DROPIDÊ TARTAROUCHE[67] *ANOCH ANOCH KATABREIMÔ[68] fearful things towards t[he?] E . . nnê katanei-kandra damastrei . . sa Most Glorious One SEROUABUOS to you I commend, because he slandered (my) intention. Let them run to/at him (?); infernal *demones,* bind the feet of those horses so that they are unable to run, those horses whose names you have here inscribed and submitted: Incletus, Nitidus, Patricius, Nauta, SIOUN . AA, Quick-Starter.[69] Bind them so that they cannot run tomorrow or the day after tomorrow in the circuses: Patricius, Nitidus, Na[ut]a, Incletus, Quick-Starter, Domina, Canpana, Lambtêras, Nitidus, Patricius, Nauta, Incletus, Quick-Starter—so that they cannot run tomorrow and the day after tomorrow, and so that at every hour they collapse in the circus. Let him perish and fall, just as you lie (here)

67. "Ruler of Tartarus," the Greek underworld. Perhaps a reference to Kronos (cf. *DT,* no. 410 and *PGM* IV, line 2242, as an attribute of the Moon).

68. KATABREIMÔ might represent either a *vox mystica* or a Greek verb meaning something like "I roar against."

69. Here the letters *tacharchên,* which Audollent (in *DT*) reads as a *vox mystica,* are taken as the name of another horse.

prematurely dead.[70] Now, now, quickly, quickly, because they drive them off, the Typhonic[71] Daimones!

12. North Africa, Carthage. Directed against Victoricus and the racing team of the Blues, with a second driver, Secundinus, and numerous horses also named. The names are all Latin although the inscription is in Greek. Found rolled up and buried with six other tablets in the grave of a Roman official, in a pagan cemetery; dated from the first to third centuries C.E. The letters are quite small. The lead tablet measures 11.5 cm. on each of its four sides. The top and bottom margins contain letters and signs; the left and right margins contain signs (Figure 8). Occasional corrections have been written in above lines. *Bibl.: DT* 241; *CIL* 8.12511; Wünsch, *Antike Fluchtafeln,* no. 3.

*SEMESILAM[72] DAMATAMENEUS[73] IÊSNNALLELAM[74] *LAIKAM[75] ERMOUBELÊ IAKOUB[76] IA[77] IÔERBÊTH IÔPAKERBÊTH[78] ÊÔMALTHABÊTH ALLASAN. A curse.[79] I invoke you (plural) by the great names so that you will bind every limb and every sinew of Victoricus—the charioteer of the Blue team, to whom the earth, mother of every living thing, gave birth—and of his horses which he is about to race[80]; under Secundinus (are) Iuvenis and Advocatus and Bubalus; under Victoricus are Pompêianus and Baianus and Victor and Eximius and also Dominator who belongs to Messala; also (bind) any others who may be yoked with them. Bind their legs, their onrush, their bounding, and their running; blind their eyes so that they cannot see and twist their soul and heart so that they cannot

70. These two words *bios thanatos* are written in Greek; they must refer to someone who had died prematurely or by violence. The translation here follows a suggestion made privately by David R. Jordan.

71. The omnipresent Typhon, widely identified with the Egyptian figure of Seth, the patron spirit of spells and charms.

72. These "words" represent the secret and powerful "great names" cited one line below.

73. A variant of DAMNAMENEUS, one of the *ephesia grammata.*

74. On the tablet, a *theta* has been inscribed directly above the first *nu.*

75. A variant, or perhaps just a misspelling, of LAILAM.

76. A possible sign of Jewish influence in the text, if the word indicates the biblical patriarch Jacob, whose name appears elsewhere in similar literature; cf. *DT* 271 (from Hadrumentum, also in North Africa); *PGM,* III, lines 1232, 1736, 1803. However, the entire formula appears in *PGM* III, line 2223; cf. also *PGM* IV, lines 277–79; XII, lines 367–70; XIVc, lines 20–23; LVIII, lines 22–25; the so-called Sethian texts from Rome (R. Wünsch, *Sethianische Verfluchungstafeln aus Rom* (Leipzig, 1898), pp. 88, 90; and several *defixiones* from Athens (no. 00). In these cases, the terms serve as an invocation of the Egyptian deity, Seth, represented as donkey-headed and normally identified with the Greek deity, Typhon; cf. Moraux, *Défixion judiciaire,* pp. 16ff.

77. Possibly a form of IAO; but see the preceding note.

78. The two preceding *voces* appear commonly with BOLCHOSETH (see glossary).

79. *katara.*

80. The names following the two charioteers (Victoricus and Secundinus) are those of their horses.

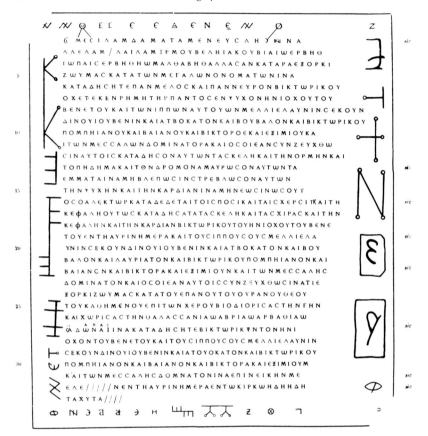

FIGURE 8. Drawing of *defixio* with *charaktêres* and numerous *voces mysticae* from Carthage (Tunisia). On this tablet, the *charaktêres* are inscribed in the margins rather than as the opening lines of the spell itself. Here again the targets are rival jockeys.

breathe. Just as this rooster[81] has been bound by its feet, hands, and head, so bind the legs and hands and head and heart of Victoricus the charioteer of the Blue team, for tomorrow; and also (bind) the horses which he is about to race; under Secundinus, Iuvenis, and Atvocatus and Bubalus and Lauriatus; under Victoricus, Pompeianos, and Baianus and Victor and Eximius and Dominator

81. There is no room on the plate for a drawing of a rooster, although Delattre indicates that the head of a rooster does appear on another plate from the same location; on this second plate there is also a list of names, including two in common with our plate—Victor and Advocatus ("Imprécations," 295–96). On the use of animals in connection with *defixiones* see Faraone, "Context," p. 21, n. 3.

who belongs to Messala and any others who are yoked with them. Also I invoke you[82] by the god above the heaven, who is seated upon the Cherubim, who divided the earth and separated the sea, ιαô, ABRIAô, *ARBATHIAô, *ADôNAI, SABAô,[83] so that you may bind Victoricus the charioteer of the Blue team and the horses which he is about to race; under Secundinus, Iuvenis, and Advocatus, and under Victoricus Pompeianus and Baianus and Victor and Eximius and Dominator who belongs to Messala; so that they may not reach victory tomorrow in the circus. Now, now, quickly, quickly.

13. Italy, Rome; on the via Appia near the Porta S. Sebastiano. Lead tablet measuring 13 × 21 cm.; inscribed on both sides. Every other line of the tablet is written upside down and backward—that is, the tablet was simply turned top to bottom at every other line as it was inscribed; here it probably manifests a deliberate attempt, through symbolic action, to "twist and turn" the intended target. The tablet was found together with approximately fifty-six lead tablets, most rolled up and pierced with nails, which were first deposited in several small terra-cotta sarcophagi and then placed inside a tomb. Of this large collection, some thirty-four were preserved more or less intact; the remaining twenty-two are quite fragmentary. Of these the editor was able to decipher writing on only forty-eight. Five are written in Latin, the rest in Greek. The language is highly formulaic and unsophisticated. The editor was able to date the tablets confidently to the end of the fourth century C.E., during the reigns of the Christian emperors Theodosius I and Honorius. The figures addressed in the spells are typically eclectic, reflecting both biblical influences (angels and archangels) from the predominantly Christian environment in Rome as well as names and titles drawn from pagan traditions, which were apparently still alive and circulating. No single figure predominates.[84] It is important to note that as with all tablets

82. Here begins a series of references to the Hebrew Bible, probably reflecting the use of an originally Jewish formula. The phrase "above the heaven" derives from passages like Isaiah 14:13 ("above the stars of heaven I will place my throne"); for "above the Cherubim," see the same words in Psalm 79:2: "Who is seated above the Cherubim." The dividing (*diorisas*) of the land and the sea refers to Genesis 1:7ff., where God separated (*chorisas*) the waters and divided the dry land from the sea, just as here. Of course, the professional who prepared this tablet probably knew none of this and simply treated the phrases and terms as traditional words of power.

83. The term SABAô is written directly above ADôNAI.

84. The original editor, Richard Wünsch, labeled all of the tablets as Sethian and regarded them as being under the influence of a known group of Christian Gnostics, called Sethians by their opponents in early Christian literature. Wünsch wrote during a period of pan-Gnostic interpretation, when everything unusual in the world of Greco-Roman culture was attributed to Gnostic influence. Wünsch's interpretation was widely criticized and is no longer accepted; cf. Preisendanz, pp. 23–37, and Moraux, *Défixion judiciaire*, p. 19, n. 3. Wünsch's edition of the texts and his commentary on them are still invaluable. I know of no other translation of these texts.

deposited in tombs or graves, the immediate agent of this spell was the spirit of the dead person in whose tomb the tablet was placed. Of particular interest is the fact that most of the tablets include drawings, usually of human figures portrayed in bonds and surrounded by a serpent about to bite or of horse heads[85] (Figures 9, 10).[86] The setting for these tablets is clear—competition among racers and charioteers in Rome. Socially, these tablets and their purchasers fall into the class of freedmen[87] and slaves. Their names are a mix of Greek and Latin and are regularly identified by maternal lineage; several are also given nicknames. *Bibl.:* R. Wünsch, *Sethianische Verfluchungstafeln aus Rom* (Leipzig, 1898),

85. These horse heads were central to Wünsch's argument for the Sethian Gnostic origin of the tablets, for he took them to be representations of the Egyptian god Seth, who was frequently portrayed with the head, not of a horse but of a donkey. More plausible is Preisendanz's view that these equine shapes depict horse-headed spirits (*daimones*), revered and feared as the presiding powers of amphitheaters and racing. Just such a horse-spirit is referred to in the *Pistis Sophia* (chap. 145), a Christian Gnostic document of the third century C.E., in connection with future punishments of murderers: "(Jesus is speaking) . . . a murderer will be bound by his feet to a great *daimon* with the face of a horse." Other elements in this chapter of *Pistis Sophia* are worth noting: the spirits will punish the murderer by whipping him; and various spirits will punish (*timôrein;* a variant of the same verb is used in our text) the murderer. One should also consider, given the fact that the horse figures appear alongside human figures who clearly represent images of the human targets of the spell, the possibility that the horse figures may also offer a visual embodiment of the target's horses, who are also named and cursed in the spells.

86. Preisendanz describes the design as follows, based on his assumption that the figures should be interpreted in close connection with the accompanying text: (1) the central figure is a horse-spirit (*daimon*), holding a whip in one hand, symbolizing both power and the charioteers' whip, and a circular object (hoop or disk?) in the other, symbolizing the wheels of the chariot; the design under the horse's left foot represents the yoke or chassis of the chariot; (2) the two figures just below the arms of the horse figure are the *paredroi* mentioned in the text as being on the right and the left; (3) the figure in the upper left is Osiris in his coffin (done in, according to legend, by Seth); the cross marks on his body, on the coffin, on the two *paredroi,* and on other figures are all symbols of binding; in addition, the lines emerging from Oriris's head and from the coffin, and the dots on the coffin depict the nails and pins that were a common feature of binding spells; the same figure shows up in nos. 20B and 29; Richard Lim makes the appealing suggestion that the figure in the coffin might better be taken as the human target of the *defixio*, depicted as dead and buried; (4) the *charaktëres,* the vowel figures and the letter formations based on EULAMO are associated with the spirits/*daimones* as indications of their cosmic power; (5) the mummy figure at the bottom, surrounded by two biting snakes and crisscrossed by lines, is the human target of the spell, dead and buried in accordance with the desires of the client; no. 49, line 54 in Wünsch's collection may contain a reference to the biting snake where the verb *akontizein,* meaning "to wound" or "to dart," may indicate the action of the snake; the star *charaktêr,* which appears under the feet of the mummy, Preisendanz takes to be a symbol of the nails and pins used in binding. Given the frequent occurrence of this symbol in many other *defixiones,* Wünsch is probably closer to the truth in arguing that it represents Osiris or the sun. In line with his general principle of locating correlations between the texts and the designs, Preisendanz offers the view that the one figure not mentioned in the text, the snake, is probably to be identified with Eulamo to whom is addressed the command, "Restrain!" The designs on another tablet, no. 20B, are virtually identical to those of our tablet, with but one exception— 20B shows several bound figures rather than the single one of no. 16.

87. DTA 3, a fragmentary item, uses the Latin word for freedmen, *collibertos.*

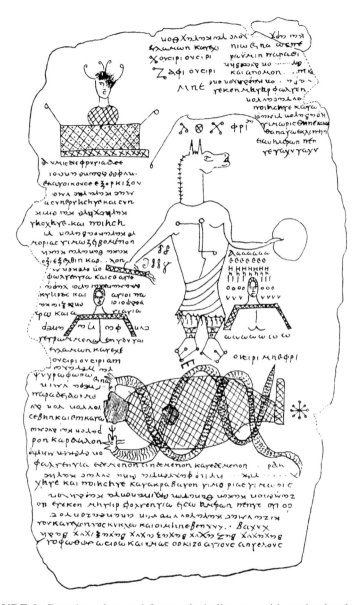

FIGURE 9. Drawing of several figures, including one with equine head, with *charaktêres* and *voces mysticae* on a *defixio* from Rome. The mummified figure at the bottom, being attacked by two snakes, probably represents the target of the binding action, in this case a rival jockey.

no. 16, pp. 14–19; *DT* 140–87; K. Preisendanz, *Akephalos. Der kopflose Gott* (Leipzig, 1926), pp. 22–41; idem, "Fluchtafel," cols. 17–18; Bonner, *Amulets*, p. 114.

(*Side A*) *EULAMÔN restrain. OUSIRI OUSIRI APHI OUSIRI MNE PHRI.[88] [I invoke you, holy angels] and archangels by the (one in the) underworld[89] in order that just as I hand over to you that impious, lawless and accursed Kardêlos, to whom his mother Pholgentia[90] gave birth, so put him on a bed of torment and make him suffer the penalty of an evil death[91] and expire within five days. Quickly, quickly! the spell[92]: To you, Phrygian goddess[93] and Nymph goddess[94] and EIDONEA[95] NEOIEKATOIKOUSE,[96] I invoke you by your [names] . . . in order that you lend a hand and restrain and hold back Kardêlos, to whom his mother Pholgentia gave birth; and make him bedridden and (make him) suffer the penalty of an evil death and come to his end in a bad condition. And you, holy EULAMÔN and holy *charaktêres* and holy assistants,[97] those on the right and on the left, and holy Symphonia,[98] who are written down on this tablet

88. This series of "words," which reappears several times in the spell, contains cryptic remains of names of Egyptian deities: OUSIRI is Osiris; MNE is Mnevis, the bull worshiped at Heliopolis; AP(H)I is Apis, the bull worshiped at Memphis; PHRI might be derived from Ra, the sun god of Egypt, whose name was pronounced P(h)re in the early Hellenistic period.

89. The phrase *tô katachthoniô* appears here and below. The reference is almost certainly to the spirit of the dead person in the grave, at the door of the underworld.

90. The target of the spell is thus Cardelus, a well-attested Latin name. His mother was Fulgentia, similarly well attested as a name.

91. This phrase is difficult. The Greek word *krab(b)aton* is not widely used, but appears to designate a bed or a couch. Thus *kata krabaton* must mean something like "while in bed" or "while lying down." Wünsch (p. 100) proposes to interpret the phrase with reference to the ladderlike figures on several tablets, which he takes to be instruments of torture. Thus the phrase here would mean that the target should suffer torments on a bed of torture.

92. Another instance where the scribe mistakenly copied instructions from the formulary, rather than beginning with the following text.

93. Preisendanz (*Akephalos. Der kopflose Gott* [Leipzig, 1926], p. 32) proposes a different interpretation. He takes the Greek *dee* as a transliteration of a Latin plural, *deae* ("goddesses") and *Phrugia* not as a reference to the well-known "Phrygian goddess" but to local spirits named Phrygia because they caused people to "dry out" (*phrugios*).

94. This translation is based on Preisendanz, who takes the second *dee* as a Latin plural for *deae*. The Nymphs thus represent the water-spirits of wells where *defixiones* were customarily deposited. Jordan (SGD, p. 167) reports an unpublished *defixio* from a pool or bathhouse in Corinth, which is addressed to "holy and powerful nymphs."

95. At the same point in the spell other texts read ADONAI, one of the names for the god of the Hebrew Bible. EIDONEA here is probably a garbled version of ADONAI. Alternatively, it might be taken as an epithet of Hekate, *aidônaia* ("infernal"); cf. *PGM* IV, line 2855.

96. Wünsch takes these letters as mysterious invocations. Preisendanz (*Akephalos*), citing parallels in other tablets (esp. no. 19, line 6ff.) from the same find, argues that they are in fact a garbled version of the phrase, *en chôrô katoikousai,* "who live in this place."

97. The Greek term here is *paredroi,* used in the papyri of the spirits or *daimones* who attended and assisted *magoi* in the course of their activities.

98. Wünsch relates Symphonia to a book by the same name attributed by the Christian author, Epiphanius, to the Sethian Gnostics. Preisendanz (*Akephalos*, pp. 34ff.) insists that the word, which he reads as *sumphôna* rather than *sumphônia*, refers clearly to the "box" of vowels,

[taken from a water conduit]—EULAMÔN restrain OUSIRI OUSIRI API OUSIRI MNE PHRI—in order that just as I hand over to you this impious, accursed, and miserable Kardelos, to whom his mother Pholgentia gave birth, bound, fully bound, and altogether bound, in order that you may in the same way restrain him—Kardêlos to whom his mother Pholgentia gave birth—and make him bedridden and (make him) suffer the penalty of an evil death and expire within five days—Kardêlos to whom his mother Pholgentia gave birth. For I invoke you by the one who grows young, under the Earth, and restrains the circles (of the zodiac)[99] and OIMÊNEBENCHUCH *BACHUCH BACHACHUCH BAZACHUCH BACHAZACHUCH BACHAXICHUCH BADÊGOPHÔTHPHTHÔSIRÔ. And I invoke you holy angels . . .[100]

14. Rome (same as no. 13); Wünsch, *Sethianische*, no. 29. Lead tablet measuring 9 × 10 cm. and written on both sides (Figure 10). Of particular interest is the scene depicted on the bottom of the tablet where two figures (one human, one birdlike) are portrayed in the act of binding the target of the spell, who lacks head and feet. Other familiar elements include a coffin with bust of the human target, depicted with pins or nails; traces of the two *paredroi* or spirit assistants on either side of the horse-headed figure; and two ladderlike drawings to the right of the horse-headed figure.[101] The human targets are racers associated with the Blue team.

(*Side A*) Column A: This is the Spell[102]: (I appeal) to you Phrygian goddess and Nymph goddess EIDÔNEA in this place that you may restrain Artemios, also called Hospês, the son of Sapêda, and make him headless, footless and powerless with the horses of the Blue colors and overturn his reputation and victory. Snatch away[103] Artemios. I ask you by the one who, under the power of Necessity, restrains the circles (of the zodiac) and OIMENE . . .

(*Column B, top*) *EULAMÔN restrain OUSIRI OUSIRI AGI OUSIRI MNE PHRI (with *charaktêres*).

(*Column B, bottom*) BENCHUCH *BACHUCH CHUCH that you may restrain . . .

(*Column C*) . . . Artemios, also called Hospês, the son of Sapêda. I ask you, holy *EULAMÔN, by your power and by the holy *charaktêres* . . . until/on the twelfth and the twenty-fourth (of the month?).

written at this point in the right column of the tablet and explicitly mentioned in the immediately following phrase, "which are written on this tablet."

99. Here occurs the common Egyptian notion that the gods renew the universe each night, in constant cyclical fashion.

100. The text continues on the other side of the tablet, largely repeating the formulas and invocations of the first side.

101. Wünsch (p. 100) regards the ladders as instruments of torture.

102. Another instance where the scribe mistakenly transcribed the label of a spell.

103. Wünsch treats the word as uncertain; Faraone has proposed to read it as *hêrpagête*, from *harpazô*, "to snatch away."

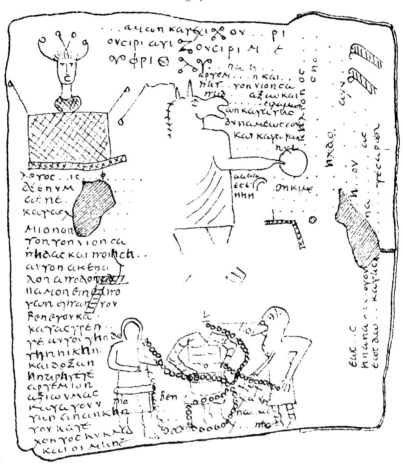

FIGURE 10. Drawing of several figures (top), virtually identical to those on Figure 9. At the bottom appears a scene of two figures in the act of binding a third; the figures probably represent, from left to right, the client, the target and the *daimon* addressed in the spell.

(*Side B*) (Restrain) . . . Restoutos the son of Restouta[104] . . . (I appeal to you) holy Nymph that you may lend a hand and restrain them completely. Quickly, quickly!

104. The same person is the target of no. 28, where he is further identified by his nickname, Artikakon. Wünsch proposes a derivation of the nickname from *artikaktos*, "artichoke."

15. Rome; discovered in an urn or vase ("anfora"), together with ashes of a dead person. Lead tablet measuring 11 × 17 cm.; originally folded several times.[105] The upper three-quarters of the tablet consists entirely of drawings and letter formations[106]; the text comes at the very bottom. The spirits addressed are "holy angels and holy names," which are clearly meant to be represented by the various letter formations depicted in the designs. Once again, the occasion is chariot racing and the effort to place a spell on one's competitors. *Bibl.:* R. Wünsch, *Sethianische Verfluchungstafeln,* no. 49; K. Preisendanz, *Akephalos. Der kopflose Gott* (Leipzig, 1926), pp. 38–41.

(*Column A*). . . . . OM . . . . ÔN . . . (a letter formation probably based on *EULAMÔN) . . .

B

AI

MO

IA

SPH

SR

MA

.

.

.

RE

IR

SG

EIZ

Z

ZÊ

ZÊR

ZÊRÔ

ZÊRÔSI

105. Preisendanz (*Akephalos*) describes the figures as follows (pp. 39ff.): (1) in the upper left-hand corner appears a horse head surrounded, as in other *defixiones,* by snakes, six in number; this horse and the second one below are the horses of Eucherios's team, depicted as bodiless and legless; (2) to the right is the lower part (foot?) of a wrapped mummy, representing Eucherios, the human target of the spell, and next to it another snake; (3) below the horse head comes a second mummy figure, with no clear shape, possibly representing Osiris; (4) in the lower right appears a second horse head resting on a triangular base (with customary cross marks indicating the act of binding); the triangle itself resembles the pegs or nails in the drawing of no. 11 in Wünsch's collection; (5) just beneath the triangle-with-horse appears a sketchy vase, probably a *hydria,* suggesting that the horse figure is to be inserted into the vase; the scene with the vase may thus duplicate the entire process of binding.

106. Other pieces from Wünsch's collection (nos. 6–8) appear to have included only drawings and mysterious words, but no spell proper.

ZÊER . SI
ZÊR . THIS
ZÊRUÊSI

. . .

E
ÊKE . .
LUMÔ

. . .

SÔMALUE[107] and TH
SÔMAUL
SÔM.AL
SÔMA
SÔM
SÔ . . . (and so on in each of the four columns).

Columns B and C consist of several successive wing formations based on EULAMÔ.

I invoke you, holy angels and holy names, join forces with this restraining spell[108] and bind, tie up, block, strike,[109] overthrow, harm, destroy, kill, and shatter Eucherios the charioteer and all his horses tomorrow in the arena of Rome.[110] Let the starting gates not [open] properly. Let him not compete quickly. Let him not pass. Let him not squeeze over. Let him not win. Let him not make the turn properly.[111] Let him not receive the honors. Let him not squeeze over and overpower. Let him not come from behind and pass but instead let him collapse, let him be bound, let him be broken up, and let him drag behind by your power. Both in the early races and in the later ones.[112] Now, now! Quickly, quickly!

16. Raraunum, Roman Gaul (modern France), on the Roman road betwween Saintes and Poitiers. Part of a cache of a dozen lead tablets found near a spring. Originally rolled up and pierced by a large nail, the tablet measures 7 × 9 cm. It is dated to the late third century C.E. The writing is difficult to decipher and consists of a mixture of cursive, capi-

107. A wing-formation based on EULAMÔS spelled backward.
108. The Greek term is *katochos,* designating the tablet and its contents.
109. The Greek verb is *akontizein,* which Preisendanz takes as indicating the striking of a snake or a javelin.
110. A precise identification of the arena is not possible. One possibility would be the Circus Maximus. No. 22, line 19, mentions an otherwise unknown "circus of (new) Babylon, that is, Rome."
111. The Greek verb is *kamptein,* a technical term for negotiating the end post (*kamptêr*) in chariot racing. All of the verbs used to describe the imagined misfortunes of Eucherios were familiar in racing circles and recur in other tablets.
112. The terms indicate the different races held in the morning (*proinas*) and in the afternoon (*aparista*).

tal, and uncial scripts; the language is a local Latin dialect, not Celtic, as Camille Jullian, a learned Celtic scholar and original publisher of the tablet, once thought. Three persons or deities, otherwise unattested, are invoked—Apecius, Aquanno, and Nana. The context seems to be professional rivalry among theatrical actors, the clue being supplied by the use of the word for "pantomime actor" (*mimus*) to describe Eumolpus and Fotius, targets of the spell. See *DT* 15 and 16 for parallel curses against theater mimes. *Bibl.:* R. Egger, "Die Fluchtafeln von Rom (Deux Sèvres): Ihre Entzifferung und ihre Sprache," *Abhandlungen der österreichischer Akademie der Wissenschaften, Phil.-hist. Klasse,* 240 (1962): 348–69; original publication by C. Jullian in *Revue celtique* 29 (1898): 168ff.; *DT* 110; Wünsch (1900) p. 268, no. 20; H. S. Versnel, " 'May he not be able to sacrifice . . . ,' Concerning a Curious Formula in Greek and Latin Curses," *ZPE* 58 (1985): 247–69, esp. 247–48, 269.

Apecius,[113] may you (singular) bind Trinemetos [and] Caticnos[114];
may you strip bare Seneciolus, Asedis,[115] Tritios,
Neocarinos, Dido.
May Sosio become delirious, may Sosio suffer from fevers,
may Sosio suffer pain everyday.
May Sosio not be able to speak.
May Sosio not triumph over Maturus and Eridunna,[116]
may Sosio not be able to offer sacrifice.[117]
May Aquanno[118] torment you.
May Nana[119] torture you.
May Sosio not be able to outshine the pantomime actor
Eumolpus.
May he not be able to play [the role of] a married woman in
a fit of drunkenness on a young horse.[120]
May he not be able to offer sacrifice.
May Sosio not be able to snatch the victory from the
pantomime actor Fotius.

---

113. One of three underworld deities invoked here.
114. Perhaps a Celtic name; possibly a Greek name, *katanikos.*
115. Another Celtic name.
116. Another Celtic name. Eridunna, Eumolpus, and Fotius are probably names of actors.
117. The phrase, *ne voteat imolare,* has been linked by Versnel ("May he not be able to sacrifice . . . ," pp. 249, 263) to other Greek and Latin curses that seek to put the victims in a dangerous position from which they could not even try to placate the wrath of the gods.
118. Perhaps one of the underworld deities, Aquaticus or "Waterman," who figured in later Carolingian rural folklore, or the name of a dead person or *daimon.*
119. Perhaps an underworld deity.
120. This is a recognized actor's role; see Festus in W. M. Lindsay, ed., *Glossaria Latina* (Paris, 1913), 281, and Egger, pp. 364–65.

17. Sicily; original location not certain (possibly near Gela), though there is some indication that it came from a grave. Lead tablet measuring 17 × 6 cm.; originally folded. The other side of the tablet reveals a text dealing with personal and financial matters. The Greek is typical of the Doric dialect found in the Greek colonies of Sicily and closely resembles that of the "Great *Defixio*" from Selinus (see pp. 139–41). The date is roughly 450 B.C.E., which places it among the very earliest of all surviving Greek *defixiones*. The text consists of fourteen lines, from which only a few letters are missing. No spirits or deities are mentioned, though they were probably invoked orally at some point during the preparation and burial of the tablet. The person who commissioned or inscribed the tablet speaks in the first person ("I curse") but the binding is intended to favor his friend Eunikos. The occasion is clear—competition and rivalry among local *chorêgoi,* directors of theatrical choruses. The purpose of the spell is to guarantee victory for Eunikos by binding his competitors, their supporters, and their families.[121] There is no reason to believe that the text of the spell was copied from an existing model, in contrast to the situation in later Roman tablets. *Bibl.:* Anne Pauline Miller, "Studies in Early Sicilian Epigraphy: An Opisthographic Lead Tablet" (Ph.D. diss., University of North Carolina at Chapel Hill, 1973), pp. 65–108; SGD 91; Jordan, personal communication of corrected text and translation based on further inspection of the tablet.[122]

> Luck/curse.[123] (I) Apelles (am writing) because of (my) love/friendship[124] for Eunikos.[125] Let no one be more successful/eager than Eunikos, or more loving/friendly, but that he should praise (Apelles?) both willingly and unwillingly and should love (him). Because of (my) love/friendship for Eunikos, I register[126]

121. The names cited in the text are typical for Greeks on Sicily.

122. David Jordan has informed us that the published versions of this text require correction—notably, L. Dubois, *Inscriptions grecques dialectales de Sicile* (Rome, 1989), no. 134. Jordan and Anne Miller plan to publish a revised text, translation, and commentary of this important tablet. In the meantime, Jordan has graciously allowed us to make use of his revised translation.

123. Miller reads *eucha;* Jordan proposes *tucha. Eucha* is not otherwise attested in *defixiones,* but is used elsewhere to describe the curse of Oedipus against his sons (Aeschylus, *Seven against Thebes* 820: "according to their father's curse"; Euripides, *Phoenician Women* 69–70: "he imprecates the most unholy curses on his sons").

124. The nature of the friendship is not specified. In the nature of things, it must have involved political and personal patronage of some sort; so Faraone, "Context," p. 31, nn. 79 and 81.

125. The name means "good at winning," which Miller takes as a possible indication that Eunikos was a professional.

126. The Greek verb is *apographô,* not otherwise attested in curse tablets, although other forms of *graphein* are common. The term is no doubt a legal one in origin and is used here in an extended sense.

all *chorêgoi*[127] for failure in word and deed[128]—and their children and fathers[129]—and to defeat both in the contests and outside the contests,[130] (all those) who would outstrip me.[131] Kaledias I curse, away from Apelles, and all those there . . . Sosias I curse, away from the shop of Alkiadas because of his love/friendship for Xanthios. Purrhias, Musskelos, Damaphantos, and the (name missing . . .) I curse away from the children and fathers, and all others who arrive here[132] so that no one be more successful with men or women than Eunikos. As this lead tablet[133] (is inscribed) so let . . . preserve victory for Eunikos everywhere . . . Because of (my) love/friendship for Eunikos I write (this).

127.   Miller suggests that the meaning of *chorêgos* here need not point to Eunikos as a wealthy patron, as at Athens, but might indicate that he was a trainer of the chorus.

128.   A common "wish" in curse tablets; cf. *DT*, nos. 68, 69, 302 (see pp. 000–000).

129.   The two references to children and fathers may indicate that there was a family tradition among *chorêgoi.*

130.   Clearly some form of public competition lies in the foreground. Such competitions involved either lyric choruses for poetic recitation or full dramatic productions. In either case, the selection and training of a chorus would have been at stake. Miller (pp. 83–85) lists a number of famous poets and dramatists (including Aeschylus) who visited Sicily and staged public performances there. These occasions must have required public competitions and no doubt generated sharp rivalries.

131.   This is close to a translation proposed by Miller in her discussion but not included in the translation proper. Her reasoning is that while this version makes better sense of the words, it raises problems of interpretation, namely, the other *chorêgoi* might have known about the *defixio* and the literary contest might thus have been subject to "forcible persuasion or even, perhaps, sabotage" (p. 89). But as we have argued, we have no choice but to assume some degree of public knowledge concerning such efforts. Furthermore, what is a *defixio* if not an attempt to change things by force, persuasion, and sabotage?

132.   This phrase raises the possibility that some chorus leaders may have come, as did the dramatists and poets themselves, from the outside, thus further intensifying the sense of competition and rivalry.

133.   At this point the text reads *bolimos,* which is used in other curse tablets instead of the Attic form, *molubdos,* to refer to the lead tablet itself.

2

Sex, Love, and Marriage

It does not require a deep commitment to Freudian psychology to recognize that the arena of sexual passion and fantasy generated the two basic circumstances associated with the use of *defixiones* in the ancient world—competition involving status and uncertainty of outcome. Thus it should come as no great surprise to discover that roughly one-quarter of all surviving tablets concern "matters of the heart."[1] While the preserved specimens—whether as recipes in the handbooks used by professional *magoi* or the tablets themselves—emerge rather late in the story (the earliest erotic tablet dates to the fourth century B.C.E., the close association between eros and the use of spells surfaces already in the earliest literary products of the Greek world, the *Iliad* and the *Odyssey*.[2] Indeed, given the fundamentally oral character of even the written spells, it is probably not too much to speculate that the marriage of eros and spells considerably antedates even the Homeric epics.

In book 14 of the *Iliad* (216ff.), Zeus's wife Hera borrows Aphrodite's embroidered girdle, "in which all her power resides, Love and Desire and the sweet bewitching words that turn a wise man into a fool." But whom does Hera wish to seduce? Her own husband, the father of men and gods! And it worked, tucked away in her bosom like many a later charm, for Zeus responds to her with unaccustomed ardor, overcome by passion and the forgetfulness of sleep. Similar procedures, no doubt much older than the epic itself, appear in book 1 of the *Odyssey,* where the more-than-human Calypso resorts to charms and spells in her unsuccessful effort to have Odysseus forget his native Ithaca and his wife Penelope in favor of remaining forever with her. Here again there is something odd about the circumstances. Why should someone more powerful than humans to begin with be forced to resort to charms and spells to work her will and, to add insult to injury, fail in her efforts? Perhaps the gods, no less than humans, suffer the afflictions of anxious

78

desire and recognize in passionate eros a power greater than all. Or perhaps male authors are no more reluctant to project their own fantasies and anxieties onto female deities than onto female humans.

However that may be, we cannot miss certain continuities between the early literary occurrences of "love" spells and later *defixiones*[3]:

(1) In Homer, Theocritus (*Idyll* 2 concerning Simaetha's efforts to regain the affections of her departed lover),[4] and Lucian of Samosata (see pp. 255–56), it is primarily women who resort to ritual means in order to charm the targets of their passion. Although in the later recipes and tablets men sometimes take the initiative in these affairs, it is worth noting that a number of women seized the opportunity to commission charms and spells. Here, then, we find one arena of ancient life where women were not only active participants in shaping their private lives but initiators of action in the public realm.

(2) The means employed—verbal spells and ritual instruments of various sorts—no doubt varied according to time and place, but we never fail to find both spells and instruments, usually together. This is not to say that *defixiones* as such, in which oral spells are recorded on strips of metal, predated Homer; it may very well be that the particular technique of using *defixiones,* attested much earlier in nonerotic affairs, was adopted by nervous lovers on the basis of its reputation as a successful device in constraining the behavior of other persons, most notably in legal matters. Indeed, one common element in *all* binding spells is constraint, through the deployment of powerful formulas, names, figures, and other materials, for the purpose of bending the actions and sentiments of others according to one's own desires. Thus an eventual transfer of *defixiones* to the realm of sex and love, if this is how things developed, must have happened quite naturally. Faraone suggests that legal *defixiones* were first "converted" into separation love spells, primarily in Greece, and only later made the further transition to attraction love spells.[5]

(3) Loss of memory plays an important role throughout—in legal *defixiones* as well—no doubt because memory was seen as the locus of social ties and obligations that interfered with erotic relationships (real or fancied) of one kind or another.[6] Just as Calypso seeks to erase Odysseus's memory of Ithaca and Penelope, so love tablets regularly insist that the object of the client's passion should forget all others.

The traditional label for these tablets as "love spells" gives a seriously misleading impression of the various types that appear even in the limited sample provided in this chapter. Faraone's basic division into *separation* spells, usually involving triangular relationships of one kind or an-

other, and *attraction* spells, designed to charm a second party, has been
subdivided further.[7] Within the first category, Petropoulos distinguishes
between attraction spells (*agôgai*) and binding tablets (*philtrokatadesmoi*).[8] Winkler offers spells to curse rivals, to divorce or separate couples, to cause a downturn in a pimp's business, and to attract a lover.[9]
Further subdivisions are no doubt possible. Our interest, however, is not
to formulate a set of categories but to underline the complexity of types
and circumstances, corresponding no doubt to the illuminating variety of
needs and fantasies experienced by the love-struck clients themselves.

We have already noted the prominence given in literary texts to
women as initiators of amatory spells and devices. In other areas involving literary representations of women as witches and purveyors of spells,
there is good reason to believe that the literary tradition has projected a
systematic distortion. Yet in this one area, relating to women's use of
erotic *defixiones,* the literary testimony finds at least partial confirmation in the tablets and recipes. Winkler asserts that these texts "are
predominantly composed by (or on behalf of) men in pursuit of
women,"[10] but he also observes that a number fall into rather different
groupings:

1. women in pursuit of men: *PGM* XV, XVI, XIXb, XXXIX; *DT*
100, 230; and no. 18 in this volume;
2. women in pursuit of women: *PGM* XXXII; SGD 151=*SuppMag*
42[11];
3. men in pursuit of men: *PGM* XXXIIa; and our no. 25(?)[12];
4. recipes to deliver men or women; *PGM* I, line 98; IV, line 2089;
and our no. 31.

In short, the professional dispensers of erotic charms were prepared to
serve all possible clients and relationships.[13]

The tablets and recipes paint a rather intimate portrait—to be sure in
the formulaic language of ritual—of the sexual appetites, anxieties, and
fantasies of ancient men and women. As such, they offer a welcome
antidote to the skewed, largely masculine perspective of the literary
tradition. Basing his observations on recent studies of Mediterranean
culture in general, Winkler asserts that women in love "are considerably
more watched and guarded and disciplined than their brothers, and
presumably had less access to male experts with their books and to
money for hiring them."[14] Yet, even more than Winkler seems prepared
to admit, both the tablets and literary figures such as Lucian demonstrate that women resorted to precisely the same ceremonies, spells, and
devices as did men. Thus it is not too much to insist that it is precisely

through *defixiones* that women emerge from their stereotyped seclusion and passivity in aggressive pursuit of their own erotic dreams.

There can be no mistaking the deeply aggressive, even violent language of the amatory *defixiones*. Various tablets specify that the object of passion is to be dragged by the hair, deprived of memory and sleep, tormented by passion, and killed with madness. In another case (no. 27), a female figurine has survived, pierced with thirteen needles.[15] Yet these items are clearly not curses in any shape or manner. Their explicit goal is not to harm the target but to constrain her. How are we to interpret the vivid language of these spells, not just in its aggressive forms but equally in its undisguised sexuality ("join belly to belly, thigh to thigh, black to black," "bring her thigh close to his, her genitals close to his in unending intercourse for all the time of her life")?

Several responses seem in order:

(1) In part, the violent language may be explained by the developmental history of *defixiones*. If, as suggested earlier, the deployment of binding spells in erotic affairs was taken over from the legal and judicial realm, where hostile techniques and formulas for harming and binding one's enemies were quite "appropriate," then it seems likely that the aggressive language simply came along with the transfer.

(2) We must exercise care not to misread a "document" like the figurine with needles. The spoken words that accompany the insertion of the needles are not "I harm you!" or "I wound you!" but "I pierce whatever part of you so that you will remember me!" In short, like much else in this and similar texts, the language is deeply symbolic and will simply not allow an overly literal interpretation. The closest modern analogy for understanding these needles is thus not the so-called "voodoo" dolls, from Haiti and elsewhere but instead the therapeutic use of needles in Chinese acupuncture. To this we must add a single caveat: the figurine is gender-specific. The penetration of the female figurine by needles probably carries sexual meaning as well.

(3) In his analysis of intense desire as a form of illness in Greek culture, requiring treatment, Winkler has indicated yet another important feature of love spells, namely, their therapeutic function.[16] Through a dual process of transference and projection, the spell reverses the actual emotional state of the two parties and thereby accomplishes a double goal—the illness is removed from the client by its projection onto the target; the target is thus made, in some fashion, to conform to the *client's* desires: "The rite assigns the role of calm and masterful control to the performer and imagines the victim's scene as one of passionate inner torment. But if we think about the reality of the situation,

the intended victim is in all likelihood sleeping peacefully, blissfully ignorant of what some lovestruck lunatic is doing on his roof."[17] In short, the "real" target of the spells is the disease, the inflamed passion of the client, whereas the spell, or rather the series of actions leading up to its procurement, functions as a cure by projecting the passion onto the other.[18]

(4) In his trenchant critique of Frazer's *Golden Bough,* Ludwig Wittgenstein reversed Frazer's understanding of ritual action by interpreting it as directed toward effecting a change not in the external world of nature but in the internal world of the actor. "Burning in effigy. Kissing the picture of the loved one. This is obviously *not* based on a belief that it will have a definite effect on the object which the picture represents. It aims at some satisfaction and it achieves it. Or rather, it does not *aim* at anything; we act in this way and then feel satisfied."[19] In other terms, the erotic *defixiones*—and the other types as well—deal primarily with the inner world of *the client's* fantasy and imagination. Their goals are largely realized in the very act of commissioning and depositing the tablets.[20] Once again, it would be a serious error to read them in an overly literal fashion. It may not be too much to propose that the chthonic powers to whom the tablet is dedicated represent the *client's* sense of domination by psychological forces beyond his or her control; the violent language expresses the turbulence of the erotic passion; and the desire to dominate the target manifests an effort to regain control of oneself.[21]

These considerations cast in a new light one of the central assumptions in virtually all traditional views of *defixiones,* their exclusively private and secret character. Indeed, the position of Frazer and his successors, which has commanded all interpretations of "magical" action—the position that all "magic" can be written off as nothing more than a series of mistaken beliefs about the world, leading to ineffectual efforts to control the natural order according to one's own desires—depends heavily on the claim that such activity is always isolated, private, and secret. But here and there in the ancient record, there are signs that the act of commissioning and deploying a *defixio* formed part of a broader amatory strategy, some of whose aspects were public. In brief, the intended target of the spell was almost certainly aware that someone had commissioned and deposited an erotic tablet because she or he was already the object of that person's attention. When Simaetha, in Theocritus's *Idyll* 2 (ca. 240 B.C.E.), intones her binding spell over her ritual apparatus (*iunx*)—"Draw my lover to my house, oh powerful iunx!"—she does not leave the matter there but adds, "And tomorrow I will go myself to

Timogetus's wrestling school to scold him for the way he has treated me." Similarly, more than six centuries later, Eunapius records an incident of erotic spell and counterspell involving the "divine philosopher" Sosipatra and her love-sick relative, Philometor.[22] He, overpowered by *eros,* had undertaken certain rituals—perhaps including the commissioning of a *defixio*—that had taken effect. The philosopher, afflicted but unaware of the source, pleads with a confidant, the well-known theurgist Maximus, to discover the cause of her ailment and to cure her. Maximus, having discovered Philometor's strategy, unleashed a powerful antidote, which brought the affair to an end. Apart from its value as testimony for the use of spells and antidotes in high places, the immediate relevance of the story is that if Philometor's amorous assault on Sosipatra has been an entirely private affair, Maximus, despite his renown as a theurgist, would not have discovered Philometor's plot.

Finally, we need to ask what these spells and tablets have to do with love—romantic style—for their traditional label has been "love spells." In some cases, as Winkler observes, "love" is not the right word.[23] Some tablets fall into the category of affairs or flings; others involve nothing more than fantasies. But a few do point to something like modern notions of romantic love, marriage, conjugal fidelity, and honor. Domitiana (no. 36) pleads to be united with Urbanus in marriage and love for the rest of their days.[24] In other cases, the issue is given a reverse spin in the sense that jealous spouses or lovers intervene with the spirits to break up an affair and induce the wandering partner to return to a stable and exclusive relationship. Here we must take note that prostitutes and concubines appear to have played a role in a number of cases (nos. 18, 21) in which a wife resorted to extraordinary measures in order to terminate her husband's involvement with a concubine (*hetaira*).[25] Such relationships were quite common throughout the ancient world and occasionally took the form of moving the *hetaira* into the husband's household. The tablets indicate that wives did not always approve. No doubt the concubines themselves made use of amatory charms, or were suspected of doing so, to win the affections of married men.[26]

Notes

1. Two excellent studies on the subject have appeared recently: see J. C. B. Petropoulos, "The Erotic Magical Papyri," in *Proceedings of the XVIII International Congress of Papyrology,* vol. 2 (Athens, 1988), pp. 215–22; and J. J. Winkler, "The Constraints of Desire: Erotic Magical Spells," in *The Con-*

straints of Desire (New York, 1990), pp. 71–98. (Winkler's essay appears also in *Magika*, pp. 216–45.)

2. On the many passages linking eros and spells, see R. Flacelière, *Love in Ancient Greece* (London, 1962), pp. 14–15 and 137–40; Petropoulos, "Erotic Papyri"; and Winkler, "Constraints," pp. 79ff.

3. Petropoulos, "Erotic Papyri," pp. 221–22, offers a particularly powerful argument to the effect that the techniques that appear later in Egypt must have existed long before Hellenistic and Roman times.

4. For a full discussion of the parallels between Theocritus's treatment of Simaetha's rituals in *Idyll* 2 and later formulas in the papyri, see A. S. F. Gow, *Theocritus*, vol. 2 (Cambridge, 1952), pp. 33–36.

5. Faraone, "Context," pp. 15–16.

6. See the full discussion in Petropoulos, "Erotic Papyri," pp. 219–20.

7. Faraone, "Context," p. 13.

8. "Erotic Papyri," p. 216. Petropoulos notes, additionally, the wide variety of terms for erotic tablets: *agôgai, philtrokatadesmoi, philtra, katochoi, potêria, diakopoi,* and *phusikleidia.*

9. Winkler, "Contraints," p. 94.

10. Ibid., p. 90.

11. This elaborate tablet, written on both sides and complete with some thirteen lines of Greek verse (iambic trimeters and choliambics), is now available in *SuppMag* 42, with full translation and commentary.

12. Jordan, "Agora," p. 223, n. 16, notes the existence of an unpublished Greek tablet from Tyre that includes the phrase, "May Juvinus lie awake in his love for me, Porphyrios."

13. For a partial listing of homosexual attraction spells, see *SuppMag,* p. 42.

14. Ibid.

15. See the insightful remarks of Winkler, "Constraints," pp. 93–98.

16. Ibid., pp. 87ff.

17. Ibid., 87.

18. The same point is made forcefully by S. J. Tambiah, "The Magical Power of Words," *Man* 3 (1968): 202: "Thus it is possible to argue that all ritual, whatever the idiom, is addressed to the human participants and uses a technique that attempts to restructure and integrate the minds and emotions of the actors."

19. *Remarks on Frazer's Golden Bough,* ed. R. Rhees (Atlantic Highlands, N.J., 1979), 4e.

20. A similar point is made by John Beattie, *Other Cultures: Aims, Methods and Achievement in Social Anthropology* (New York, 1964), chap. 12 ("The Field of Ritual: Magic"), p. 204: "Once the essentially expressive, symbolic character of ritual, and therefore of magic, has been understood. . . ."

21. So also Winkler, "Constraints," pp. 87–88.

22. *Lives of the Philosophers* 468–69 (pp. 410–15 in the translation of W. C. Wright, Loeb Classical Library [Cambridge, Mass., 1922]).

23. "Constraints," p. 72.

24. A recently discovered and very early (375–359 B.C.E.) Greek tablet from ancient Pella in Macedonia speaks in similar terms. The client, a woman named Thetima, appeals to the *daimones* as follows: "May he indeed not take another wife than myself but let me grow old by the side of Dionusophon." We are indebted to the editor of this tablet, Dr. Emmanuel Voutiras, for granting us permission to cite from his still unpublished edition.

25. On courtesans and prostitutes, not always properly distinguished, see V. Ehrenberg, *The People of Aristophanes* (London, 1947), pp. 194–98; Sarah B. Pomeroy, *Goddesses, Whores, Wives, and Slaves: Women in Classical Antiquity* (New York, 1975), pp. 88–92, 114–17, 139–41, 201–2; M. R. Lefkowitz and M. B. Fant, eds., *Women's Life in Greece and Rome* (Baltimore, 1982), s.v. "concubines," "courtesans," "prostitutes"; A. Rousselle, *Porneia: On Desire and the Body in Antiquity* (Oxford, 1988), chaps. 5 ("Adultery and Illicit Love") and 6 ("Separation, Divorce and Prostitution"); and Winkler, *Constraints*, pp. 199–202 ("The Laughter of the Oppressed").

26. The connection between concubines and love spells remained strong well into the Middle Ages; see M. Rouche, "The Early Middle Ages in the West," in *A History of Private Life from Pagan Rome to Byzantium,* ed. P. Veyne (Cambridge, 1987), who speaks of the notion that "concubines used spells, potions, amulets, and magic of all sorts to inflame passion and hold onto their lovers" (p. 418).

18. Greece, Boeotia; original location not known. Lead tablet measuring 8 × 7 cm.; written on both sides. No date given by the editor. The figures invoked are Earth and Hermes. The form of the spell is simple—the target is handed over to the deities for appropriate action. The action is not specified but clearly presupposes constraint rather than punishment. The occasion would appear to involve a triangular love affair. If so, the client was "the other woman," seeking to steal the affections of Kabeira from his wife. From the various aspects of Zois's personality and character, as listed in the curse, it is obvious that she presented the client with a formidable challenge. *Bibl.:* Wünsch (1900), p. 71; *DT* 86; Ziebarth (1934), no. 22; Faraone, "Context," p. 14.

(*Side A*) I assign Zois the Eretrian, wife of Kabeira, to Earth and to Hermes— her food, her drink, her sleep, her laughter, her intercourse,[1] her playing of

---

1. The Greek term *sunousia* could be used of social or sexual intercourse.

the kithara,[2] and her entrance,[3] her pleasure, her little buttocks,[4] her think-ing, her eyes . . .

(*Side B*)[5] and to Hermes (I consign) her wretched walk, her words, deeds, and evil talk. . .

19. Greece, Karystos on the island of Euboea; exact place of origin not known. Flat lead figurine, measuring 5 × 9 cm.; 5 mm. thick (Figure 11). There are no sexual features to indicate whether the figure is meant to represent a man or a woman; the arms and legs are mere stumps. In-scribed on both sides; only faint traces of letters on side B are visible. There are two inscriptions on side A, one above (beginning at the right arm and covering the upper part of the figurine) and one below and at right angles (beginning at the left leg and covering the lower part of the body).[6] The letters are very small, 3mm. high. Dated by Guarducci to the fourth century B.C.E. In both texts, the god invoked is Hermes the Restrainer, a common figure in binding spells. The target of the spell is a woman, Isias, identified by her mother. No occasion is cited, but judicial proceedings and matters of love and sex frequently made use of figu-rines. If so, the figurine probably represents the "target," that is, Isias herself. *Bibl.:* Robert, *Froehner,* 17–18 (text); M. Guarducci, *Epigrafia greca* IV: *Epigrafi sacre pagane e cristiane* (Rome, 1978), pp. 248–49; SGD 64; Faraone, "Context," p. 3.

I record[7] Isias, the daughter of Autoclea, with Hermes the Restrainer. Re-strain her near you.

I bind[8] Isias before Hermes the Restrainer—the hands, the feet of Isias, the whole body.[9]

2. A common musical instrument, related to the zither.

3. The Greek term *parodos* might mean "entrance" or "passage," thus designating a particu-lar way of entering a room. But it was also used as a technical term in Greek theater and could refer to public recitation. Here it may also have sexual overtones.

4. The Greek term *pugeôn* generally referred to the buttocks but might also be used of certain kinds of dancing, which seems to fit well here where other aspects of performance or entertainment are in focus.

5. The writing on Side B is quite fragmentary.

6. Other figurines have survived with writing, but none of them flat like ours.

7. The Greek is *katagraphein* which may be used in various other contexts: "to enroll," "to summon by written order," or "to convey" (as with property via a written deed).

8. Here the Greek is *katadesmeuein,* an uncommon word, which appears in *PGM* V, line 321, as part of a recipe for producing a binding spell on a sheet of papyrus or lead.

9. Robert notes that another binding spell, from Boeotia in roughly the same period, relating to love and sex specifies the same elements to be bound—hands, feet, and body (no. 20). It is tempting to suggest that the two may represent products of one local Boeotian professional.

FIGURE 11. Flat lead figurine with inscription from island of Euboea (Greece). Although this figurine is without sexual features, it typifies the use of three-dimensional objects in Greek spells of the same period.

20. Greece, Boeotia. A round lead tablet, 9.5 cm. in diameter and inscribed on both sides. Side A is inscribed in eleven circular and concentric lines; side B has twenty-one parallel lines. The Greek is written in the Boeotian dialect, with awkward grammatical constructions. The presence of many seemingly non-Greek and perhaps even nonsensical words and the poor condition of the inscription (especially on side B) make a full and adequate translation difficult; it is possible that some of the undecipherable words represent *voces mysticae*. The date is perhaps second or third century C.E. This binding spell—or as the author of the tablet calls it, a blocking spell—plays on the analogy between the dead person, Theonnastos; the similarly lifeless and buried lead on which this spell was inscribed; and the desired effect on Zoilos against whom the spell was directed. The stated purpose of the spell is to keep Zoilos and Antheira, Zoilos's female lover, apart from one another. From the fact that Zoilos is cited as the primary target of the spell, we may suppose that the person who commissioned the spell was a rival suitor competing for Antheira's favor. *Bibl.:* Wünsch (1900), no. 70, p. 55; *DT* 85; Ziebarth (1934), no. 23 (text); Faraone, "Context," pp. 13–14.

> (*Side A*) Just as you, Theonnastos,[10] are powerless in any act or exercise of (your) hands, feet, body . . . to love and see maidens (?) . . . so too may Zôilos remain powerless to screw[11] Antheira and Antheira (remain powerless toward) Zôilos in the same way, of beloved Hermes (?) . . . the bed and the chitchat and the love[12] of Antheira and Zôilos . . . and just as this lead is in some place separate from humans, so also may Zôilos be separated from Antheira with the body and touch and kisses of Antheira and the love-makings of Zôilos and Antheira . . . the fear of Zôilos (?)
> I inscribe even this blocking (spell)[13] with a seal.
>
> (*Side B*) . . . may you not catch, O god, Antheira and Zôilos [together] tonight and may they not . . . with one another and . . . Timoklês . . . binding spell . . . thus also Zôilos . . . this binding spell . . . just as this lead (tablet) has been completely buried, deeply buried and . . . thus also bury for Zôilos . . . his business and household affairs and friendships and all the rest.

21. Greece, Athens; from a well in the Agora. Lead tablet measuring 10.1 × 6.7 cm. In this case, the person who produced the tablet has applied the formulas not to athletics but to love spells, specifically to the goal of breaking up a relationship between a woman and one or two men; three tablets of this sort were found together. Who was the client—another jealous patron, a relative? One understands that in such relationships the potential for anger and jealousy is always present. A

---

10. The name of the dead person near which this curse tablet was deposited.

11. The text reads *bainimen.* Faraone translates the word based on *bainein* ("to come") but also suggests that the intended verb might be *binein*, "to screw (in a sexual sense)."

12. There is an attempt at wordplay in the Greek by the duplication of endings in this and the previous line: in the first line, -ata, -ta, -ata, and -ta; in the second, -an, -an, -esin, and -esin.

13. The term is *aporia,* meaning something which hinders the ease of passage; in this context, the blocking refers to efforts to prevent Zoilos and Antheira from coming together.

parallel tablet by the same hand, but from a separate location in Athens, portrays sharp enmity between the unnamed client and a woman, Tyche, the daughter of Sophia.[14] Most of the *voces mysticae* on this tablet are identical or closely similar to other *defixiones* from the same cache, suggesting that they were copied from written recipes in a formulary. *Bibl.:* Jordan, "Agora," no. 8, pp. 225–27; SGD 31.

*BÔRPHÔRBABARPHORBARBARBARPHORBABARPHORBABAIÊ Oh powerful BEPTU,[15] I deliver to you Leosthenês and Peios, who frequent[16] Juliana,[17] to whom Marcia gave birth, so that you may chill them and their intentions, in order that they may not be able to speak or walk with one another, nor sit[18] in Juliana's place of business,[19] nor may Leosthenês and Peios be able to send messages to Juliana. And also (chill) in your gloomy air those who bring them together.[20] Bind (them) in the darkened air of forgetfulness and chill and do not allow Proklos (?)[21] and Leosthenês and Peios to have sexual/social intercourse with (her). MONZOUNÊ ALCHEINÊ PERPERTHARÔNA IAIA, I deliver to you Leosthenês and Peios. Powerful Tuphon KOLCHLO PONTONON Seth SACHAÔCH EA, Lord APOMX PHRIOURIGX who are in charge of disappearing and chilling, KOLCHOICHEILÔPS,[22] may Leosthenês and Peios cool off, so that they are unable to talk with Juliana. Just as these names are cooling off,[23] so may the names of Leosthenês and Peios cool off for Juliana and also their soul, their passion,[24] their knowledge, their passion, their charm,[25] their mind, their

14. Jordan, pp. 251–55. Of special interest in this tablet are clear imprints left by pieces of hair; hair is also mentioned in the spell ("I give you Tuche . . . whose hairs these are, here rolled up"). No reason for the enmity is indicated.

15. The opening invocation is virtually identical in all ten of the tablets that Jordan identifies as coming from the same hand.

16. The verb *proserchesthai* might designate visits in a general sense, but in Xenophon, *Symposium* 4.38, it is used of sexual visits.

17. Tablet no. 9 in Jordan's series also mentions a Juliana, the daughter of Marcia, along with Polynikos (a man), as one of the two main characters. There is every reason to suppose that they are one and the same. The issue in no. 9 is to separate Juliana and Polynikos "so that their affection and their sexual intercourse (*sunêtheia*) and their sleeping together may cool off."

18. Jordan comments (p. 227) that the Greek verb *kathizein* might be understood here by analogy with the Latin verb *sedere*, both meaning "to sit." In Latin, the use of *sedere* in such a setting would have a clear sexual implication.

19. The word *ergastêrion* could be used of any place of business, including a brothel; so by the orator Demosthenes, 59.67. Together, this and the earlier remarks merely indicate, without demonstrating, that Juliana was a professional prostitute.

20. Here, as in no. 7 of the same series published by Jordan, the spell includes those who were responsible for bringing man and woman together.

21. Here a third man appears in the text. Is this perhaps the go-between?

22. This series of mysterious names appears in virtually identical form in all ten tablets published by Jordan.

23. The cooling off, while a possible indication of the sexual nature of the relationship, clearly presupposes that the tablet was deposited in a well where conditions would indeed have been quite cool.

24. Twice here the text uses *orgê*, which Jordan translates as "impulse." The last line uses the

knowledge, and their reasoning. May they stand deaf, voiceless, mindless, harmless, with Juliana hearing nothing about Leosthenês and Peios and they feeling no passion or speaking with Juliana.

22. Greece, Attica; original location unknown. A thick tablet inscribed lightly on both sides, broken at left, and measuring 12 × 8 cm. Fourth century B.C.E. The tablet must have been deposited in a grave. The Greek contains many abbreviations. The author hopes that Theodora will cut off relations with Charias, her lover. The curse invokes first Hekate; then an ill-defined category of "the unmarried," or "unfulfilled"[26]; and later Hermes and Tethys (the wife of Oceanos the sea god). *Bibl.: DT* 68; J. C. B. Petropoulos, "The Erotic Magical Papyri," in *Proceedings of the XVIII International Congress of Papyrology*, vol. 2 (Athens, 1988), pp. 219–20.

(*Side A*) I bind Theodôra in the presence of the one (female) at Persephone's side[27] and in the presence of those who are unmarried. May she be unmarried and whenever she is about to chat with Kallias and with Charias— whenever she is about to discuss deeds and words and business . . . words, whatever he indeed says. I bind Theodôra to remain unmarried to Charias and (I bind) Charias to forget Theodôra, and (I bind) Charias to forget . . . Theodôra and sex[28] with Theodôra.

(*Side B*) [And just as] this corpse lies useless, [so] may all the words and deeds of Theodôra be useless with regard to Charias and to the other people. I bind Theodôra before Hermes of the underworld and before the unmarried and before Tethys. (I bind) everything, both (her) words and deeds toward Charias and toward other people, and (her) sex with Charias. And may Charias forget sex. May Charias forget the girl,[29] Theodôra, the very one whom he loves.

verbal form *orgizesthai;* again Jordan translates as "having no impulse toward." Although the more "neutral" meaning appears in early texts for the noun, the verb is almost always used to indicate "anger." In the Greek of our period, the noun too generally means "anger." One plausible target of anger associated with our tablet would be its client, whose anger is already proved by the tablet itself.

25. The term here is *epipompê*, possibly referring to a counterspell; so Jordan, p. 247.

26. The meaning of *atelestos* and *atelês* is usually "without issue." LSJ conjectures that they mean "unmarried" in this instance. However the literal meaning is "someone who has not reached fulfillment"—perhaps a pun for one untimely dead.

27. That is, her daughter, Hekate.

28. *Koitê*, literally means "marriage bed" but certainly with a more explicit connotation here.

29. Wünsch ([1900], p. 65) unnecessarily complicates the scenario by reading *paidion* as Theodora's child rather than as a term of endearment for Theodora herself.

23. Greece, Attica; original location not known. Lead tablet measuring 27 × 3 cm. and originally folded. Wünsch and Wilhelm agree in dating it to the fourth century B.C.E. Like other curse tablets of this period, this one contains no verb but names the target in the accusative case as the direct object of an implied action of binding. Nor is there any spirit or deity explicitly cited in the written text. The occasion is "romantic" jealousy regarding Aristokudes's relationships with other women. If so, the client must have been a woman, perhaps his wife or fiancée. *Bibl.:* DTA 78; Wilhelm, p. 113; Faraone, "Context," p. 14.

(I bind?) Aristokudês and the women who will be seen with him. May he not marry any other woman or young maiden.

24. Greece, Attica; original location not known. The tablet measures 17 × 8 cm. and appears to have been folded. Inscribed on both sides. Here the occasion seems to be competition and jealousy regarding erotic matters.[30] The names in parentheses were deliberately scrambled by the person who prepared the tablet; curiously, the names also appear in their normal form. This unusual procedure suggests that the scrambling represents a symbolic attempt to scramble the persons themselves, rather than an effort to conceal the names from potential human readers. *Bibl.:* DTA 77.

(*Side A*)
We bind (*Kallistratê),[31]
the wife of (*Theophêmos) and
Theophilos, son of Kallistratê,
and the children/slaves of (*Kalli)stratê
both Theophêmos and (*Eustratos) the brother . . . I bind
their souls and their deeds
and their entire selves and all
their belongings.

(*Side B*)
and their penis and their vagina
and Kantharis and Dionusios, son of (*Kantharis)
both themselves and their soul and deeds and
all their entire selves and (their) penis[32] and
unholy vagina. (*Tlesia)

30. C. Faraone has suggested that the binding of penis and vagina may be intended to cause infertility, in which case the tablet would not be erotic in the strict sense.
31. Again in this inscription personal names are often scrambled on purpose.
32. The Greek term is *psôlê*, a rare word used to describe the penis with foreskin pulled back.

(be) cursed.[33] (*Theophêmos Euergos
Kantharis Dionusios).

25. Greece, Nemea, approximately twelve kilometers southwest of Corinth. Found in a pit within a large building.[34] The pit dates from the late
fourth century B.C.E. Its varied contents, including this tablet, must be
earlier; thus second half of fourth century B.C.E. Five additional tablets,
probably inscribed by the same person, were recovered at the same site;
cf. SGD, p. 167. This tablet seeks to separate one man from another. No
spirit or deity is invoked. Particularly interesting is the enumeration of
parts of the body, which gives a symbolical portrait of the appropriate
human anatomy. *Bibl.:* S. Miller, "Excavations at Nemea, 1979,"
Hesperia 49 (1980): 196–97; *SEG* 30.353; SGD 57.

I turn away Eubolês[35]
from Aineas, from his
face, from his eyes,
from his mouth,
from his breasts,
from his soul,
from his belly, from
his penis, from
his anus,
from his entire body. I
turn away Eubolês
from Aineas.

26. Palestine, Horvat Rimon, thirteen kilometers north of Beer Sheva.
Discovered in excavations near the ancient synagogue. A clay potsherd,
now in several fragments, which was deliberately cut and inscribed before being fired. There are traces of black marks, probably caused by
flames when it was subsequently thrown into a fire.[36] Original size approximately 9 × 9 cm., narrower at top (Figure 12). There are nine lines
of text in Aramaic, of which the first two contain six words, each one

---

33. The Greek term is *kataratos.*

34. In *Hesperia* 50 (1981): 64–65, S. Miller announced the discovery of five additional tablets
whose date, writing, and language indicate that they stem from the same person who produced
our tablet.

35. The name is taken here as that of a man, in the Doric form, Eubolês or Eubolas.

36. The editors note that the use of new sherds of pottery or unbaked sherds is prescribed in
various recipes for the preparation of love curses on clay; cf. *The Sword of Moses,* ed. M. Gaster
(New York, 1970), p. XV, lines 17–18, and several fragments from recipe books for love and hate
spells from the Cairo Geniza.

FIGURE 12. Clay pot inscribed with attraction spell and "boxed" *voces mysticae,* from Horvat Rimon (Palestine). The boxed names address the angels by their secret and powerful names. (J. Naveh and S. Shaked, *Amulets and Magic Bowls: Aramaic Incantations of Late Antiquity* [Jerusalem: Magnes Press, Hebrew University, 1985], p. 86 [Fig. 12]. By permission.)

"boxed" by a solid line; the last line contains traces of markings, which the editors call "magic characters." The text is heavily reconstructed. The date is fifth or fourth century C.E., possibly earlier. The curse is clearly Jewish and the occasion is love-longing. The agents invoked are angels. *Bibl.:* Naveh and Shaked, amulet no. 10, pp. 84–89.

HR'WT 'TB'WT QWLHWM SPTWN SWSGR . . .[37] You ho[ly (and mighty)] angels [I adjure] you, just as [this sherd] [burns,[38] so shall] burn the heart of R[. . .] [Mar]ian after me, I. . [. . and turn] [his/her heart and mi]nd and kidney, so [that she/he will do] my desire in this . . . (*charaktêres*).

27. Egypt; the various collections of recipes, called formularies, in *PGM* have preserved models on which specific charms, spells, and *defixiones* were patterned and copied. Until recently, we had only the models and few copies.[39] Now we have several examples of a single, richly complex love spell based on a recipe, or variants of it, preserved in *PGM* IV, lines 296–466: two lead tablets from Oxyrhynchus; one lead tablet from the region of Antinoöpolis (see next item); a lead tablet from Hawara; and a lead tablet at the University of Michigan (Papyrus Michigan 6925).[40] *PGM* IV contains an extensive collection of fifty-three separate recipes dating from the fourth century C.E. The special value of this extensive recipe is that it reveals the full complexity of ritual actions associated with the use of *defixiones*. In other words, it was never a simple matter of buying a piece of inscribed lead and tossing it into an open grave. Our translation largely follows the version of E. N. O'Neil in *GMP*, pp. 44–47.

Marvelous binding spell[41]: Take wax (or clay) from a potter's wheel and form two figures, male and female. Make the male like Ares, fully armed, bearing a sword in his left hand and striking down on her right collarbone; make her with her arms behind her back and resting on her knees. Attach the "stuff"[42] on her head or neck. Write (the following) on this representation of the woman being attracted—on her head, ISEÊ IAÔ ITHI OUNE BRIDÔ LÔTHIÔN NEBOUTO-

37. These words are the secret names of the angels invoked in the spell. Similar names of angels appear in *The Sword of Moses* and in fragments from the Cairo Geniza; see Naveh and Shaked, p. 89.

38. A clear case of symbolic action in which the desired human effect is acted out through a physical object. We must assume that the spell was thrown into a fire.

39. For a full list of duplicate spells, copied from masters preserved in *PGM*, see Wortmann, p. 58, nn. 4 and 5.

40. See D. G. Martinez, *P. Michigan XVI. A Greek Love Charm from Egypt* (*P. Mich. 757*) (Atlanta, 1991).

41. The title or label of the spell, not part of the spell itself.

42. The Greek term is *ousia;* something from or belonging to the "victim" was regularly required for a binding spell. Often this was hair.

SOUALÊTH; on her right ear—OUER MÊCHAN; on her left ear—LIBABA ÔIMATHOTHO; on her forehead—AMOUNABREÔ; on her right eye—ÔRORMOTHIO AÊTH; on the other—CHOBOUE; on her right shoulder—ADETA MEROU; on her right arm—ENE PSA ENESGAPH; on the other—MELCHIOU MELCHIEDIA; on her hands—MELCHAMELCHOU AÊL; on her breast, the name of the woman being attracted, on the mother's side; on the heart—BALAMIN THÔOUTH; and below her stomach—AOBÊS AÔBAR; on her sexual organs—BLICHIANEOI OUÊIA; on her buttocks—PISSADARA; on the sole of her right foot—ELÔ; on the other—ELÔAIOE. Take thirteen copper needles and stick one in her brain, saying "I pierce your brain, so and so." Then stick two in her ears, two in her eyes, one in her mouth, two in her stomach, one in her hands, one in her sexual organs, two in the soles of her feet, all the while saying, "I pierce whatever part of so and so, in order that she may remember no one but me alone, so and so." Then take a lead tablet and write the same spell and recite it. Then fasten the tablet to the figures with a piece of thread from a loom and tie 365 knots while you speak, as you have learned, "*ABRASAX, you are restraining her." While the sun is setting, place it near the grave of one who has died prematurely or by violence, putting some seasonal flowers along with it. (This is) the spell to be written and spoken: "I entrust this spell to you, gods of the underworld . . . (from this point the text runs parallel to the text of no. 28, until line 384: "I will quickly give you rest.") . . . For I am BARBAR *ADÔNAI who hides the stars and governs the shining heaven; I am the lord of the COSMOS. ATHTHOUIN IATHOUIN SELBIOUÔTH AÔTH SARBATHIOUTH IATHTHIERATH ADÔNAI IA ROURA BIA BI BIOTHÊ ATHÔTH *SABAÔTH ÊA NIAPHA AMARACHTHI CATAMA ZAUATHTHEIÊ CERPHÔ IALADA IALÊ CBÊSI IATHTHA MARADTHA ACHILTHTHEE CHOÔÔ OÊ ÊACHÔ KANSAOSA ALKMOURI THUR THAÔOS SIECHÊ I am THÔTH OCÔMAI, bring and bind so and so, loving, desiring and longing for so and so (the usual thing), for I invoke you, spirit of the dead, by the fearful and great IAEÔ BAPHRENEMOUN OTHI LARIKRIPHIA EUEAI PHIRKIRALITHON UOMEN ER PHABÔEAI, so that you will bring to me so and so, and join head to head, make lips touch lips, join belly to belly, bring thigh close to thigh, and fit black to black[43]; so that so and so will consummate her sexual pleasure with me, so and so, for all time. Next write on another part of the tablet the heart and the *charaktêres*, as below:

IAEÔBAPHRENEMOUNOTHILARIKRIPHIAEUEAIPHIRKIRALITHONUOMENERPHABÔEAI[44]
AEÔBAPHRENEMOUNOTHILARIKRIPHIAEUEAIPHIRKIRALITHONUOMENERPHABÔEA
EÔBAPHRENEMOUNOTHILARIKRIPHIAEUEAIPHIRKIRALITHONUOMENERPHABÔE
ÔBAPHRENEMOUNOTHILARIKRIPHIAEUEAIPHIRKIRALITHONUOMENERPHABÔ

43. The Greek here, *to melan*, refers to pubic hair.
44. This "heart," mentioned just above, is a lengthy and frequently cited palindrome; it is used in the individual spells copied from this recipe (Wortmann, no. 1, lines 1–2). It appears in *PGM* I, line 141; III, lines 59–60; XIXa, lines 16–45, where it is copied out line by line until only the middle letter remains. Sections of the palindrome show up also in *PGM* XXXVI, lines 115ff. Elsewhere it is referred to simply as the IAOE word—for example, *PGM* V, line 366, and so on.

BAPHRENEMOUNOTHILARIKRIPHIAEUEAIPHIRKIRALITHONUOMENERPHAB
APHRENEMOUNOTHILARIKRIPHIAEUEAIPHIRKIRALITHONUOMENERPHA
PHRENEMOUNOTHILARIKRIPHIAEUEAIPHIRKIRALITHONUOMENERPH

The prayer which goes with the ritual: at sunset, while holding the "stuff" from the tomb, say the following[45]:

> Borne on the breezes of the wandering winds,[46]
> Golden-haired Helios, who wield the flame's
> Unresting fire, who turn in lofty paths
> Around the great pole, who create all things
> Yourself which you again reduce to nothing,
> From whom, indeed, all elements have been
> Arranged to suit your laws which nourish all
> The world with its four yearly turning points.
> Hear, blessed one, for I call you who rule
> Heaven and Earth, Chaos and Hades, where
> Men's spirits dwell who once gazed on the light,
> And even now I beg you, blessed one,[47]
> Unfailing one, the master of the world,
> If you go to the depths of earth and search
> The regions of the dead, send this spirit
> From whose body I hold this remnant in my hands,[48]
> To her, so and so, at midnight hours,
> To move by night to order beneath your force,
> That all I want within my heart he may
> Perform for me; and send him gentle, gracious
> And pondering no hostile thoughts toward me,
> And be not angry at my potent chants,
> For you yourself arranged these things among

45. Once again, a label and a set of instructions.

46. The following lines are written in Greek verse, specifically dactylic hexameters. Different versions and parts of this hymn to Helios appear in *PGM* I, lines 315–27; IV, lines 1957–89; VIII, lines 74–80. In his analysis of this and other hymns in *PGM,* the great student of Greek religion, Martin Nilsson, argued that these hymns were not the products of the professional *magoi* who authored and sold the spells but were taken over from a much earlier tradition of Greek magic, which survived largely in these poetic texts; cf. "Die Religion in den griechischen Zauberpapyri," *Opuscula Selecta* 3 (Lund, 1960): 129–30. Although Nilsson's view is possible, it is based on his view that "magic" and religion represent antithetical phenomena, a view not shared by the authors of this book. Thus, we must hold open the possibility that it was precisely the *magoi* who produced not only the spells but the hymns as well.

47. This and the following lines make it apparent that the hymn in its original form was written for precisely the sort of spell that it accompanies here. It is not, as some have suggested, a "purely religious" cult hymn, borrowed from the "religious" sphere and adapted secondarily for "magical" purposes.

48. A reference to the requirement in some spells that they must be accompanied by some material—hair, clothing, shoes—belonging to the individual in question. In this case, the individual is not the intended target but the dead person who is being used as the agent of the spell.

Mankind for them to learn about the threads
Of the Fates, and this with your advice.[49]
I call your name, Horus,[50] which is in number
Equivalent to those of the Fates.
ACHAIPHÔ *THÔTHÔ PHIACHA AIÊ ÊIA IAÊ ÊIA THÔTHÔ PHIACHA
Be kind to me, forefather, scion of
The world, self-gendered, fire-bringer, aglow
Like gold, shining on mortals, master of
The world, spirit of restless fire, unfailing,
With gold disk, sending earth pure light in beams.·

28. Egypt, possibly Antionoöpolis, above Oxyrhynchus on the Nile. The original location is not known, though the objects were certainly deposited in a cemetery. Dated by the editor to the third or fourth century C.E. Lead tablet measuring 11 × 11 cm., found together with a female figurine of unbaked clay, pierced with needles still in place (Figure 13), and a clay vase that contained both objects. The sheet was rolled up. The text covers twenty-eight lines and was inscribed by an expert hand. The content of the text closely follows the "marvelous binding spell," given in *PGM* IV, lines 335–84 (preceding text, no. 27). The spell as given in the recipe from *PGM* IV, or ones closely related to it, has now turned up on several metal tablets from Egypt: our tablet from Antionöpolis; two lead tablets from Oxyrhynchus (published by Wortmann, nos. 1–2)[51]; a lead tablet from Hawara, in the Fayoum, published in 1925 by C. C. Edgar[52]; and a lead tablet at the University of Michigan (Papyrus Michigan 6925). The figurine conforms closely to the

49. In this and the preceding lines, the hymn expresses a common fear that the power of the spell or, in this case, the power of the spirit might turn against the client. The source of the fear was no doubt twofold: first, the effect of the spell was to disturb the dead person's spirit, which might quite plausibly make it angry at the client; and second, as the spell itself prescribes, the spirit belonged to someone who had died early or by violence, under circumstances that rendered it angry and vengeful.

50. The Egyptian god Horus is here identified with Helios, as also in line 989; in Egyptian religion of our period, Horus was worshiped at Heliopolis and revered as the god of the morning sun.

51. On Wortmann, no. 1, see D. R. Jòrdan, "A Love Charm with Verses," *ZPE* 72 (1988): 245–59.

52. Edgar, "A Love Charm from the Fayoum," *Bulletin de la Société Archéologique d'Alexandrie* 21 (1925): 42–47; *SEG* 8.574; cf. Wortmann, pp. 58–75. This lead tablet measures 22 × 16 cm. and shows two sets of holes both about 1 cm. apart, one set between lines 10 and 11, the other at line 19; the holes seem to have been made before the text was carved in. Edgar proposes that the holes served to attach two figurines, as prescribed in *PGM* IV, lines 330–31. Two such figurines have been preserved in the Cairo museum but are of uncertain provenance (Edgar, p. 43); on this charm see also, K. Preisendanz, "Eine neue Zaubertafel aus Aegypten," *Gnomon* 2 (1926): 191–92, and A. D. Nock, "Greek Magical Papyri," *JEA* 15 (1929): 233–34.

FIGURE 13. Elegant female figurine pierced by thirteen needles and found with *defixio* in a clay pot from Egypt. The figurine is made according to directions in a recipe preserved in *PGM* IV, lines 296–329. (Musée du Louvre. By permission.)

directions for producing figurines in the same *PGM* recipe, lines 296–329, except that there are no signs of writing on our figurine. As the recipe instructs, the figurine has her arms behind her back and rests on her knees (lines 301–2); she is also pierced with thirteen nails, as prescribed (lines 320ff.). Whereas the model in *PGM* calls for the relevant names to be introduced (*deina,* "so and so"), our tablet names the client as Sarapammon, son of Area, and the desired woman as Ptolemais, daughter of Aias and Origenes. The spell invokes a number of deities (or one deity under a long series of names?), but the immediate agent is the spirit of a dead man, whose name is Antinous (Hadrian's friend?) and whose task it is to carry out the charge. In return for the successful completion of its charge, the client promises to release the spirit of Antinous from its restlessness. The occasion is love-longing; whether marriage is the ultimate goal is not certain, though Sarapammon does envisage a long-standing relationship with Ptolemais. The italicized portions of the text represent diversions from the *PGM* recipe. The translation here largely follows that of *GMP*, pp. 44–45. *Bibl.:* Sophie Kambitsis, "Une nouvelle tablette magique d'Égypte," *Bulletin de l'Institut Français d'Archéologie Orientale* 76 (Cairo, 1976): 213–23 (plates); *SEG* 26.1717; G. H. R. Horsley, *New Documents Illustrating Early Christianity,* vol. 1 (North Ryde, New South Wales, 1981), no. 8; SGD 152; *SuppMag* 47 (text).

I entrust this binding spell[53] to you, gods of the underworld, Plutôn and *Korê *Persephonê *Ereschigal and Adônis[54] and *BARBARITHA and Hermes of the underworld and *Thôoth PHÔKENSEPSEU EREKTATHOU MISONKTAIK and to mighty Anoubis[55] PSÊRIPHTHA who holds the keys to (the gates of) Hades,[56] to infernal gods, to men and women who have died untimely deaths, to youths and maidens, from year to year, month to month, day to day, hour to hour, *night to night.* I conjure all spirits in this place to stand as assistants to this spirit, *Antinoos.* And arouse yourself for me and go to every place and into every quarter and to every house and bind *Ptolemais, to whom Aias gave birth, the daughter of Ôrigenês, in order that* she may not be had in a promiscuous way, let her not be had anally, nor let her do anything for pleasure with another man, just with me alone, *Sarapammôn, to whom Area gave birth, and do not let her* drink or eat, that she not show any affection, nor *go out,* nor find

---

53. The term is *katadesmos.*
54. A legendary figure of Greek mythology and later a deity associated with Aphrodite and Persephone; his role in this love spell probably derives from his own erotic connections with Aphrodite. He is not a prominent figure in charms and spells.
55. An Egyptian god, normally represented as a jackal or dog; he was associated primarily with the afterlife and the underworld and thus sometimes identified with Hermes.
56. Anubis, the dog-headed god, is often represented as holding keys to the underworld; cf. Wortmann, p. 70.

sleep without me, *Sarapammôn, to whom Area gave birth.* I conjure you, spirit of the dead man, *Antinoos,* by the name that causes fear and trembling, the name at whose sound the earth opens, the name at whose terrifying sound the spirits are terrified, the name at whose sound rivers and rocks burst asunder. I conjure you, spirit of the dead man, *Antinoos,* by *BARBARATHAM CHELOUMBRA *BAROUCH *ADÔNAI and by *ABRASAX and by *IAÔ *PAKEPTÔTH PAKEBRAÔTH[57] SABARBAPHAEI and by *MARMARAOUÔTH and by *MARMARACHTHA[58] MAMAZAGAR. Do not fail, spirit of the dead man, *Antinoos,* but arouse yourself *for me* and go to every place, into every quarter, into every house and *draw* to me *Ptolemais, to whom Area gave birth, the daughter of Ôrigenês* and with a spell keep her from eating and drinking, *until* she comes to me, *Sarapammôn, to whom Area gave birth,* and do not allow her to accept for pleasure the attempt of any man, just that of me, *Sarapammôn.* Drag her by the hair and her heart *until she no longer stands aloof from me, Sarapammôn, to whom Area gave birth, and I hold Ptolemais herself, to whom Aias gave birth, the daughter of Ôrigenês, obedient* for all the time of my life, filled with love for me, desiring me, *speaking to me the things she has on her mind.* If you accomplish this for me, I will set you free.

29. Egypt, Oxyrhynchus; original location uncertain. A small clay pot, now reconstructed from numerous fragments, measuring roughly 11 × 11 cm. The editor dates the pot to the third or fourth century C.E. The pot belongs with two lead tablets, also from Oxyrhynchus (Wortmann, nos. 1–2, pp. 57–80). The writing on the pot comes from the same person who inscribed the tablets; the text is largely derived (in much abbreviated form) from the text recorded on the tablets; and the three items concern the same "problem"—a certain Theodoros undertakes by various means to win the affections of a woman. The spirits addressed here include none of the pagan figures (Kore, Persephone, Adonis, Thoth, Anubis, Hekate, and so on) cited in the tablets. Instead, the only names, of the secret and mysterious variety, are those associated commonly with Jewish texts (Adonaios, Baruch, Sabaoth, Abrasax, Iao, and so forth). As in all of the spells derived from *PGM* IV (lines 296ff.), the immediate agent is the spirit of a dead person, here not named, in whose grave the tablets and pot were deposited. While unusual, the use of a pot for writing a binding spell is not without precedent (Wortmann, p. 81; cf. no. 26). *Bibl.:* Wortmann, no. 3, pp. 80–84; cf. SGD 155–56; *SuppMag* 51 (text).

---

57. The same *voces* appear in the parallel passage of the tablet from Hawara; these two *voces* appear also in *PGM* XII, line 186, a spell that shows other parallels with our tablet.

58. *Voces mysticae* built on the basic form of *marmar,* with numerous suffixes, are quite common; cf. *PGM* XXXV, line 2, and VII, line 572.

Let Matrôna, to whom Tagenê gave birth, whose "stuff"[59] you have, including the hairs of her head, love Theodôros, to whom Techôsis gave birth. I
invoke you, spirit of the dead, by *BARBARATHAM *BAROUCH BAROUCHA
*ADÔNAIOS god and by . . . SESENGEN PHARANGÊS IAÔ IAÔ[60] and by . . . Do
not ignore me, whoever you are, but awaken yourself for me and go off to
Matrôna, so that she may freely give me everything that is hers,[61] and carry
out this binding spell.[62] Now, now. Quickly, because I invoke you, spirit of
the dead, by . . . NÔPHRIS SAXA BAPHAR. Do it quickly, quickly. Just as Isis
loved Osiris,[63] so may Matrôna love Theodôros for all the time of her life.
Now, now. Quickly, quickly. Today . . . IAÔ SABAÔTH ADÔNAI BARBARATHAM
BAROUCHA BAROU[BA]CH . . . (the following appears on the bottom of the
pot) I invoke you by the name of ABRASAX.

30. Upper Egypt, north of Assiut; original location not known. From
the text it is clear that the pot and its contents were originally deposited
in a cemetery. Four items: (1) two wax figurines portrayed in an embrace
and wrapped inside two folded sheets of papyrus (Figure 14); (2) large papyrus sheet measuring 22.5 × 55 cm., with fifty-three lines of writing; (3)
blank sheet of papyrus used to protect the sheet with writing; (4) fragments of a clay pot into which the folded sheets and wax figurines were
stuffed; the pot was then sealed with chalk or plaster. The editor assigns
a date in the fifth century C.E. Our couple is the only Greco-Roman
example portraying such an embrace. There is, however, a close parallel
in an Arabic text, called *Picatrix,* where two wax figures are to be placed
face-to-face if the goal is union or back-to-back if it is separation.[64] The

59. The Greek word *ousia* is used here to designate something associated with the target,
such as hair or threads of clothing. Jordan, "Agora," p. 251, discusses other instances where such
ousia is prescribed or has actually survived.

60. The two occurrences of IAO are written with a short line above each letter, probably as an
indication that the words were to be regarded as *nomina sacra,* much like the "boxed" names in
Shaked and Naveh, Amulet 1. As such, they indicated names of special power; see C. H.
Roberts, *Manuscript, Society and Belief in Early Christian Egypt* (London, 1979), pp. 26ff.
("*Nomina Sacra:* Origins and Significance").

61. The reference here is not to personal property but to the woman herself and her sexual
favors.

62. *Katadesmos* here refers both to the text of the spell and to its written form on the pot—to
the entire operation and its intended results.

63. Cf. *PGM* XXXVI (a collection of spells for various purposes), lines 288–89: "Let so-and-
so love me for all the time of her life, as Isis loved Osiris." The love of Isis for her husband Osiris
was legendary and formed the heart of the legends and rites associated with the cult of Isis in the
Greco-Roman world; cf. Plutarch, *On Isis and Osiris.*

64. *Picatrix. Das Ziel des Weisen von Pseudo-Magriti,* trans. H. Ritter and M. Plessner
(London, 1962), pp. 267ff. The *Picatrix,* a collection of magical recipes and astrological lore,
circulated in Arabic throughout the Middle Ages. There are numerous late Latin translations,
recently edited by D. Pingree, *Picatrix: The Latin Version of the Ghayat al-Hakim* (London,
1986). The ultimate sources of its contents are undoubtedly Greco-Roman.

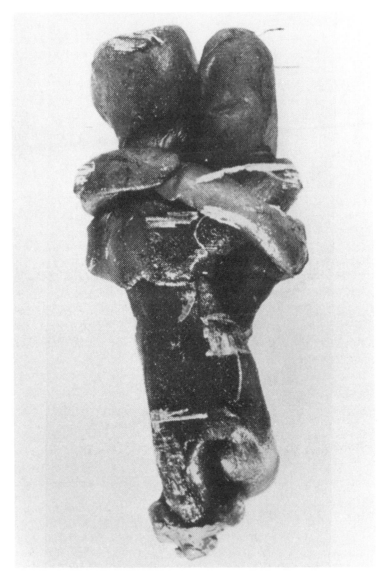

FIGURE 14. Wax couple in embrace found carefully wrapped in a papyrus sheet inscribed with an attraction spell; the couple was then wrapped inside an additional sheet of papyrus and placed in a clay pot; finally, the ensemble was deposited in a cemetery. (D. Wortmann, "Neue magische Texte," *Bonner Jahrbücher* 168 [Bonn: Rheinisches Landesmuseum, 1968], p. 87 [Fig. 9]. By permission.)

figures addressed in the text include a broad array of names. But here again the primary and immediate agents are the *daimones* and spirits of dead people in and around the cemetery. Otherwise, the deities through whom these "lower" entities are commanded include the following: IAO, the unconquerable god; various spirits with mysterious names; ADONAI; a series of twenty-three mysterious names, each beginning, in order, with the letters of the Greek alphabet; Anankê or Fate; Chnum-Horus; ABRASAX; the seven thrones of heaven. In summing up, the text refers to them collectively as "these holy names and powers." The occasion is a relatively simple one: Theon seeks to bring Euphemia to him, in love and passion, for a period of ten months.[65] The spell is unusual both for its length and for the complexity of its invocations which include elements whose origins are Greek, Egyptian, and Jewish, although Egyptian elements clearly predominate. Despite its late date, however, there are no signs of Christian influence. As the many parallels with other texts indicate, almost every line of our spell was copied from recipes in reference works much like those preserved in the large collections of *PGM*. Overall, the spell consists of the following elements: five separate invocations of the spirits and deities ("I invoke . . ."), a series of threats and promises directed to the spirits of the dead people in the cemetery, and several repetitions of the request. *Bibl.:* Wortmann, no. 4, pp. 85–102; *GMP* CI; *SuppMag* 45 (text).

> I bind you with the unbreakable bonds of the Fates in the underworld and powerful Necessity.[66] For I invoke you *daimones* who lie here, who are continually nourished here and who reside here and also you young ones who have died prematurely. I invoke you by the unconquerable god IAÔ *BARBATHIAÔ BRIMIAÔ*[67] CHERMARI. Rouse yourselves, you *daimones* who lie here and seek out Euphêmia, to whom Dôrothea gave birth, for Theôn, to whom Proechia gave birth. Let her not be able to sleep for the entire night, but lead her until she comes to his feet, loving him with a frenzied love, with affection and with sexual intercourse. For I have bound her brain and hands and viscera and genitals and heart[68] for the love of me, Thêon. If you ignore me and fail to carry out quickly what I tell you, the sun will not sink below the

65. The ten months may represent a "fling" or, as suggested by J. Winkler, "The Constraints of Desire: Erotic Magical Spells," in *Magika*, p. 245, n. 108, a marriage confirmed by pregnancy.

66. As with the first lines of Wortmann's nos. 1 and 2, which were copied out from the recipe prescribed in *PGM* IV but to which the *magos* added a first line, the first line of our spell seems to have been prefixed to the required formula, which begins in the following line, "I invoke you . . ."

67. A combination of Brimo, a common epithet of Hekate, and Iao. *Voces mysticae* with combinations of BRIM- are quite common.

68. This series of binding verbs might mean that Theon had already peformed some other symbolic act of binding, such as planting nails in a clay figurine, before preparing this spell.

earth, nor will Hades and Earth continue to exist.[69] But if you bring Euphêmia, to whom Dôrothea gave birth, to me, Theon, to whom Proechia gave birth, I will give you Osiris NOPHRIÔTH,[70] the brother of Isis, who brings cool water[71] and will give rest to your soul. But if you fail to do what I tell you, EÔNEBUÔTH will burn you up.[72] I invoke you, *daimones* who lie here: IEÔ IIIAIA ÊIA IAÔ IAÊ IAÔ ALILAMPS.[73] I hand over (this spell) to you in the land of the dogs.[74] Bind Euphêmia for love of me, Theôn. *Daimones,* I place an oath on you in/by the stele of the gods.[75] I place an oath on you by those (gods) in the inner sanctuary. I place an oath on you by the names of the all-seeing god: IA IA IA IÔ IÔ IÔ IE IE IE OUÔA ADÔNAI. I invoke you who are content in the temple and (are content with) the blood seized/drunk by the great god IÔTHATH (in the temple).[76] I invoke you by the one who sits upon the four points of the winds.[77] Do not ignore me, but act very quickly, for I have commanded you— *AKRAMMACHAMARI, BOULOMENTHOREB, GENIOMOUTHIG, DÊMOGENÊD, ENKUK-LIE,[78] ZÊNOBIÔTHIZ, ÊSKÔTHÔRÊ, *THÔTHOUTHÔTH, IAEOUÔI, KORKOUNOÔK, LOULOENÊL, MOROTHOÊPNAM, NERXIARXIN, XONOPHOÊNAX, ORNEOPHAO,

69. Such threats are not uncommon in the magical papyri, especially where Egyptian influence is strong. *PGM* XXXIV threatens that "the sun will stand still; and should I order the moon, it will come down and should I wish to delay the day, the night will remain for me." Conversely, *PGM* LVII consists of a spell in which the *magos* promises to withhold similar threats if certain commands are fulfilled: "Accomplish this for him, whatever I have written here for you, and I will leave the east and the west." For a discussion of the ideas and practices in traditional Egyptian religion that underlie these threats, particularly the extent to which the gods were thought to be utterly dependent on daily rituals, see Wortmann, pp. 92–93.

70. This term, together with Osiris, is akin to Osoronnophris, a common name for Osiris meaning "Osiris the beautiful being."

71. In general, the dead were thought to suffer from thirst in the underworld. Osiris in particular brought water, as in a common grave inscription, "May Osiris bring you cool water."

72. The figure named here is not otherwise attested. The idea of burning as a form of punishment in the underworld was widespread in antiquity, reaching from Plato's myth of "the vast region of blazing fire" in the *Phaedo* (113–15) to later Jewish and Christian texts. Here it serves as the counterpart to the refreshing water of Osiris.

73. The suffix -*lamps* is common among *voces mysticae.*

74. Probably a reference to the cemetery where the tablet was deposited; normally, cemeteries in Egypt were thought to be ruled by Anubis, represented with the head of a jackal or dog.

75. The reference here is to steles or official inscriptions set up in temples. Such monuments sometimes contained lists of instructions or secret names of the gods. *PGM* VII, lines 41–42, for instance, speaks as follows: "Your true name is inscribed on the holy stele in the sanctuary at Hermoupolis, where you were born."

76. The reference is to the blood of the enemies of the gods, notably Seth as the archenemy. In turn, this blood was traditionally identified with the flood waters of the Nile, especially at the moment of creation (see D. Wortmann, "Kosmogonie und Nilflut," *Bonner Jahrbücher* 166 [1966]: 87–88). These combined powers are here invoked against the spirits in case they fail to cooperate in carrying out the spell. Behind the name, Iothath, lies Thoth, the great Egyptian god of wisdom and ritual power. The word translated here as "temple" appears as *ôap* and is not a Greek word. It probably represented a transliteration of the Coptic word for temple or holy place; cf. Wortmann, p. 97.

77. A common attribute of various Egyptian gods, including Isis. Control of the winds expresses great cosmic power.

78. Greek for "circular, round."

PUROBORUP,[79] REROUTOÊR, SESENMENOURES, TAUROPOLIT,[80] UPERPHENOUPU, PHIMEMAMEPH, CHENNEOPHEOCH, PSUCHOPOMPOIAPS,[81] Orion[82] the true! Let me not be forced to say the same things again ΙΟÊ ΙΟÊ. Lead Euphêmia, to whom Dôrothea gave birth, to me, Theôn, to whom Proechia gave birth, loving me with love, desire, affection, sexual intercourse, and a frenzied love. Cause her limbs, her liver, and her genitals to burn until she comes to me, loving me and not ignoring me. For I invoke you by powerful Necessity—*MASKELLI MASKELLÔ PHNOUKENTABAÔTH OREOBAZAGRA RÊXITHÔN HIPPOCHTHÔN PURICHTHÔN PURIPÊGANUX LEPETAN LEPETAN MANTOUNOBOÊL—so that you may bind Euphêmia to me, Theôn, in love and longing and desire for a period of ten months from today, which is the twenty-fifth of Hathur in the second year of the indiction.[83] Once again I invoke you by the one who rules over you, so that you do not ignore me; and again I invoke you by the one who governs the air; and again I invoke you by the seven thrones—ACHLAL LALOPHENOURPHEN BALEÔ BOLBEÔ BOLBEÔCH BOLBESRÔ UUPHTHÔ,[84] and by the implacable god CHMOUÔR[85] *ABRASAX IPSENTHANCHOUCHAINCHOUCHEÔCH.[86] Grab Euphêmia and lead her to me, Theôn, loving me with a frenzied love, and bind her with bonds that are unbreakable, strong and adamantine, so that she loves me, Theôn; and do not allow her to eat, drink, sleep, or joke or laugh but make (her) rush out of every place and dwelling, abandon father, mother, brothers, and sisters, until she comes to me, Theôn, loving me, wanting me (with a) divine, unceasing, and a wild love. And if she holds someone else to her bosom, let her put him out, forget him, and hate him, but love, desire, and want me; may she give herself to me freely and do nothing contrary to my

79. Greek for "fire-eater."

80. Possibly derived from an epithet of Artemis, "the one worshiped at Tauris" or "related to the bull."

81. Related to the Greek "guide of souls," a traditional epithet of Hermes.

82. Another instance of "wordplay." Each name begins with a letter of the Greek alphabet, in order. Except for the first and last items in this series, the names also begin and end with the same letter. A number of them contain recognizable elements of Greek and Egyptian words and names, such as *zenobiothiz* as a combination of a form of the name of Zeus and *bios* ("life"). Some are also palindromes, such as *puroborup*.

83. The issues raised here are quite interesting. The date is November 21. The year (indiction, a term used only after the reign of the emperor Diocletian, 287 c.e.; the term designated the cycles of tax assessments, counted in cycles of five or fifteen years) is not known. Theon seems to have in mind neither marriage nor a permanent relationship but instead a ten-month affair. Alternatively, the ten months may designate the institution of "trial marriages," attested in Egypt of the Roman period; see the discussion of S. Eitrem in *Papyri Osloenses*, vol. 2 (Oslo, 1931), p. 33, n. 1. The normal period for such marriages seems to have been five rather than ten months.

84. This is an unusually complete listing of secret names of the rulers of the seven heavenly thrones. They appear elsewhere in curse tablets from North Africa (*DT* 240, lines 2ff.; from Carthage), from Syria (*DT* 15, lines 51ff., and no. 16, fragment II, line 8; from Apheca in Syria). The idea of seven heavenly thrones appears also in Jewish texts; cf. Ascension of Isaiah 7:13ff. (seven heavens, each with a throne in the middle).

85. Probably formed by combining the names of two Egyptian gods, Chnum and Horus.

86. The same series of letters appears in *PGM* IV, line 4 (a primarily Coptic spell).

will.[87] You holy names and powers, be strong and carry out this perfect spell. Now, now. Quickly, quickly.

31. Probably from Egypt, although Palestine is a possibility. This Jewish document entitled *Sepher ha-Razim,* or "The Book of Mysteries," was edited from various sources by M. Margalioth and translated into English by Michael Morgan. It contains numerous spells and recipes for a wide variety of purposes. They are embedded in a detailed description of six heavenly firmaments, to each of which are assigned groups of angels. The text provides their secret and powerful names along with their functions and specialties. The document itself dates most likely from the third or fourth centuries C.E.; the individual spells and recipes are no doubt earlier. While the descriptions of the heavenly firmaments and their angelic inhabitants are distinctively Jewish, the spells are not. The book provides numerous recipes for producing *defixiones* from various metals—copper, gold, iron, lead, tin, and silver. This particular spell concerns love and marriage.[88] *Bibl.: Sepher ha-Razim,* pp. 45–46.

> If you wish to put love of a man into the heart of a woman, or to arrange for a poor man to wed a rich woman, take two copper *lamellae* and write upon them, on both sides,[89] the names of these angels, and the name of the man and the name of the woman and say thus: "I ask of you, angels who rule the fates of the children of Adam and Eve, that you do my will and bring in conjunction the planet of N son of N[90] into conjunction with[91] (the planet of) the woman N daughter of N. Let him find favor and affection in her eyes and do not let her belong to any man except him." Place one *lamella* in a fiery furnace[92] and the other where she bathes.[93] Do this on the twenty-ninth of the month when the moon has completely waned. Take care to keep yourself from intercourse, from wine, and from all (kinds of) meat.

87. At this point it becomes clear that Theon's interest is primarily sexual; "nothing but my desire" most likely indicates Theon's own sexual fantasies.

88. Lines 30ff. of the second firmament; *Sepher ha-Razim,* pp. 45–46.

89. Here the instructions confirm the physical evidence, that the metal sheets could sometimes be inscribed on both sides.

90. In the recipes of the *Sepher,* it is not possible to determine whether the names are designated by paternal or maternal lineage. However, the common practice of identifying persons by maternal descent is well attested even in the circles of Rabbinic Jews; in the Babylonian Talmud (Shabbath 66b), Abaye states that all spells repeated several times must include the name of the person's mother.

91. The means specified here for bringing man and woman together also involved astrological techniques of some sort.

92. Here again the action is symbolic; as the fire burns, so should the heart of the desired woman. In this case, the "real" fire must refer to the furnaces (hypocaust) of the bathhouse.

93. Morgan comments that according to Rabbinic law (which may or may not be relevant here) women who use the ritual bath (*mikvah*) are either married or about to be married. If so, the formula would only be useful to husbands, bridegrooms, or adulterers.

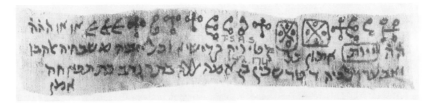

FIGURE 15. *Charaktêres* on a Jewish amulet from the Cairo Geniza; the material is cloth. The amulet also shows the technique of placing both *charaktêres* and powerful names inside boxes. (J. Naveh and S. Shaked, *Amulets and Magic Bowls: Aramaic Incantations of Late Antiquity* [Jerusalem: Magnes Press, Hebrew University, 1985], p. 216. By permission.)

32. Egypt, Cairo; from the Geniza (treasure room for storing worn out manuscripts) of the Ben Ezra synagogue. Along with texts of many other sorts, the Geniza preserved numerous charms, spells, and amulets as well as extensive fragments from larger collections of recipes, much like *PGM, Sepher ha-Razim,* and the *Sword of Moses.* The materials from the Geniza reveal how widespread such beliefs and practices were in Jewish communities of the ancient world and how broadly this material circulated, crossing linguistic, chronological, cultural, and religious boundaries. This spell is written on a piece of cloth[94] (Figure 15). The language is Aramaic. The editors give no specific indication of date, although most of the Geniza materials come from the Middle Ages. It should be noted, however, that their contents are much older. As in other spells, the figures invoked here are a combination of "holy *charaktêres*" and "all the revered letters." No other deities or spirits are cited. The text, imprecisely designed as an amulet by the editors, is actually a binding spell of the common variety designed to secure the affections of another person. In this case, the client appears to be a woman. *Bibl.:* Naveh and Shaked, p. 216 (Geniza 1).

> (*charaktêres* and letters) You holy *charaktêres* and all the revered letters, kindle and burn the heart of ṬRŠKYN, son of Amat-Allah,[95] for GDB, daughter of Tuffaḥa. Amen.

94. Geniza fragment no. 6, p. 1, line 15 (Naveh and Shaked, pp. 230–31) prescribes that a love spell should be written on a new piece of cloth, placed in a reed, and buried on a river bank.

95. If this is to be taken as a reference to Allah, the name of God in Islam, the spell would have to be dated no earlier than the seventh century C.E. From the structure of many similar texts, it would appear that GDB, daughter of Tuffaha, is the client who commissioned the spell.

33. Egypt; from the Cairo Geniza (for details, see p. 107). Among the finds in the Geniza were scattered pages from collections of recipes and spells. It is worth noting that in comparison with the complex and detailed instructions that accompany similar recipes in *PGM*, there are no explicit instructions here concerning prayers, fasting, sexual abstinence, and the like. *Bibl.:* Naveh and Shaked, pp. 230–36 (Geniza 6).

> (*Page 1*) Another (spell). It should be written on an unbaked potsherd and thrown into the fire. This is what should be written: "In the name of Him who says and does. This writing is designated for X, son/daughter of Y, that he should love A, son/daughter of B, and that this heart should burn. Just as this piece of pottery burns, so should the heart of X, son/daughter of Y, burn after A, son/daughter of B. In the name of Nuriel the great angel who is appointed over grace and loveliness. Bring down a light from your light and a fire from your fire and kindle the heart of X, son/daughter of Y. In the name of Abrasax the great angel who overturned Sodom and Gomorrah,[96] so should you turn the heart, the mind, the kidney of X, son/daughter of Y, after A, son/daughter of B. A(men) A(men) S(elah) H(allelujah) . . .

> Another (spell). Write on the hide of a deer with sukk and saffron and hang it on yourself: "In the name of God we shall do this and succeed. This mystery is designated for the endearment of X, son/daughter of Y, that he should love A, son/daughter of B. In the name of klbw bdw blhw ryswdws dz hwwh wzh b' gtyt you praiseworthy letters, kindle and burn the heart of X, son/daughter of Y, after A, son/daughter of B with great love. A(men) A(men) A(men) S(elah) H(alleljuah).

34. Egypt; original location not known. From the letters, which the editor describes as "rather carefully inscribed," the date would appear to be the second or third century C.E. The tablet measures 19.4 × 11 cm. The figures depicted on the tablet may derive from the animal-gods of traditional Egyptian religion. Beyond this, the tablet makes use of two further techniques found in the spells and charms: first, the spell is said to have been used by no one less than Isis herself and thus claims her authority; and second, the client identifies himself, through a series of mysterious names, with the unknown spirit or deity addressed in the spell. At the bottom, following the words "And here are the figures,"

---

96. There is a mention of Sodom and Gomorrah in a fragmentary *defixio* from Carthage, directed at a rival charioteer and his horses; it dates from the mid-third century C.E. The tablet from Carthage shows close similarities to *DT* 252 and 253; see the discussion by D. R. Jordan, "New Defixiones from Carthage," in *The Circus and a Byzantine Cemetery at Carthage*, vol. 1, ed. J. H. Humphrey (Ann Arbor, Mich., 1988), pp. 118–20.

FIGURE 16. Drawing of elaborate designs at the bottom of a *defixio* designed to attract a woman to the client, a man. The significance of some of the drawings remains obscure; they include *charaktêres*, at least two human figures (kissing?), a crocodile, and perhaps a deity (left).

several *charaktêres* and designs were inscribed (Figure 16). These figures probably depict the series of animals cited in the spell. One appears to depict a crocodile; another shows two people kissing; still another may be a schematic effort to represent a penis entering a vagina. The occasion is a familiar one: the client, a man, seeks to win the affections of a woman and to prevent her from having sexual relations with anyone but himself. There is no indication that the relationship was meant to involve marriage. Although the language of the tablet is Greek, the pervading atmosphere is Greco-Egyptian. Like other love spells of a similar kind, this one expresses primarily a set of male fantasies about sexual pleasures and female subservience. *Bibl.*: V. Martin, "Une tablette magique de la Bibliothèque de Genève," *Genava* 6 (1928): 56–64; SGD 161; *SuppMag* 38.

I bind you, Theodotis, daughter of Eus, by[97] the tail of the snake, the mouth of the crocodile, the horns of the ram, the poison of the asp, the hairs of the cat, and the penis of the god[98] so that you may never be able to sleep with any other man, nor be screwed, nor be taken anally, nor fellate,[99] nor find pleasure with any other man but me, Ammonion, son of Hermitaris. For I

97. The preposition is *eis*.

98. The client here invokes the most powerful parts of various animals associated with traditional Egyptian cults: the crocodile was honored at Kom Ombo and in the Fayoum, where numerous crocodile mummies have been excavated; the ram was associated with several Egyptian gods, including Osiris; the cat, identified with Bastet, was widely worshiped and numerous mummies of cats have survived; finally, the ithyphallic god Min was worshiped at Coptos. The Greek word translated as "penis" is *prosthema*.

99. This series of graphic verbs covers various forms of sexual behavior whose meaning cannot be specified with precision. *Binein* and *pugizein* appear in *PGM* IV, lines 350–51, a text with which our tablet shows other similarities. These two verbs appear also in SGD 155, the same as Wortmann 1 (line 21), which is based largely on *PGM* IV.

alone am[100] LAMPSOURÊ'[101] OTHIKALAK' AIPHNÔSABAÔ' STÊSEÔN' UELLAPHONTA' SANKISTÊ'[102] CHPHURIS'[103] ÔN.[104] Make use of this binding spell,[105] employed by Isis, so that Theodotis, daughter of Eus, may no longer try anything with any other man but me alone, Ammôniôn, and may be subservient, obedient, eager, flying through the air seeking after Ammôniôn, son of Hermitaris and bring her thigh close to his, her genitals close to his, in unending intercourse for all the time of her life. And here are the figures[106]: (The figure follows the text.)[107]

35. Egypt, Oxyrhynchus; original location not known. Ostracon measuring 12.5 × 8 cm., part of "the upper part and rim of a big open bowl"; written on both sides. Dated to the second century C.E. The spell uses the imagery of fevers and quarrels, abetted by the powers of Greek vowels and *voces mysticae*, to separate a husband and wife. The focal verb, *apallassô*, derives from the legal word for divorce, while the heat imagery suggests sexual motivations on the part of the client. The client (unnamed) desires to break up a marriage and (presumably) attain the wife for himself. Yet this second intent is not spelled out. *Bibl.:* L. Amundsen, "Magical Text on an Oslo Ostracon," *Symbolae Osloenses* 7 (1928): 36–37; S. Eitrem and L. Amundsen, *Papyri Osloenses* 2 (Oslo, 1931), pp. 29–33 (Papyrus Oslo 15); *PGM,* vol. 2, pp. 209–10 (Ostrakon 2).

100. At this point the spell follows a common procedure of identifying the speaker/client with a more-than-human authority; in this case the authority, an otherwise unnamed god or spirit, is invoked by its secret, mystical name.

101. This name appears beside the figure of a *daimon* in *PGM* II, line 167. It should be noted that the text as given in *PGM* (cf. *GMP,* p. 18) reads CHAMPSOURÊ; Martin comments that the papyrus itself clearly shows the first letter as a lambda; thus the word should read LAMPSOURÊ, a common term in *PGM* (e.g., V, line 62).

102. The term appears several times in the papyri; cf. *PGM* II, line 234; V, line 425; VII, lines 680–81.

103. The same term appears in *PGM* XXXVI, line 170. It refers to the Egyptian god, Khepri, embodied in the scarab.

104. At this point the text reads *ôn,* whose meaning remains unclear. It might represent *ôon* (Greek for "egg") or possibly the present participle of the verb "to be," as in "I who am . . . (the preceding *voces*)."

105. The Greek term is *philtrokatadesmos;* it also appears in the lengthy love charm whose preparation is detailed in *PGM* IV, lines 296ff.

106. Here the scribe mistakenly copied onto the tablet a label that was intended only as a set of instructions; such labels are common in *PGM* (e.g., VII, line 477).

107. *SuppMag,* p. 122, describes the figures as follows: "to the left, the god holding a staff; at his feet, the snake; to the upper right of the snake, the crocodile; at the extreme right, the cat (?); above the crocodile, two figures, presumably the ram (though we do not recognize it) and the woman; yet more magical signs and letters, and drawings that remain obscure."

(*Outside*)

OOOOOOAOÊÊÊÊÊAÔOÔÔÔÊIÊ
. IEIEEEÊÊÊERÔEIÔTAÔÔ
. . UÊÊE AAA ÊIÊÔAIAA ÔOÔÔ
. . UEÔÔÔOOE LATH ARMATRÔAEA
. OCHUSOIOIO NUCHIE NARAEEAEAA
. . . OS BAL SABAIÔTH* Ô MAÔSAIO[108]
UEÔAÔUÔUEÔAEÔAEÔA
ARITHOSAAAAA SKIRBEU MITHREU
MITHRAÔ[109] ARUBIBAÔ THUMÔ ÊOAU
EAAUUEEAUEAE MOULA A . . .
. . . . . IMSIU OULATSILA MOULA
. . AAIEÊIIII AI EÔ EÔ IEE ÔAIÊ
AAAÊ . EÊ IÊ OAOAOA
AA ALO ALARÔ ARÔ ARÔ UUU IÊÊÊ
ÊÊÊÊÊÊÊ IIIIIII AÊAÊAÊÊIA
AIÊ ÔUOÔUOOÔUOOU IEOUIEIEIE
IEIE OAÔAÔÔ BAAAAAAAAAA
AAAAAA OOOOOO UUUUUUUU
EEEEE ÔÔÔÔÔÔÔÔÔÔÔÔ
ÔÔOÊ ÔOÔO AEÊIOUÔ AEÊIOUÔ
AEÊIOUÔ SOUMARTA MAX AKARBA[110]
MIUCHTHAN SALAAM ATHIASKIRTHO
DABATHAA ZAAS OUACH KOL MOL
PHRÊ ZÔCHRAIE ZANEKMÊT SATRA
PEIN[111] EBLARATHA ARNAIAUSAIA
EAE AEÊIOUÔ ÊIÊÊIÊ O AOAOA
ÔOAOAÔÊAÔ Let burning heat con-
sume the sexual parts[112] of Allous,
(her) vulva, (her) mem-
bers, until she leaves the house-
hold of Apollonios. Lay
Allous low with
fever, with sick-
ness unceasing,
starvation—Allous—
(and) madness!
Allous.

108. Perhaps a version of the name Moses.
109. *Mithreu mithrao:* a rendering of the god Mithras, apparently to exploit the sound of his name rather than evoke his power.
110. *Soumartamaxakarba:* reversed, *abrakaxa-matramous.*
111. *Satrapein:* from Persian *satrap,* meaning viceroy.
112. The Greek word is *psyche.*

(*Inside*)
Remove Allous from A-
pollonios her husband;
give Allous insolence, hatred, ob-
noxiousness, until she departs the household of Apollonios. Now, quickly.

36. North Africa, Hadrumentum. Found rolled up, with holes, probably made by a nail. Lead tablet measuring 25 cm. on each side, deposited originally in the Roman cemetery of Hadrumentum. The language and script point to a date in the third century C.E. Similar mélanges of Greek and Latin appear in other tablets from Hadrumentum: *DT* 267, 269, and 270 contain lengthy passages of Latin text in Greek letters; 295 shows a mix of Greek and Latin; a tablet from Augustodunum in Gaul, dating to the second century C.E., includes a list of Latin names and a series of *voces mysticae* in Greek letters (SGD 132). The spell is addressed, as was frequently the case, to the spirit of a dead person in the cemetery, whose role is to carry out the immediate task of the spell. But the greater power invoked, whose function was to command the local spirit, is the god of Israel. No other spirits or deities are invoked. The language of the invocation is strongly reminiscent of the Septuagint (LXX) and other Greek Jewish texts.[113] The spell was commissioned by a woman, Domitiana, for the purpose of inducing a man, Urbanus, to take her as his wife. Both the client and her "target" were probably freedmen. Despite the numerous echoes of terms and phrases from the Greek Bible, there is no need to assume that either the client or the professional was Jewish, though both may have been. The language of the spell is certainly Jewish in its origins but had probably become part of the common culture of professional *magoi* by the time it was copied from a reference book in Hadrumentum in the third century C.E. *Bibl.:* G. Maspero, "Sur deux *tabellae devotionis* de la nécropole romaine d'Hadrumète," *Bibliothèque égyptologique* 2 (1893): 303–11; *DT* 271; Wünsch, *Antike Fluchtafeln,* no. 5; a full discussion and commentary by A. Deissmann in *Bible Studies* (Edinburgh, 1901), pp. 269–300 ("An Epigraphic Memorial of the Septuagint").

---

113. Deissmann draws attention to a Jewish text of the first century C.E., the Prayer of Manasseh, which shows striking similarities to the language of our spell: "Almighty Lord, God of our fathers Abraham, Isaac, Jacob and of their righteous seed, who made the heaven and the earth with all their beauty, who bound the sea by the word of thy command, who closed up the abyss and sailed it with thy fearful and glorious name, which causes all things to shudder and tremble at thy power." On this important text, see *The Old Testament Pseudepigrapha,* ed. J. H. Charlesworth, vol. 2 (New York, 1985), pp. 625–37.

I invoke you *daimonion* spirit[114] who lies here, by the holy name[115] ΑΟΤΗ
ΑΒΑΟΤΗ,[116] the god of Abraham[117] and *ΙΑΟ, the god of Jacob,[118] ΙΑΟ ΑΟΤΗ
ΑΒΑΟΤΗ, god of Israma,[119] hear the honored, dreadful and great name,[120] go
away to Urbanus, to whom Urbana gave birth, and bring him to Domitiana,
to whom Candida gave birth, (so that) loving, frantic, and sleepless with love
and desire for her, he may beg her to return to his house and become his
wife.[121] I invoke you, the great god, eternal[122] and more than eternal, almighty[123] and exalted above the exalted ones. I invoke you, who created the
heaven and the sea.[124] I invoke you, who set aside the righteous.[125] I invoke
you, who divided the staff in the sea,[126] to bring Urbanus, to whom Urbana
gave birth, and unite him with Domitiana, to whom Candida gave birth,
loving, tormented, and sleepless with desire and love for her, so that he may
take her into his house as his wife. I invoke you, who made the mule unable

114. That is, the spirit of the dead person, in whose grave the tablet was deposited.
115. Up to this point in the first line, the text is Greek written in Roman letters. This line,
along with the Greek in Roman letters of lines 4–5 (in fact this is just one line inasmuch as it
begins on line 5 and slopes upward into a blank space in the middle of line 4), was added at a
later time. Maspero explains the confusion of the opening lines as follows: the original scribe left
the first line blank, with the intention of later inserting the name of the local spirit to be invoked;
the scribe also left a blank space in line 4, for the special names of the deity, and drew a vertical
line toward the end, just before the *kai* of "and bring him," indicating the limit of the space
available for them; the scribe's eye then skipped from one *kai* to the next, a common scribal
error, and in the process omitted the first part of the command and the name of the target; at this
point, a second hand, perhaps the chief *magos* in the shop or possibly even Domitiana herself,
discovered the omission and added the missing phrase in Roman letters, beginning on the next
line and sloping upward to fill the blank space in line 4 (the text reads, in Roman letters, *cae
apelthe pros ton orbanon hon ethecen urbana*).
116. The identical words appear in *PGM* V, line 135. Abaoth is a variant of Sabaoth.
117. The text reads *Abraan*. This mistaken spelling of the name of Abraham is but one of
several errors that suggest that the immediate writer of the tablet was probably not a Jew or, if
so, not a highly literate one.
118. Here, as in line 38, the word is *Iakou*, a corruption of Jacob.
119. Here, as in line 39, *Israma* is written instead of Israel. The double occurrence points to
the likelihood that the corruption existed already in the reference book from which this spell was
copied.
120. The name of God is frequently described in the LXX as *entimon*/"honored" (Deuteronomy 28:58), *phoberon*/"dreadful" (Psalm 110:9); the combination of *megas* ("*great*") *kai
phoberos* is common (Deuteronomy 10:17; Sirach 43:29).
121. The Greek term *sumbios* is used regularly in papyri for husband or wife.
122. The identical phrase appears in Isaiah 26:4.
123. The Greek term *pantokratôr* is commonly applied to God in the LXX.
124. A clear echo of Genesis 1:1; 14:19, 22.
125. The phrase means that the righteous are separated from the ungodly. The text refers to
Sirach (LXX) 33:11ff. ("In his great wisdom the Lord separated them . . . and the sinner is
opposite the ungodly").
126. A curious reversal whereby the undoubted original read "divided the sea with the
staff," referring to the staff of Moses at the crossing of the Red Sea in Exodus 14:15–16, and has
now been transformed into its current meaningless version.

to bear offspring.[127] I invoke you, who separated light from darkness.[128] I invoke you, who crushes rocks.[129] I invoke you, who breaks apart mountains.[130] I invoke you, who hardened the earth on its foundations.[131] I invoke you, by the holy name which is not spoken.[132] . . . I will mention it[133] by a word with the same numerical equivalent[134] and the *daimones* will be awakened, startled, terrified, to bring Urbanus, to whom Urbana gave birth, and unite him with Domitiana, to whom Candida gave birth, loving and begging for her. Now! Quickly! I invoke you, who made the heavenly lights and stars[135] by the command of your voice,[136] so that they should shine on all men.[137] I invoke you, who shook the entire world, who breaks the back of mountains and casts them up out of the water, who causes the whole earth to tremble and then renews all its inhabitants.[138] I invoke you, who made signs in the heaven, on earth and on sea, to bring Urbanus, to whom Urbana gave birth, and unite him as husband with Domitiana, to whom Candida gave birth, loving her, sleepless with desire for her, begging for her, and asking that she return to his house and become his wife. I invoke you, great, everlasting and almighty god, whom the heavens and the valleys fear throughout the whole earth,[139] through whom the lion gives up its spoil[140]

127. No such claim is attributed to God in the Hebrew Bible, although later Jewish texts make mention of the mule as the sterile offspring of a horse and an ass; see L. Ginzberg, *Legends of the Jews,* vol. 7 (Philadelphia, 1946), s.v. The reproductive incapacity of the mule was much discussed among ancient authors; as such, its wondrous character made it attractive to those who produced charms; cf. Deissman, pp. 285–86.

128. A reference to Genesis 1:4 ("and God separated the light from the darkness").

129. An echo of 1 Kings 19:11 ("the Lord was passing by: a great wind came shattering rocks . . .").

130. Compare Psalm 77:15 (LXX): "He shattered the rock in the wilderness."

131. Proverbs 8:29 speaks of God as "making firm the foundations of the earth."

132. The idea that the name of the God of Israel could not be spoken outside the Temple and thus possessed enormous potency was widespread in ancient Judaism.

133. The name(s) of the God of Israel recur frequently in magical papyri. Here the *magos* uses the ultimate threat, the pronouncing of the divine name, to motivate the spirit. Josephus, *Bellum Judaicum* 5.438, speaks of the "terrible name of God" (*to phrikton onoma tou theou*).

134. Wünsch reconstructs the word *isarithmô;* this would refer to the common practice of calculating the numerical sum of letters in words in order to determine their mystical significance. Here the issue is not simply to know their special power but to invoke it for one's own purposes.

135. An almost exact citation of Genesis 1:16ff.

136. A common theme in the Hebrew Bible; cf. Psalm 32:9 (LXX): "For he spoke, and it was; he commanded, and it was created."

137. Cf. Genesis 1:17, which lacks the reference to all men.

138. The Greek phrase, *kainizonta pantas tous katoikountas,* echoes passages like Psalm 33:14 ("He oversees all the inhabitants of the earth . . ."). Wisdom of Solomon 7.27 speaks of the divine *sophia* as "renewing all things."

139. Compare Psalm 33:8 ("Let the whole world fear the Lord").

140. A curious statement, somewhat reminiscent of the earlier reference to mules, in that both affirm divine control over every aspect of the created order. The idea that God can change the behavior of the terrible lion appears also in Isaiah 11:6–7 ("the calf and the young lion shall grow up together . . . and the lion shall eat straw like cattle").

and the mountains tremble with earth and sea, and (through whom) each becomes wise who possesses fear of the Lord[141] who is eternal, immortal, vigilant, hater of evil,[142] who knows all things that have happened, good and evil, in the sea and rivers, on earth and mountain, AÔTH, ABAÔTH, the god of Abraham and IAÔ of Jacob, IAÔ AÔTH, ABAÔTH, god of Israma, bring Urbanus, to whom Urbana gave birth, and unite him with Domitiana, to whom Candida gave birth, loving, frantic, tormented with love, passion, and desire for Domitiana, whom Candida bore; unite them in marriage and as spouses in love for all the time of their lives. Make him as her obedient slave, so that he will desire no other woman or maiden apart from Domitiana alone, to whom Candida gave birth, and will keep her as his spouse for all the time of their lives. Now, now! Quickly, quickly!

141. Cf. Proverbs 1:7 ("The fear of the Lord is the beginning of wisdom").

142. Cf. Psalm 96:10 ("Those who love the Lord hate evil"); the term in our text is *misoponêros*, which appears in 2 Maccabees 4:49; 8.4.

3

Tongue-Tied in Court:
Legal and Political Disputes

From the writings of ancient historians and public orators, it has long been recognized that lawsuits and public trials constituted a fundamental feature of life in the Greco-Roman world. With the possible exception of modern America, no society has been more notorious for litigation than classical Athens. A character in Aristophanes' *Peace* comments that Athenians never do anything but try cases.

We know much about the formal aspects of ancient legal culture—how charges were filed, how evidence and witnesses were presented, how juries were selected, and how verdicts were rendered.[1] But for our purposes, we may safely ignore these formal matters, including the numerous important differences between Greek, Roman, and other ancient legal systems, in order to focus instead on one particular aspect of the legal process, namely, how prospective litigants prepared to defend themselves against their accusers. According to the standard treatments, such preparations covered the following items: preliminary sparring before magistrates over the legitimacy of the charges and the competence of the court; the gathering of evidence and witnesses in support of one's case; and, once a trial seemed inevitable, the decision to seek professional assistance in the form of an orator or speech writer.[2] Once again, however, these were strictly formal matters, and we know from our own experience that involvement in legal proceedings evokes and engages powerful emotions—fear, shame, guilt, panic—that require preparation and treatment of an entirely different kind. In antiquity, how were these emotions recognized and handled, if at all? To be honest, most scholarly discussions have paid them no heed whatsoever, so that we may perhaps excuse them for having also ignored a substantial body of relevant material whose precise if unstated function was to deal with the nonlegal side, the emotional dimension, of lawsuits and public trials.[3]

This body of material comprises curse tablets, directed against judicial opponents and accusers, from classical Athens to the late Roman Empire. Altogether some sixty-seven Greek examples have been published, along with a lesser number (approximately forty-six) in Latin; a Hebrew collection of spells and curses, the *Sepher ha-Razim,* includes a recipe for preparing a *defixio* to reverse bad fortune in a trial.[4] Indeed, among the published Greek *defixiones,* judicial or legal types constitute the second largest subgroup. In addition, this group contains some of the earliest and best preserved of all ancient curse tablets. The very earliest date from the Greek colony at Selinus, on Sicily, around the year 500 B.C.E. (no. 49), while a somewhat later tablet (no. 50) from the same site has been described as "the largest complete *defixio* of the fifth century."[5] From the Greek mainland, several well-preserved tablets have survived from the fifth and fourth centuries B.C.E., while curses in Greek still occur as late as the third century C.E. in Palestine and the fifth century in Egypt. Among these later examples, the most intriguing is a large cache of more than two hundred tablets, discovered in a well on Cyprus.[6] Dating from the second or third century C.E., these tablets—of which only seventeen have yet been published—employ the same formulas throughout and concern a variety of different trials. In each instance, the tablets name the defendant as well as the plaintiff, but only one of the published curses gives any details about the case (no. 46).

The procedures and the circumstances involved in commissioning and deploying *defixiones* of a judicial character are relatively straightforward—and common. Indeed, we must now begin to consider the likelihood that commissioning a curse tablet against prospective judicial opponents was a regular feature of the legal process in the Greco-Roman world. Two facts seem certain: first, and contrary to earlier views, which held that judicial tablets represented efforts to seek revenge against one's accusers *following* a trial,[7] it is now clear that the deployment of the *defixio* belongs to the preliminary phase of preparations for an anticipated trial; and second, by the very nature of the case, those who commissioned the tablets were prospective defendants.[8] Exceptions to those two observations are rare: first, a spell prescribed in the Jewish collection, *Sepher ha-Razim,* promises relief for a friend who has already suffered an unfavorable ruling; and second, a number of cases involve disputes over inheritances and ownership of property that do not lend themselves to the category of criminal law (defender versus prosecutor) but fall instead into the category of civil cases in which the emphasis lay on efforts at mediation and the "loser" was not found guilty of any crime. While it is not always easy to distinguish between criminal and civil trials in the tablets, one thing is clear in every case—the clients who commissioned the *defixio* saw themselves as threatened by the legal proceedings and consequently took extralegal measures in order to guarantee a favorable outcome.[9]

Overall, the language and structure of judicial *defixiones* follow the developmental pattern of curse tablets in general. The earliest examples name the opponents; they sometimes invoke local deities by their familiar names, though some examples omit both verbs of binding and the names of the deities; and they mention the physical and mental faculties, usually tongue and mind, to be bound or tied up so that the targets will be unable to pursue their case. Later examples, by contrast, tend to be longer, to use secret names and mystical terms (the *voces mysticae*), to invoke spirits and deities from many traditions, and to provide specific details for every aspect of the binding process.[10]

In general, the potential defendant in a legal proceeding seems to have resorted to a local professional, perhaps to a *magos,* in order to commission a *defixio.*[11] Very little is known about this stage of the transaction, but it seems fair to guess that the professional played a leading role: he (most seem to have been men) supplied the text and the metal tablet, perhaps offering the client a range of options according to a sliding scale of fees; he inscribed the tablet, or in some cases selected one from a pile of preinscribed sheets, filling in the client's name, the names of the opponents, and sometimes providing a few details related to the case; and in all likelihood, he arranged for the placement of the tablet, in a well but more often in a grave or tomb, where its binding action was entrusted to the restless spirit, the "ghost," of the dead person.[12] At least with the later tablets, the role of the deities and *daimones* was not to carry out the curse itself but rather to see to it that the spirit of the dead person executed the spell as commanded.

For the most part, the binding action of judicial *defixiones* was carried out through the symbolic medium of written and spoken words understood as "things" that could change the world. The written form of words is obvious in the inscribed text on the metal sheet; the spoken word is less obvious but no less important and, as the recipes in the Greek papyri make clear, took the form of oral prayers and invocations delivered at various stages in the preparation and placement of the *defixio.* As for the binding action itself, it is sometimes made explicit, as in cases where the opponents are to be rendered speechless, like the corpse in the grave, or where they are to become cold and useless, like the lead of the tablet itself. In other cases, the symbolic character of the binding action is extended from words-as-things to other physical objects, like dolls, figurines, and dead animals.[13] Several examples of such figurines have been recovered in Greece and elsewhere. One (no. 41) is clearly judicial and bears the name of the target inscribed on its leg. Other examples, some of which are also judicial, represent figures in various suggestive postures—laid out in miniature coffins, hands bound behind their backs, heads sometimes turned sideways or completely removed, and so on.

The first figurine just mentioned (no. 41), from the Kerameikos in Athens and dating to fourth century B.C.E., raises a number of further issues. First, in a tablet whose judicial character is unmistakable, the text names not just one but nine accusers and, by way of caution, incorporates "any one else with them as an advocate or witness." In short, the client's enemies constituted, in his mind, a group, perhaps even a party. Second, multiple accusers are not at all uncommon in Greece: in the *Apology* Socrates always refers to his accusers in the plural. What this suggests, as is patently manifest in the case of Socrates, is that many of the tablets with multiple accusers—in one case the total approaches seventy-seven—lead us from the realm of the law into the world of political strife and competition. In other words, the Greek public was neither the first nor the last to use courts of law as a cover for advancing their own political causes and for crushing those of their opponents.

Whether or not one chooses to distinguish a separate category of political *defixiones* is largely a matter of taste.[14] Here we have chosen to recognize a distinction while treating them in the same chapter (legal, nos. 37–54; political, nos. 55–59). More important is to recognize that the tortured political history of Greece from the end of the Peloponnesian War to the Macedonian conquest under Philip, Alexander, and their successors witnessed an era of unparalleled political strife and that much of the drama was played out in the courts.[15] As the testimony of the curse tablets clearly reveals, the confusion of political and legal affairs generated a need on the part of many to call upon higher powers to sustain their cause. Third and last, the Greek and Latin *defixiones* demonstrate conclusively that the use of curse tablets was by no means limited to "unlettered and superstitious" members of the lower classes. In classical Greece as in imperial Rome, their power was accepted and employed by all, including the wealthy and powerful Athenian aristocrats cited on numerous Greek tablets.[16] Similarly, just as *defixiones* cut across all social categories, they were no respecters of gender. Several tablets, including some from Greece and at least one from Sicily, mention women as potential witnesses. Their presence contradicts the traditional view that women had no legal standing in Greek courts, but the evidence of the *defixiones* as well as other considerations must now call this view into question.[17] Women must also have been among those who commissioned judicial tablets, as is even more clearly the case in tablets relating to matters of sex and pleas for justice.

The persistent and pervasive belief in the binding power of judicial *defixiones* may be further illustrated by the evidence of literary sources. Generally speaking, the Greeks of classical Athens were fully convinced that words—or rather, correct words skillfully arrayed—could shape human behavior. Words, in the form of rhetorical and philosophical discourse, were frequently likened to charms and spells. Socrates toys with a young man in the *Charmides* who is looking for a spell to cure headaches; what he really needs, answers Socrates, is a spell to cure the soul, meaning, of course, that philosophical training alone can provide the ultimate cure. At almost exactly the same date, Aristophanes, in the *Wasps,* mentions the case of a well-known orator whose jaws suddenly and inexplicably froze as he was defending himself before an Athenian jury. Two competing explanations of the orator's dilemma circulated at the time: one, no doubt advanced by his enemies, held that he was simply outargued by his rhetorically gifted opponents, while the other, certainly preferred by the orator himself, insisted that his tongue had been tied up, certainly by a judicial *defixio*.[18] At a much later date, in the first century B.C.E., Cicero tells of a similar case in which a lawyer, in midspeech, forgot which case he was pleading, subsequently blaming his lapse on spells and curses (*veneficiis et cantionibus*).[19] Later still, in the mid-second century, the physician-philosopher Galen indicates his disdain for all who believed in the power of spells and charms, citing their claim as follows: "I will bind my opponents so that they will be incapable of saying anything during the trial."[20] But of course, Galen's very disdain for such beliefs and practices merely confirms their broad circulation in his time.

Can we take one further step, beyond the now incontrovertible belief in the effectiveness of curse tablets? Dare we ask whether these curses actually worked? The answer must be that they did. Or, at the very least, that they were widely believed to work. Not, perhaps, precisely as the *magos* and his clients believed, through the agency of *daimones,* deities, and ghosts. But work they did, more likely, as literary figures argued, through the coercive power of words, a power much exploited by practitioners of rhetoric and philosophy.[21] If two essential components in understanding how judicial curses worked were, first, the general belief in the power of words, and second, the related belief that this power could extend beyond the human realm to coerce unseen but no less real personalities via the medium of *defixiones*, a third element certainly lay in the tensions and anxieties experienced by all public speakers, whether orators or defendants at trial, especially when they "knew" that they were likely targets of binding spells. Under these circumstances, it is no wonder that they worked.

Two famous orators who suffered bouts of verbal paralysis aptly reflect Euripides's description of how it must have felt to speak before a large public audience with one's reputation or even one's life on the line. First, Euripides: "Whenever anyone stands opposite in the debate and is about to speak at a trial for homicide, fear paralyzes the mouth of men and prevents the mind from saying what it wants to say."[22] Aelius Aristides, the most noted orator of the second century c.e., found himself frequently incapable of delivering his speeches due to respiratory disorders, no doubt of a psychosomatic nature. Although Aristides gives no hint of attributing his difficulties to spells or curses, perhaps because his problems were chronic in nature, the case is quite different with Libanius, one of the leading orators of the fourth century c.e. Numerous references in his autobiographical speech (*Oratio* 1) attest the frequency with which orators resorted to charges that their enemies had used binding spells, but one incident in particular paints a vivid portrait of how effective these spells must have been.[23] Libanius was devastated by physical maladies in his later years, especially by migraine headaches, which made it impossible for him to lecture before his students. Doctors proved unable to provide a cure. But a dream revealed to him that he had been the target of hostile rituals; the truth of the dream was soon confirmed when a mutilated chameleon turned up in his lecture hall, its head stuffed behind its rear legs and one of its front legs placed over its mouth. Discovered and removed from the building, the obvious source of Libanius's illness was eliminated and the orator soon returned to full health. Later he complained to his students about their indifference to his sufferings, in a passage that clearly demonstrates his conviction that he had been done in by a curse tablet: "When you believe that a charioteer or a horse has been hobbled in this manner (by a *defixio*), everything is in an uproar, as if the city itself had been destroyed; but I am treated with indifference when the same things happen to me."

One final word about the commonly held view that the beliefs about curse tablets and binding spells so fully displayed by Libanius belong exclusively to "the superstitious age in which he lived,"[24] an age once caricatured by Gilbert Murray for its "failure of nerve."[25] The evidence of the *defixiones* themselves, stretching from the sixth century B.C.E. to the sixth century C.E., proves, as Peter Brown has noted, that "it is far from certain that there was any absolute increase in fear of sorcery or in sorcery practices in the late Roman period."[26] Instead, as Brown observes, such beliefs function like X-rays, revealing "pockets of uncertainty and competition."[27] Such pockets, we have good reason to believe, belong to every age and to every place.

Notes

1. On Greek law and trials, see R. J. Bonner's still useful *Lawyers and Litigants in Ancient Athens* (Chicago, 1927); D. M. MacDowell, *The Law in Classical Athens* (Ithaca, 1978); A. R. W. Harrison, *The Law of Athens* (Oxford, 1968–1971). On Roman legal affairs, see A. Berger, *Encyclopedic Dictionary of Roman Law* (Philadelphia, 1953); J. A. Crook, *Law and Life of Rome, 90 b.c.– a.d. 212* (Ithaca, 1967); Alan Watson, *The Evolution of Law* (Baltimore, 1985) and *Roman Slave Law* (Baltimore, 1987).

2. Generally speaking, Greek defendants memorized or read aloud the speeches written for them by trained orators, whereas in Rome the orator-lawyers regularly spoke on behalf of their clients.

3. J. Ober, *Mass and Elite in Democratic Athens* (Princeton, 1989), offers a most satisfactory treatment of the personal and social aspects—the broader and deeper roots of legal proceedings of lawsuits and public trials in classical Athens. He is also one of the very few historians of Greece to consider the evidence of *defixiones* (p. 149).

4. The recipe appears in the second firmament, lines 145–54. The second firmament also contains a spell (lines 21–25), without mentioning any tablet, for guaranteeing a favorable outcome in a pending lawsuit.

5. Jeffery, p. 73.

6. See the discussion in Faraone, "Context," n. 11.

7. The older view was represented by Ziebarth (1899), p. 122. Others, among them Wünsch, argued against this position. P. Moraux, *Une défixion judiciaire au Musée d'Istanbul* (Brussels, 1960), pp. 42–44, argues convincingly that the all of the judicial tablets must have been commissioned as preventive measures, not as acts of vengeance. For a more recent treatment, see Faraone, "Context," n. 67.

8. Greek judicial tablets use a variety of technical legal terms—judge (*dikastês*), accuser (*sunêgoros* and *katêgoros*), witness (*martus*), and prosecuting opponent (*antidikos* and *sundikos*); see E. Kagarow, *Griechische Fluchtafeln* (Leopoli, 1929), p. 54, for a catalogue of technical legal language in the Greek tablets. The Latin tablets generally employ the term *inimicus*.

9. Part of the tension in these trials was certainly due to class antagonism. Ober (*Mass and Elite*, p. 18), without mentioning judicial *defixiones*, constructs a scenario in which their use would seem both rational and inevitable: "the elite litigant retained functional advantages over his ordinary opponent, and the poorer Athenian's envy and resentment of the social privileges enjoyed by the wealthy man were far from eliminated." It is not too much to speak of *defixiones* in general as tokens of envy and resentment.

10. So Moraux, *Défixion*, pp. 5–10, though we need not follow him in describing the later period as a time of delirious syncretism (p.7).

11. C. Faraone has suggested that in the earlier *defixiones*, those of the sixth to fourth centuries B.C.E., the great variety of formulas and vocabulary points to the likelihood that many individuals simply prepared their own tablets. Still, Plato (*Republic* 364C) speaks of wandering specialists who offered their services for a fee. Certainly the clear evidence for professional hands on *defixiones* comes from a later time, the first to sixth centuries C.E.

12. Recently, D. R. Jordan has presented evidence from Athens that the inscribers and dispensers of *defixiones* may have been professional scribes, "moonlighting," rather than *magoi*. His evidence is of two sorts: a large horde of tablets was turned up in a well near the stoa where professional scribes worked in the city's civic offices; and curse tablets of the period in question "have scripts comparable in quality with those of the Civic Office well, and most show evidence for the use of formularies." See Jordan, "New Evidence for the Activity of Scribes in Roman Athens," *Abstracts of the American Philological Association—120th Annual Meeting* (*Baltimore*) (Atlanta, 1989), p. 55.

13. The most thorough treatment of such objects is by David Jordan, "New Archaeological Evidence for the Practice of Magic in Classical Athens," in *Praktika tou XII diethnous synedriou klasikês archaiologias* (Athens 1988), pp. 273–77, which includes a number of recent and yet unpublished finds from Athens; see also Jordan's brief description of these objects in SGD, p. 157.

14. Preisendanz (1972), cols. 9–10, among others, isolated a separate category of political curses. Faraone, "Context," pp. 16–17, is inclined to do away with the distinction. It is worth noting, however, that there is little substantive difference between the two positions. Both agree that political disputes in Greece were fought out in lawsuits and public trials.

15. On the politics of this period, see W. R. Connor, *The New Politicians of Fifth-Century Athens* (Princeton, 1971); C. Mossé, *Athens in Decline: 404–86 B.C.* (London, 1973); and Ober, *Mass and Elite*, esp. pp. 43–52.

16. Ober, *Mass and Elite*, p. 149, mentions curse tablets (the tablet is treated here as no. 56) in his treatment of Athenian social groupings. Stressing the point that wealthy and poor citizens seem regularly to have associated with one another, despite differences of class, he cites a *defixio* from the fourth century B.C.E. that names politicians of high standing together with prostitutes; presumably they belonged to the same circles.

17. See the discussion on women as witnesses in Harrison, *The Law of Ath-*

ens, pp. 136–37. It must also be remembered that the Greek term *sundikos,* normally translated as "(legal) witness" might also be rendered more broadly as "supporter."

18. See Faraone, "An Accusation of Magic in Classical Athens (Ar. *Wasps* 946–48)" *TAPA* 119 (1989): 151–53.

19. In Cicero's *Brutus* 217, dated to 46 B.C.E. At this point in the essay Cicero is discussing orators with weak memories.

20. The essay is Galen's "On the powers of all drugs," *Opera omnia,* ed. C. G. Kühn (Hildesheim, 1965). The passage appears in XII, p. 251.

21. See Jacqueline de Romilly, *Magic and Rhetoric in Ancient Greece* (Cambridge, 1973), and G. E. R. Lloyd, *Magic, Reason and Experience: Studies in the Origins and Development of Greek Science* (Cambridge, 1979), esp. pp. 10–58.

22. The passage appears in a fragment (*CAF,* no. 67) of Euripides' lost play, *Alcmeon;* see the discussion in Faraone, "Accusation of Magic," p. 152.

23. See the following instances: *Oratio* 1.43 (Libanius's bitter rival attributes Libanius's success to his use of spells); 1.62ff. (another rival accuses him of resorting to charms and thereby causing the death of his wife); 1.98 (a rival bribes a young boy to accuse Libanius of cutting off the heads of two young girls—real or fabricated?—for use in preparing *defixiones.* The account of the chameleon appears in 1.243–50. See the discussions in Campbell Bonner, "Witchcraft in the Lecture Room of Libanius," *TAPA* 58 (1932): 34–44; A. F. Norman, *Libanius' Autobiography (Oration 1)* (New York, 1965); and P. Brown, "Sorcery, Demons and the Rise of Christianity: From Late Antiquity into the Middle Ages," in *Religion and Society in the Age of Saint Augustine* (New York 1972), pp. 127–28.

24. So Bonner, "Witchcraft," p. 34.

25. The phrase appears as the title of chapter 4 in G. Murray's *Five Stages of Greek Religion* (Garden City, N.Y., 1925).

26. Brown, "Sorcery," p. 122.

27. Ibid., p. 128.

37. Greece, Peiraeus; original location not known. Lead tablet measuring 13 × 9 cm. Found in a grave, rolled up and pierced by a nail. From the names and the shape of the letters, the editor dates the tablet to the early part of the fourth century B.C.E. This tablet is typical of many Attic judiciary curses in which prospective defendants seek to bind their accusers in advance of the formal legal proceedings. There is no verb of binding and no mention of any deities. The names appear in the nominative case. *Bibl.:* DTA 38; Wilhelm, p. 120 (text).

Philippidês
Euthukritos
Kleagoros
Menetimos[1]
and all the others, however many are advocates for them.[2]

38. Greece, Peiraeus; original location not known. Lead tablet measuring 13 × 6 cm., folded and fixed with a nail. The tablet dates from the mid-fourth century B.C.E.; close to 323 according to Wilhelm. This is a judicial spell deposited in anticipation of an impending lawsuit involving the people named. Most of the names are well known and appear in inscriptions of the period. All were associated with the naval affairs of Athens, mostly as trierarchs, wealthy citizens chosen to maintain and command a war vessel (trireme). In 323 B.C.E., Athens revolted against the rule of the Macedonians; the revolt failed, in part because of the poor condition of the Athenian navy. Lawsuits involving trierarchs and naval affairs were rampant during these years. Sending a letter to the gods probably involved using a dead person, in whose grave this tablet would have been placed, as an emissary to bring the tablet to the attention of Hermes and Persephone. *Bibl.:* DTA 103; Wilhelm, pp. 122–25 (text).

I am sending this letter to Hermes and Persephone, since I am presenting wicked people to them, for it is fitting for them to obtain the final penalty, O Justice/*Dikê: Kallikratês son of Anaxikratês,[3] Eudidaktos,[4] Olympiodôros[5] . . . Theophilos . . . Zôpuros Pasiôn[6] Charinos, Kallenikos,[7] Kineias[8] . . . Apollodôrus,[9] Lusimachos, Philoklês,[10] Dêmophilos[11] and their

1. All of the names appear among well-attested Attic figures; cf. *PA*, s.v. For Kleagoros, Kirchner records only Kleagoras; cf. nos. 8453–55.
2. The Greek term is *sunêgoros*.
3. Kallikratês was trierarch in 342 B.C.E. (*PA* 7953).
4. Eudidaktos was priest of Asclepius at Delphi in 352/351 B.C.E. and a prominent figure at the time (*PA* 5414).
5. Olympiodoros (of Peiraeus) was trierarch in 323 B.C.E. (*PA* 11407).
6. Wilhelm calls him the financial backer of the party; he was the grandson of the famous Athenian banker of the same name (*PA* 11673).
7. Kallenikos was trierarch in 323 B.C.E. (*PA* 7769).
8. Kineas Lampreus was trierarch in 323 B.C.E. and earlier (*PA* 8436).
9. Apollodôros was trierarch in 323 B.C.E. and earlier (*PA* 1413).
10. There were two trierarchs named Philoklês in 323 B.C.E. (*PA* 14541 and 14546).
11. Dêmophilos, son of Dêmophilos, was involved in decrees concerning naval matters in 323 B.C.E.; he also accused the philosopher, Aristotle, of impiety (*PA* 3675).

associates[12] and any other friend of theirs. Dêmokratês,[13] the one going to court for the case: Mnêsimachos,[14] Antiphilos.

39. Greece, Attica; original location not known. Lead tablet measuring 7 × 23 cm. and written on both sides; originally folded and pierced with a nail. From names given in the tablet, it is possible to date it to the late fourth century B.C.E. Although no deity is named, the formulas used indicate that Hermes was probably the one invoked. Some of those targeted are women. The spelling in some of the names appears to have been deliberately scrambled. No clear occasion is given; court proceedings seem most likely. *Bibl.:* DTA 95; Wilhelm, pp. 119–20 (text).

(*Side A*) I will bind . . . Aristias and Euaristês and Kalliadês and . . . the friend . . . and I bind those . . . me . . . (the rest is lost).

(*Side B*) and Menôn the son of Aristoklês, both him and the actions of Menôn and his tongue and words and actions; and that he may prove useless to the authorities, and also Pithios and Eukolinê and her life(?) . . . and (name lost) Anaphlustios[15] and Xenokritos and Sôsinomos (?); and Aris[. . .], Nikias (?), Charisios (?), the sons of Diophan(tês?) . . . and (Lusi)machos, of Phula; Lusimachidês, the son of Philinos from (the deme of) Peiraeus.[16] The god who restrains[17] holds the advocates[18] with Nikios and Hêdulê the daughter of Timokratês.

40. Greece, Attica; original location unknown. Lead tablet measuring 17 × 6 cm.; originally folded. Wünsch dates it to the fourth century B.C.E., but is uncertain whether the author of the tablet is Attic, since he employs spellings that can best be explained as Doric. The names Hermes and Hekate appear on the verso as if the tablet was addressed as a letter to the gods of the underworld. The context appears to be legal. Galênê was a known *hetaira*, mentioned in DTA 102. Social intercourse

12. The Greek is *sundikoi,* meaning the associates in the same party to a legal suit.
13. Dêmokratês is known as a treasurer (*tamias*) in charge of trireme construction in 323 B.C.E. and earlier (*PA* 3525).
14. Mnêsimachos was trierarch in 323 B.C.E. (*PA* 10335). A Mnêsimachos is also mentioned on a tablet from Athens (see no. 41).
15. Probably a demotic, "of the deme of Anaphlustos."
16. The same Lysimachidês appears in an Athenian inscription of the fourth century B.C.E. (*Sylloge Inscriptionum Graecorum,* 2d ed., no. 725), which praises him and his brother for his devotion to the gods and to the affairs of the *orgeones,* formal groups or associations dedicated to the worship of specific gods. For their piety, the brothers are awarded gold crowns and exemption from taxes; cf. A. Koerte, "Die Ausgrabungen am Westabhange der Akropolis," *AM* 21 (1896): 298–302. See also *PA* 9482.
17. Though not named, the deity must be Hermes, who is commonly called by the epithet *katochos* used here.
18. The common word *sunêgoros* found in numerous texts relating to court trials.

between leading citizens and courtesans or prostitutes was not uncommon. *Bibl.:* DTA 107; cf. Wilhelm, p. 112.

(*Side A*) Hermes of the underworld and Hekate of the underworld.[19]

(*Side B*) Let Pherenikos be bound before Hermes of the underworld and Hekate of the underworld. I bind Pherenikos's (girl) Galênê to Hermes of the underworld and to Hekate of the underworld I bind (her). And just as this lead is worthless[20] and cold, so let that man and his property be worthless and cold, and those who are with him who have spoken and counseled concerning me.

Let Thersilochos, Oinophilos, Philôtios, and any other supporter of Pherenikos be bound before Hermes of the underworld and Hekate of the underworld. Also Pherenikos' soul and mind and tongue and plans and the things that he is doing and the things that he is planning concerning me. May everything be contrary for him and for those counseling and acting with . . .

41. Greece, Athens; in the Kerameikos, just beyond the city walls. Three lead objects (two plates and a figurine; Figure 17). (1) A crude human figure, 6 cm. tall, with phallus. The arms are folded behind the back, indicating that they were meant to be tied or bound; the feet show no signs of having been bound. On the right leg of the figurine is inscribed a name, Mnesimachus.[21] (2) The first of two oblong plates shaped like saucers and measuring approximately 11 × 6 cm. and 2 cm. deep. Together, the two sheets formed the top and bottom of a miniature sarcophagus for the figurine. On the first sheet, there is no writing. (3) The second sheet contains the curse proper, where the name of Mnesimachus reappears. Beneath the inscription, and to the right, there are two holes, presumably made by a nail. These objects are of particular interest in that they were found precisely where they were originally deposited. The objects were found in a grave, along with the partially disturbed and mutilated remains of a human skeleton and a red-figure vase (*lekythos*). The figurine was placed at the pelvis of the skeleton and the inscribed plate was discovered just above it. The date of the grave itself lies close to 400 B.C.E.; the names on the lead objects belong to well-known public figures from the late fifth century.[22]

19. These phrases, on the outside of the scrolled tablet, were meant to function as the address to which the "letter" was posted.

20. *Atimos,* "cheap."

21. The editor mistakenly claims that this type of figurine is the only example. For other figurines bearing the name of the cursed person, see Preisendanz (1933), pp. 163–64; cf. a find of eight clay figurines from Pozzuoli (Italy), each bearing a name written twice (*DT* 200–7).

22. In a recent article ("Archaeological Evidence"), Jordan presents a notable find from two graves not more than a few yards from our site. The find includes three lead sheets formed into

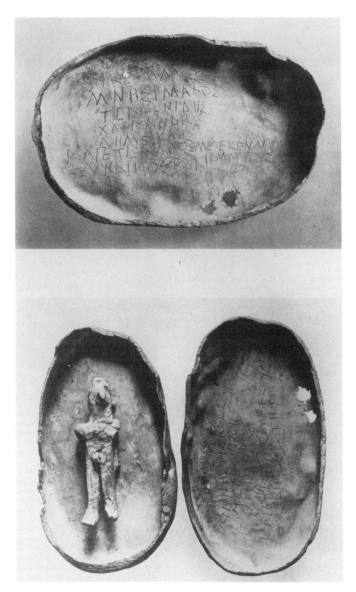

FIGURE 17. Lead figurine and miniature coffin from the Kerameikos (Athens); similar to those of Figure 3 and found quite close to them. These four figurines are quite crude in design. The chief target of the spell was probably Mnesimachus, whose name was inscribed a second time on the right leg of the figurine. (J. Trumpf, "Fluchtafel und Rachpuppe," *Athenische Mitteilungen* [Berlin: Verlag Gebr. Mann, 1958], Beilage 72.1 and 2. By permission.)

The text of the inscription places it in the category of curses associated with public trials. No verb is used and no deities or spirits mentioned. *Bibl.:* J. Trumpf, "Fluchtafel und Rachpuppe," *AM* 73 (1958): 94–102; *BE* (1963): no. 32, p. 125; SGD 9; Jordan, "New Archaeological Evidence for the Practice of Magic in Classical Times," in *Praktika tou XII diethnous synedriou klasikês archaiologias* (Athens, 1988), pp. 275–76.

> Barburtidês Xôphugos
> Nikomachos[23] Oinoklês
> Mnêsimachos[24]
> Chamaios Teisônidês
> Charisandros[25]
> Dêmoklês[26]
> And if there is any one else with them[27]
> as an advocate or witness.[28]

42. Greece, Attica; original location not known. The last line indicates that the tablet belongs to the category of judicial curses. Several well-known figures are cited here, notably Demosthenes and Lycurgus, probably to be identified as the important political figures in Athens of the late fourth century B.C.E. All of the names listed and the first line of the request are written backward. The final two lines are written from

boxes, each one containing a lead figurine with large genitals; two have their hands tied behind their backs. Each of the boxes and two of the figurines have men's names inscribed on them. The third box and its figurine reveal several names, among them Theozotides, whom Jordan identifies as a well-known Athenian politician; his son, Nikostratos, was a follower of Socrates (Plato, *Apology* 33e).

23. Nikomachos may be the controversial and prominent commissioner (*anagrapheus*) who in 411 B.C.E. and again in 403 was appointed to draw up a revised copy of the laws of Solon. On both occasions he took much more time than expected. The orator Lysias, along with others, sought to impeach him; Lysias's oration against Nikomachos has survived (*Against Nikomachos*). In the *Frogs* (1506), Aristophanes mentions Nikomachos among the fools of Athens. A Nikomachos also appears in another curse tablet, from approximately the same period; cf. Wilhelm, p. 122. The name also appears on a lead figurine from the Kerameikos (no. 41).

24. A common name of the period; cf. Trumpf, p. 101. The name appears on DTA 103. There is evidence of a lost oration of Lysias against a Mnêsimachos.

25. A figure with this name served as *prutanis* (annually chosen officials who presided over a variety of public events); his date falls at the beginning of the fourth century; cf. *PA* 15482 and Trumpf, p. 101.

26. Several figures with this name occur in inscriptions from around the year 400 B.C.E., all of them prominent public figures; cf. Trumpf, p. 101.

27. A common formula in curses, following the list of proper names so as to cover any individuals not cited by name.

28. The words here, *xundikos* and *martus,* are technical terms for coprosecutors and witnesses in Athenian judicial proceedings.

left to right.[29] The person who commissioned this tablet must have been the target of an imminent trial; those mentioned on his list would have been his accusers, a number of whom seem to have belonged to the circle of Demosthenes. *Bibl.:* Wünsch (1900), p. 63; *DT* 60; Faraone, "Context," n. 76.

Nêreidês
Dêmosthenês[30]
Sôklês
Lukourgos[31]
Euthukratês
Epiklês[32]
Charisios[33]
Boêthos
Poluokos
and all of the others who are accusers with Nereidês.

43. Greece; possibly Megara. Probably found in a tomb. Two 7 × 3 cm. lead tablets inscribed with letters on one side, and partially folded. Written in Doric Greek. Both tablets are intended to forestall lawsuits through the intermediary of a corpse, named Pasianax (in whose grave, presumably, the tablets were buried). The first tablet (A) protects Aristandros from Neophanês; the second (B) protects Eratophaenea from both Akestor and Timandridas. A curious set of assumptions underlies the words addressed to the corpse: the first thought is that the curse will become effective as soon as the corpse reads the letter ("whenever you read . . ."); the second thought reflects a sudden awareness that the corpse cannot read anything[34] ("but neither shall you ever read . . ."); thus the third and final thought takes the spell in an entirely new direction, no longer based on the assumption that the corpse will read the spell and carry it out but instead that the target should become as idle in

29. No. 7 in Ziebarth's collection (1899) (*DT* 61) seems to have been produced by the same hand. Both include nine names on the initial list; both write the names and the first two lines of the request backward; and both write the final two lines left to right. The request in no. 7 reads as follows: "all those with Plathanê, men and women."

30. There seems little doubt that this is the same person as the great Athenian orator who died in his sixties in 322 B.C.E.

31. Almost certainly to be identified with Lycurgus, the famous Athenian politician and orator, who died ca. 324 B.C.E.

32. In the much later writing called the *Lives of the Ten Orators* (848C), falsely attributed to Plutarch, the account of Demosthenes mentions an Epiklês who once chided the orator.

33. In his speech *Against Euboulidês* (57.2), Demosthenes mentions a Charisios, perhaps the father or uncle of the figure listed here.

34. Opinions were divided as to whether corpses could read the contents of *defixiones;* no. 73 appears to assume that the spell will be read by spirits or possibly by a corpse.

his actions as the body is idle in the grave. *Bibl.:* Ziebarth (1899), pp. 120–25; Wünsch (1900), pp. 67–68 (nos. 21–22); *DT* 43–44.

> (*Tablet A*) Whenever you, O Pasianax,[35] read this letter[36]—but neither will you, O Pasianax, ever read this letter, nor will Neophanês ever direct a lawsuit against Aristandros (?). But just as you, O Pasianax, lie here idle, so also let Neophanês be idle and nothing.

> (*Tablet B*) Whenever you, O Pasianax, read this letter—but neither will you ever read this (letter), nor will Akestôr direct a lawsuit against Eratophanês— and not Timandridas either.[37] But just as you lie here idle and nothing, so also let Akestôr and Timandridas become idle.

44. Greece, Athens; original location not known. Lead tablet measuring 20 × 16 cm., with five nail holes. The date is approximately 300 B.C.E. While no god is mentioned, other tablets from Athens of the same period (notably *DT* 50) indicate that Hermes or Persephone or both are the most likely candidates. Several figures cited among the targets are known from other sources. The precise occasion is not clear, but the pending court case must have involved a dispute between cooks or butchers (the Greek term *mageiros* can mean both), some of whom appear to have been famous public figures. Most if not all of the figures involved were probably slaves. *Bibl.:* Ziebarth (1899), no. 10; Wünsch (1900), no. 10 (pp. 63–64); *DT* 49.

> Theagenês,[38] the butcher/cook, I bind the tongue and soul and speech that he is practicing. Purrias,[39] I bind the hands and feet and tongue and soul and speech that he is practicing. I bind the wife of Purrias, her tongue and soul. Also Kerkiôn, the butcher/cook I bind and Dokimos the butcher/cook, the tongue and soul and speech that they are practicing. I bind Kineas, his tongue and soul and speech that he is practicing with Theagenês. And Phereklês, I bind the tongue and soul and evidence that he gives for

35. Pasianax appears to be the name given to the corpse. Wünsch notes that Pasianax may be an old name for Pluto, ruler of the underworld, and suggests that the name belonged to the person in the grave only as a corpse, not during his lifetime. The name itself is once attested as an epithet of Zeus (LSJ, s.v.).

36. In both spells this conditional clause is left hanging. Presumably it might have been followed by "accomplish for me what lies herein."

37. The addition of Timandridas is clearly an afterthought.

38. Theagenes, mentioned first on the list and cited again several times below, was clearly the client's chief antagonist.

39. Lucian, the Greek satirist of the second century C.E., mentions a Pyrrias, a butcher/cook (*mageiros*), in his *Menippus* 15. Surveying the underworld, which consisted of nothing but skeletons, the speaker comments that it was no longer possible to distinguish famous men of the past from ordinary types: "I could not distinguish the cook Pyrrias from Agamemnon." Despite the difference in spelling, there is some possibility that Lucian's Pyrrias, clearly chosen for his fame as a chef, is identical with our own.

Theagenês. Seuthês,[40] I bind the tongue and soul and speech that he is practicing and his feet and hands and eyes and mouth. Lamprias,[41] I bind the tongue and soul and speech that he is practicing, and his hands and feet and eyes and mouth. All of these I bind, I hide, I bury, I nail down.[42] If they lay any counterclaim before the arbitrator or the court,[43] let them seem to be of no account, either in word or in deed.

45. Cyprus, Amathous, near Paphos.[44] Discovered by workmen "at the bottom of a disused shaft, under a quantity of human bones."[45] From the language of the tablet it seems possible that the shaft served as a common grave. D. R. Jordan notes that the original find consisted of more than two hundred tablets on lead and an additional sixty on sheets of selenite. Altogether, seventeen tablets were published by L. Macdonald in 1891. The rest are in the British Museum in London and the Biblio-

---

40. According to Athenaeus (*The Learned Banquet* 377bc), written ca. 200 C.E., the Attic comedian Poseidippus (ca. 289 B.C.E.) mentions the same Seuthes in one of his comedies. Poseidippus was noted for having introduced slaves as cooks into his dramas.

41. According to Athenaeus (379e), Euphron, another new comedian and a contemporary of Poseidippus, mentioned a butcher/cook, with the name of Lamprias, in one of his plays. Lamprias is credited with having first created black broth (*zômos melas*). The speaker goes on to rank Lamprias and six other cooks as the second group of "Seven Sages" in Greek history.

42. In other words, the client "wishes" for his enemies the same fate as that of the tablet itself.

43. The client here anticipates two possible stages in legal proceedings—the preliminary hearing before an arbitrator (*diaitêtês*) or a full trial in the public court (*dikastêrion*).

44. According to the report (p. 184) of P. Aupert and D. R. Jordan, this and the other tablets were in fact found at Amathous, not at Kourion, as reported by all previous scholars. Cyprus was known in antiquity as a center of "magical" activity. In his *Natural History,* Pliny the Elder writes of the various "schools" or traditions of the *magi:* "There is yet another branch of magic, derived from Moses, Jannes, Lopates, and the Jews, but living many thousand years after Zoroaster. And much more recent is the branch in Cyprus" (30.11). Jews were known as *magoi* on Cyprus. Josephus (*Antiquities* 20.142–44) reports efforts by the Roman procurator of Judea, Felix, to win the affections of Drusilla, great-granddaughter of Herod and sister of Agrippa, by using the services of a Jew from Cyprus, named Atomus, who had advertised himself as a *magos.* Atomus was summoned to provide Felix with a binding spell, of the sort well known and attested among the finds from Amathous. With this must be compared the tale in Acts 13:4–12. Paul sails to Cyprus and preaches his message in the synagogues. Arriving at Paphos, near Amathous, he encounters a Jewish *magos* (or possibly two figures, conflated by the author) by the name of Bar-Jesus or Elymas; presumably Bar-Jesus called himself *magos.* This Elymas was a friend of the Roman proconsul, Sergius Paulus; in other words he was a figure of considerable standing. Paul confronts Elymas with his own superior power and renders him temporarily blind. Sergius Paulus embraces Christianity on the spot. Assuming, roughly, a late second century date for the tablets, we are thus able to locate two important moments in the history of "magic" on Cyprus, reaching from the mid-first century to the late second or early third century. On the story in Acts, see A. D. Nock, "Paul and the Magus," in *The Beginnings of Christianity,* ed. F. J. Foakes Jackson and K. Lake, vol. 5 (London, 1933), pp. 164–88, and on the considerable evidence for the Jewish community on Cyprus, see T. B. Mitford, "New Inscriptions from Early Christian Cyprus," *Byzantion* 20 (1950): 110–16.

45. So Macdonald, p. 162.

thèque Nationale in Paris. According to Jordan, "those [unpublished tablets] whose texts have been read follow the formula of *DT* 22–37 and evidently issue from the same atelier."[46] All were found rolled up, with the writing on the inside. The one chosen for translation here is a lead sheet, measuring 14.7 × 25.9 cm. The right and left edges are worn and the bottom is missing portions of a line or two; the full text covered some sixty lines. The letter shapes and the postclassical Greek point to a date in the late second or the mid-third century c.e. Except for a "rudely drawn figure of a bird" (a rooster?) on the fragmentary no. 141 in Mitford's edition (*DT* 36), there are no designs on the tablets. The action of the spell or curse consists of two parts, each repeated several times and not always in the same manner: first, the deities and spirits are invoked; next they are bidden to take over the named targets. The bulk of this massive find consists of brief spells directed against opponents, the prosecutors, in legal proceedings. Their intended goal is not to kill or injure these opponents but to render them incapable of speaking at the trial. The clients were thus prospective defendants. Among the tablets from Amathous, only one gives any information concerning the substance of the dispute; no. 134, line 18, at the place where all of the parallel texts say simply "so that he may be unable to oppose (me) in any matter," adds "concerning the *thremmata* (young animals/slaves?)." The names of the litigants, both defendant/client and prosecutor/target, appear in every text.[47] The deities and spirits invoked include a wide variety of name types from diverse sources (Greek, Jewish, and Egyptian), among them several with mysterious combinations of letters of the sort known from other spells and tablets. With very few exceptions, the formulaic portions of the tablets from Cyprus are identical, word for

46. SGD, p. 193; in a subsequent communication, Jordan comments that the tablets clearly show the work of several hands.

47. No. 127: Sotêrianos also called Limbaros versus Aristôn (the name of the defendant will be given first); 128: Ari . . . versus Aphrodisios and Nesotrios; 129: Kalokeros versus Zotê (a woman) or Zotas (a man); 130: Alexandros the son of Matidia also called Makedonios versus Theodoros governor (*hêgemôn*) of Cyprus and Timôn the son of Markia (whose case was presumably being supported by the governor); 131: Alexandros also called Makedonios versus Theodôros (probably the same persons as named in no. 130); 132: Alexandros also called . . . (the same as in nos. 130 and 131?) versus Metrodoros Asbolios the banker, Alexandros also called Louskinios, Timôn, Philodemos, Eumenês, Makarios, Demokratês, Markos, Demokratês, Dôrothes, and Neôn; 133: Artemidôros versus Aphrodisianos; 134: Eutuchês versus Sozomenos; 135: Kallis versus Krateros; 136: Serapias (a woman) versus Mariôn (her husband; line 27: *ton andra*); 137: the names of both parties are written in cryptic letters (Mitford likens them to a kind of Latin shorthand; p. 272, n. 1); 138: Zoilos versus Soteria (a woman), Truphôn, Demetrios, and Demetria (a woman); 139: Didumos versus Mormulos; 140: Eirênes, Aristôn, and Timôn (same as in no. 132?) versus Onasas and Demetrios (same as in no. 138?); 142: Mariôn versus Euanthios and Demetrios (same as in no. 138 and 140?).

word and, in the case of the mysterious names, letter for letter. It is clear that they were copied from the same master version and produced by the same professional. According to Jordan and Aupert, "we now have more texts by this scribe than by any other ancient magician."[48] The tablet translated here is *DT* 22, which is MacDonald 1 and Mitford 127 (text). *Bibl.:* L. Macdonald, "Inscriptions Relating to Sorcery in Cyprus," *Proceedings of the Society of Biblical Archaeology* 13 (1891): 160–90; DTA, pp. xviii–xix; *DT* 22–27; Robert, *Froehner,* pp. 106–7; T. B. Mitford, *The Inscriptions of Kourion* (Philadelphia, 1971), pp. 246–83; T. Drew-Bear, "Imprecations from Kourion," *Bulletin of the American Society of Papyrologists* 9 (1972): 85–107; P. Aupert and D. R. Jordan, "Magical Inscriptions on Talc Tablets from Amathous," *AJA* 85 (1981): 184; SGD, p. 193; C. Harrauer, *Meliouchos. Studien zur Entwicklung religiöser Vorstellungen in griechischen synkretistischen Zaubertexten* (Vienna, 1987), pp. 58–63.

Daimones under the earth and daimones whoever you may be; fathers of fathers and mothers (who are a) match (for men), you who lie here and you who sit here, since you take men's grievous passion from their heart,[49] take over[50] the passion of Aristôn which he has toward me, Sotêrianos also called Limbaros, and his anger; and take away from him his strength and power and make him cold[51] and speechless and breathless, cold toward me, Sotêrianos also called Limbaros. I invoke you by the great gods,[52] MASÔM-ASIMABLABOIÔ MAMAXÔ EUMAZÔ ENDENEKOPTOURA MELOPHTHÊMARAR AKOU RASRÔEEKAMADÔR MACHTHOUDOURAS KITHÔRASA KÊPHOZÔN goddess ACHTHA-MODOIRALAL AKOU RAENT AKOU RALAR hear ALAR OUECHEARMALAR KARA-MEPHTHÊ SISOCHÔR ADÔNEIA of the earth CHOUCHMATHERPHES THERMÔMASMAR ASMACHOUCHIMANOU PHILAESÔSI gods of the underworld, take over from Aristôn and his son the passion and anger they hold toward me, Sotêrianos also called Limbaros, and hand him over to the doorkeeper in Hades, MATHUREUPHRAMENOS, and (to/of?) the one who is appointed over the gate to

48. Jordan and Aupert, p. 184.
49. The sense of the phrase is not that the spirits should kill the targets but that as the spirits have died, that is, given up their passions (*thumos*), so they should take away the passion (*thumos*) of the opponents. Originally, the opening of the tablet consisted of four lines in dactylic hexameter, now metrically corrupt. The lines contain numerous Homeric words and phrases; see Drew-Bear, p. 89.
50. The Greek verb is *paralambanein,* used four times in the text.
51. Chilling one's opponent is a common request in judicial curses; see Moraux, *Défixion judiciaire,* pp. 49–52.
52. What follows are the secret, authentic, and mysterious names by which the "great gods" are to be addressed. These names do not derive from the stock of *voces mysticae* characteristic of *PGM,* the amulets, or other *defixiones* from other parts of the Greco-Roman world.

Hades and the door bolts of heaven, STERXERX ÊRÊR[XA][53] earthshaker[54] ARDAMACHTHOUR PRISGEU LAMPADEU.[55] And bury in this mournful (grave) the one whose name is written on this curse tablet[56] which brings about silence.[57] I invoke you the king of the deaf/voiceless *daimones*. Hear the great name, for the great SISOCHÔR rules over you, the ruler of the gates to Hades. Of my opponent Aristôn bind and put to sleep the tongue, the passion and the anger he holds toward me, Sotêrianos also called Limbaros, lest he oppose me in any (legal) matter.[58] I invoke you *daimones*—buried in a single grave, violently slain, untimely dead, not properly buried—by her who bursts forth from the earth and carries down (into the grave) the limbs of *MELIOUCHOS and MELIOUCHOS himself.[59] I invoke you by ACHALEMORPHÔPH who is the only god of the earth OSOUS *OISÔRNOPHRIS OUSRAPIÔ,[60] do whatever is written herein. O much lamented tomb and gods of the underworld and Hekate of the underworld and Hermes of the underworld and Ploutôn and the infernal Erinues[61] and you who lie here below, untimely dead and unnamed, EUMAZÔN, take away the voice (s) of Aristôn who is opposing me, Sotêrianos also called Limbaros, MASÔMACHÔ. I deposit with you this charge/ spell[62] to make Aristôn silent, and (you) give over his name to the infernal gods. ALLA ALKÊ KE ALKEÔ LALATHANATÔ, three-named Kore. These shall always carry out (my wishes) for me and silence Aristôn the opponent of me,[63] Sotêrianos also called Limbaros. Awaken yourself for me, you who hold (?)

---

53. Audollent (*DT*, p. 42) suggests that behind these letters may lie the Greek word for "falcon," *ierax/irex*.

54. The Greek is *rêsichthôn*, a common epithet of Hekate and Dionysus; here it appears to modify STERXERX.

55. Audollent interprets these two words as the description of a spirit gnashing its teeth with a face on its hand (*DT*, p. 42). The reading of PRISTEU is taken from Audollent in preference to Mitford's PRISSGEU.

56. The Greek term is *katathema*. Here it clearly designates the tablet itself.

57. Compare *PGM* VII, lines 396–404: "An excellent spell for silencing (*phimotikon*), for subjecting, and for restraining: Take lead from a cold-water pipe and make a *lamella* and inscribe it with a bronze stylus, as shown below, and set it with a person who has died prematurely." *PGM* XLVI, line 4, also prescribes a spell for silencing (*phimotikon*).

58. The Greek term *pragma* probably refers to pending legal actions.

59. An allusion to the story of Adonis entering and leaving the underworld; Adonis is mentioned explicitly in line 14 ("ADÔNEIA of the underworld"). An association with Osiris is also evident from the immediately following lines. Harrauer, p. 61, notes that the cult of Adonis was prominent on the island of Cyprus and that he was also identified with Osiris, most notably at Amathous.

60. An early name for Serapis, built by combining elements from Osiris and Apis.

61. The traditional Greek spirits of vengeance and punishment, attested as early as Homer (*Iliad* 9.454, 571; *Odyssey* 17.457) and regularly associated with the underworld. They do not appear frequently in the papyri or tablets; cf. *PGM* IV, lines 2339, 2860; V, line 191.

62. The Greek term is *parathêkê*. Here it refers either to the task or to the tablet—possibly to both.

63. The Greek term is *antidikos*, a common word for designating an opponent in judicial proceedings.

the infernal kingdom of all the Erinues. I invoke you by the gods in Hades, OUCHITOU, the dispenser of tombs, AÔTH IÔMOS TIÔIE IÔEGOÔEOIPHRI, who in heaven rule the upper kingdom, MIÔTHILAMPS, in heaven, IAÔ, and the (kingdom) under the earth, SABLÊNIA IAÔ SABLÊPHDAUBÊN THANATOPOUTÔÊR. I invoke you, BATHUMIA CHTHAORÔOKORBRA ADIANAKÔ KAKIABANÊ THENNANKRA. I invoke you gods who were exposed by Kronos, ABLANAIANABLA[64] SISIPETRON, take over Aristôn the opponent of me, Sotêrianos also called Limbaros, ÔÊANTICHERECHER BEBALLOSALAKAMÊTHÊ, and you, earthshaker, who holds the keys of Hades. Carry out for me, you . . . Provide . . .[65]

46. Same as no. 45. This tablet (*DT* 25, which is Macdonald 4 and Mitford 130) is only partially preserved. Its text follows some of the formulas from the other tablets from the same deposit but not all. At places its scribe seems to have abbreviated (corrupted?) the master copy, while at others he has used different formulas. Was this a shortened, less expensive version? The client, Alexandros also called Makedonios, is probably the same as the client of Mitford, nos. 131 and 132. The targets are Timon and Theodoros, the latter identified as governor of Cyprus.[66] *Bibl.:* same as no. 45.

Daimones under the earth and *daimones* whoever you may be; fathers of fathers and mothers (who are a) match (for men), whether male or female,[67] *daimones* whoever you may be and who lie here, having left grievous life,[68] whether violently slain or foreign or local or unburied, whether you (plural) are borne away from the boundaries of (the) cities[69] or wander somewhere in the air,[70] and you (singular) who lie under here,[71] take over the voice(s) of my

64. A variant of the common palindrome ABLANATHANALBA.

65. The bottom of the tablet is damaged and impossible to reconstruct. This tablet, like several others in the group from Amathous (128, 134, 138, 140, 142), appears to include four lines (57–60) of *voces mysticae* below the text proper. Mitford reports "a horizontal line, not shown by previous editors, . . . cut from margin to margin, with line 57 to 60 beneath it" (p. 248).

66. The term translated here as governor (*hêgemôn*) is restored in the text. Under Roman rule until the reforms of Diocletian, Cyprus was governed by a proconsul (Greek *anthupatos*, the title of Sergius Paulus in Acts 13:7). Mitford notes, however, that other Roman governors in the same region, such as in Cilicia, were called *hêgemôn*.

67. A new word (*andrioi*) not found in the other tablets, unless *polu-* has dropped out, in which case the phrase would refer again to the collective grave site (so Jordan in a private communication).

68. At this point the text skips some twenty-six lines found in the other tablets (lines 5–31 in no. 127).

69. Jordan (in SGD) has proposed reading *asteôn* instead of *astrôn*, assuming a scribal mistranscription. In this case, the text refers to the practice of burying the dead beyond the city limits.

70. The phrase may refer to the practice of cremation and to the resultant floating of ashes in the air.

71. The text echoes no. 127, line 31 (*biothanatoi*), but adds a new set of phrases for addressing the unknown dead.

opponents, (I) Alexandros also called Makedonios, to whom Matidia gave birth, namely, Theodôros the governor and Timôn to whom Markia gave birth[72] NÊTHIMAZ . . . MASÔLABEÔ MAMAXÔMAXÔ ENKOPTÔDIT . . . ENOUOUMAR AKNEU MELOPHTHÊLAR AKN . . . ruler of *daimones* beneath the earth. . . . And give a muzzle to Theodôros the governor . . . of Cyprus and to Timôn, so that they will be unable to do anything against me, Alexandros MAZO . . . also called Makedonios. But just as you are . . . wordless and speechless . . . so also let my opponents be speechless and voiceless. Theodôros the governor and Timôn . . .

47. Asia Minor, Claudiopolis, in the province of Bithynia; original location not known. Lead tablet, broken into four sections; 35 × 15 cm. Breakage lines suggest that the tablet was originally folded in quarters. Third or fourth century C.E. The spell includes common *voces mysticae* in addressing primarily "gods" and "angels." The occasion seems to be judicial: a man, Capetolinus, seeks to bind (*katadein*) a large number of men and women to prevent them from giving any information about him. It can be deduced from the names that both client and opponents belong to a servile class. This spell employs the same series of *voces mysticae* as a horse-racing *defixio* from Hadrumetum (no. 11) and several Greco-Egyptian spells (for example, Wortmann, no. 1). *Bibl.:* J. M. R. Cormack, "A *Tabella Defixionis* in the Museum of the University of Reading, England," *HTR* 44 (1951): 25–34; SGD 169.

Granilla, Kapitôn, Granilla, Rouphas, Philôninus, Stalianus, Agathêmerus, Eutuchês, Mercourios, Eunous, Bassus, Primos, Thalamus, Helenus, Capitôlinus, Dêmêtras, Apoll[i]naris, D[êm]êtras, Dêmêtras(?), Parth . . . , Hilaros, Auxiochos, Ariarathês, Kastôr, Hermês, Eutuchês, Philônin[o]s, Bassos, Euagros, Pêgasos, Nigellos, Oualeria, Ma, Lamura, Plokion, Loukia, Dêmêtras, [A]thênais, [An]tônia, [Ath]ênais, [Hill]aros, Huakinthos—let all these (people) cease from speaking ill, from gossiping, from spying; rather, let them be silent, dumb, making no accusation against Kapetôlinos, to whom Danaê gave birth, also called Beautiful, through the power of the names:

Lord Gods, restrain all those inscribed (herein)! *HUESEMMIGADÔN ORTHÔ BAUBÔ NOÊR ODÊRE SOIRE SAN KANTHARA *ERESCHIGAL SANKISTÊ *DÔDEKATKISTÊ *AKROUROBORE *KODÊRE IÔ IÔ IÔ ARBETHE IÔ APERBETHE IÔ ARBATH IAÔ IÔ IÔ ARBETHE IÔ APERBETHE IÔ ABRATH IAÔ IÔ ABRATH IAÔ IÔ *ABRASAX, Lord Gods, angels, restrain all those inscribed (herein)—every bit of their strength (which they might use) against Kapetôlinos to whom Danaê gave birth, also called Beautiful.

72. Only in this tablet are the participants defined with reference to their mothers, though the practice is well attested elsewhere.

48. Olbia Pontica near the estuary of the Dneiper, a trading town on the Black Sea. Found with two other lead tablets in three contiguous tombs, it measures 13 × 8 cm., with a vertical fracture roughly down the middle. The Ionian Greek is not polished; although the syntax and language are elliptical, the general sense of the binding spell can be established with fair certainty. Dated between the third and first century B.C.E.[73] The setting is most probably judicial; the client seeks to initiate a preemptive strike against the people named, who were about to give testimony in court. Of interest is the switch from "we" to "I/me," which may point to the ritual context of the spell-casting act. This spell is important to the broader discussion of curse tablets in that here the dead person in the tomb is directly invoked and promised an offering in return; that is to say, he is not just a messenger to the deities below. *Bibl.:* Benedetto Bravo, "Une tablette magique d'Olbia pontique, les morts, les héros et les démons," in *Poikilia. Études offertes à Jean-Pierre Vernant* (Paris, 1987), pp. 185–218; SGD 173.

> [Just as][74] (it is a matter of fact that) we do not know you, in the same manner (it is also true that) Eupolis and Dionusios, Makareos, Aristokratês and Dêmopolis, [K]ômaios, Heragorês are coming (to court) in order to do a terrible deed,[75] and Leptinas, Epikratês, Hestiaios. (We do not know) for what deed they are coming (to court), (we do not know) upon what testimony those men have agreed, just as we do not know you. If you restrain[76] and constrain them for me, I will honor you and prepare a most agreeable gift[77] for you.

49. Sicily, Selinus (modern Selinunte). Lead tablet roughly 4.0 × 3.7 cm., inscribed on both sides; broken into three parts. This tablet belongs to a group of five tablets discovered near the cemetery at Buffa (SGD 94–98). The Greek is the archaic Doric dialect of Sicily. The letters are small and corrosion has made the script difficult to decipher. The reconstruction is discussed in detail by A. Brugnone, who dates the tablet to the end of the sixth century B.C.E.; this date must be treated as uncertain. The occasion was a trial. The word *sundikos* is mentioned, which in Athens means the accuser's advocate. The author seeks to bind the tongues of at least seven men, presumably his accusers and their wit-

---

73. See the discussion in Bravo, pp. 192–94.

74. *Hôsper . . . houtôs.* This corresponds to the "persuasive analogy" category of spell as defined by Faraone, "Context," pp. 4–10.

75. For *deinon* used in a judicial context, see Demosthenes, *Against Pantainetus* 38.39. In Herodotus, 3.14 and 5.41, *deina poiein,* is to make complaints.

76. The verb *katechô* is used.

77. The Greek is *ariston dôron.*

nesses. *Bibl.: SEG* 26.1113; A. Brugnone, *Studi di storia antica offerti dagli allievi a Eugenio Manni* (Rome, 1976), pp. 73–79 (no. 2); SGD 95; Dubois, no. 31.

> (*Side A*) The tongue of Euklês and that of Aristophanis and that of Angeilis and that of Alkiphrôn and that of Hagestratos. Of the advocates, of Euklês and Aristophanis,[78] their tongues. And the tongue of . . .
>
> (*Side B*) and that of Oinotheos and that of . . . the tongue.

50. Same as no. 49. Discovered in a group of eight tablets near the sanctuary of the local goddess, Malophoros. Lead sheet measuring 17.2 × 9.9 cm., originally folded down the middle but unusually well preserved. Indeed, along with other tablets from Sicily, this is one of the oldest surviving tablets. The language is the normal Doric Greek of Sicily. The tablet is given a tentative date by L. Jeffery of 475–450 B.C.E. The text contains "misspellings, corrections, clumsy spacing, and variation in letter forms and sizes."[79] In line 16, a solid line was etched, dividing the sheet into two parts, one (A) of sixteen lines[80] and the other (B) of just three lines.[81] The deity invoked is never named, but referred to simply as "the holy goddess." The verbs of binding are compounds of *graphô* ("to write"), three times with the prefix *kata-* and once with *enkata-*. No specific occasion is given in the spell, but the fact that the seventeen persons cursed are all male, that they appear to reflect family groups (Calder identifies seven families)[82] and that line 8 mentions

---

78. It is not clear whether Euklês and Aristophanês are the names of the advocates or additional figures.

79. Calder, p. 172.

80. Part A consists of eight subsections, each with a formula of cursing and the names of those cursed. The names in A appear in the accusative case, as direct objects of the binding verbs.

81. In B, the names all appear in the nominative case. There are other significant differences between the two sections, described by Calder, pp. 164, 169–71; for example, of the twelve individuals named in B, nine are repeated from A. Calder formulates the following hypothesis, not accepted by Masson (p. 378): the author of A, perhaps the son of the author of B, wrote first and produced the longer text; between the writing of A and B, the names omitted by B had ceased to be a threat. Despite the differences signaled by Calder, Masson prefers to see B as a simple recapitulation of A. Calder proposes a division of A into eight subsections, each with "a formula of consecration" and each naming from one to four persons (these sections are indicated—roughly following Calder's divisions—by periods in the translation). The formulas range from a simple "with the holy goddess" to the fuller "I record/register with the holy goddess the life and the strength. . . ." B again differs from A, this time in excluding any curse formula and giving only personal names.

82. Calder cautiously defines the families as follows: (1) Pykeleios = father (?) of Halos; Halos = father of Lykinos (= father of Apelos) and Nauertos (= father of Atos ?; only if one assumes, with Calder, that Nauertos and Naueridas are the same person); (2) Tamiras = father of Rotylos = father of Saris and Apelos; (3) Haiaios (our Kailios, following Masson) = father of

"tongues" leads naturally to the assumption that the occasion was a pending lawsuit. The client thus endeavors in the tablet to curse the prosecution and witnesses. Furthermore, as Calder suggests, the fact that the tablet targets family groups points in the direction of a dispute about testaments or inheritance. The ethnic and national diversity of names in the text illustrates the social diversity of Selinus itself.[83] The translation below generally follows the readings and interpretations of Jeffery and Masson. Where Calder differs with Jeffery and Masson on points relevant to the text as a whole, indications will be found in the notes. *Bibl.*: Jeffery, p. 73, no. 10; W. M. Calder, "The Great Defixio from Selinus," *Philologus* 107 (1963): 163–72; O. Masson, "La grande imprécation de Sélinonte (SEG XVI, 573)," *BCH* 96 (1972): 375–88; L. H. Jeffery, "The Great Defixio from Selinus: A Reply," *Philologus* 108 (1964): 211–16; SGD 107; Dubois, no. 38.

> (*Part A*) I record[84] Apelos, (son) of Lukinos with the holy goddess, along with his life[85] and power/strength[86]; and also Lukinos, the son of Halos, and his brother. And (I record) with the holy goddess this one, Nauerotos,[87] the son of Halos, and . . . otulos (the son) of Tamiras and their sons. And Saris and Apelos and Romis (son) of Kailios, (I record) with the holy goddess, and his/ their (?) sons and Saris, the son of Purinos and (also) Puros. With the holy goddess (I record) Puros and the sons of Rotulos (the son) of Puros—with the holy goddess—both their power/strength and their tongues.[88] Plakitas (son) of Nannelaios and Halos (son) of Pukeleios, I recórd their life with the holy

Romis = father of Saris and Pyrrhos; (4) Nannelaios = father of Titelos; (5) Matylaios = father of Kadosis; (6) Magon = father of Ekotis; (7) Kaiaios (whom Calder takes to be different from Haiaios in no. 3; Masson reads both as Kailios) = father of Phoinix = father of Apelos and Titelos. Not all of these relations are certain. For a somewhat different organization of family groups, see Masson, p. 388.

83. Masson classifies the names as follows: (1) Greek: definitely, Pyrrhos, Pyrrhinos, Phoinix, Plakitas; and possibly, Nauerotos, Naueridas, Halos, and Atos; (2) Semitic (Punic): Magon; (3) Sicilian or Italic or Etruscan: probably, Apelos, Titelos; and possibly, Rotylos, Kailios, Romis, Matylaios, Pykeleios; (4) Asiatic: Tamiras, Nannelaios; (5) uncertain: Ekotis, Kadosis, Saris.

84. The verb here is *katagraphein*, a common binding verb in spells. It is used here with the preposition *par*, a shortened form of *para*. The sense here as elsewhere is that the target is to be transferred to the realm or to the authority of the holy goddess.

85. The Greek word is *psucha*, the Doric variant of *psuchê*. As Calder notes, it connotes the idea of "essential life-force."

86. The Greek here is *dunamis*, which is frequently paired with *psuchê* on other tablets (e.g., DT 234, line 16; 237, lines 9 and 30–31). They may be read either as approximate synonyms or, more plausibly, as embracing respectively the inner and outer aspects of human life in its totality.

87. Nauerotos seems to be identical with the brother named just previously. Perhaps, as Calder suggests, the author of the text recalled his name and introduced it only after mentioning him anonymously.

88. Calder reads *glosas* ("tongues") and the following *plakitan* as "the tongue's flat surface." More likely, *plakitan* is to be taken as a proper name; so Masson, pp. 382–84.

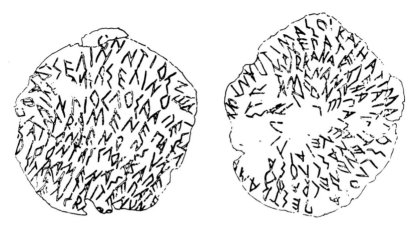

FIGURE 18. Round *defixio* from Selinus (Sicily). The use of round tablet for curses was not common, although a round *defixio* from Athens (no. 20) has been preserved. A round lead tablet from Nemea (still unpublished), inscribed with various *voces mysticae*, was probably intended as a *defixio* or an amulet of some sort.

goddess and also their power/strength. Kadosis (son) of Matulaios and Ekotis (son) of Magôn, I record their life with the holy goddess. And the son of Phoinix (son) of Kailios, I record/register with the holy goddess . . .

(*Part B*) Apelos (son) of Lukinos, Lukinos (son) of Puros, Nannelaios, Ekotis (son) of Magôn, Halos (son) of Pukeleios, Romis (son) of Kailios, Apelos (son) of Phoinix, Titelos (son) of Phoinix, Atos (son) of Naueraidas, Titelos (son) of Nannelaios, Saris (son) of Romis.

51. Same as no. 49; from the sanctuary of the goddess Malophoros (a local equivalent of Demeter). Round lead tablet measuring approximately 4.3 cm. in diameter (Figure 18). Inscribed on both sides; on Side A the writing appears in rough lines, with some letters at various angles on the right and left; on Side B the writing proceeds in concentric circles. Jeffery dates the tablet to the early fifth century B.C.E., thus placing it among the earliest of all surviving *defixiones*.[89] Although no deity is invoked explicitly, we must assume that an oral address to Malophoros accompanied the deposition of the tablet. The occasion is judicial, to judge from the reference to "witnesses." The tablet mentions both men

89. No. 13 in E. Gabrici, "Il santuario della Malaphoros a Selinunte," *Monumenti Antichi* 32 (1927), cols. 388–90 (SGD 108), only fragmentarily preserved, resembles our tablet closely; it is round and uses the same formulas, but with different names.

and women. *Bibl.:* E. Schwyzer, "Zu griechischen Inschriften," *RM* 73 (1924): 426–29; *SEG* 4.37 and 38 (text); Jeffery, no. 1, p. 72; Anne P. Miller, "Studies in Early Sicilian Epigraphy: An Opisthographic Lead Tablet" (Ph.D. diss., University of North Carolina, Chapel Hill, 1973), no. 36; J. Heurgon, *Kokalos* 18–19 (1972–1973): 70–72; SGD 99; Dubois, no. 37.

> (*Side A*) I inscribe[90] Selinontios[91] and the tongue of Selinontios, twisted to the point of uselessness for them. And I inscribe, twisted to the point of uselessness, the tongues of the foreign witnesses.[92]
>
> (*Side B*) I inscribe Timasoi and the tongue of Timasoi, twisted to the point of uselessness. I inscribe Turrana and the tongue of Turrana, twisted to the point of uselessness for all of them.

52. Emporia, near Gerona in the eastern foothills of the Pyrenees (modern Spain). Three lead tablets, each inscribed in Latin on two sides and found inside three separate funerary vases. The Latin is corrupt, with no regard for cases other than nominative and genitive. The tablets can be dated precisely to 78 C.E., on the basis of external evidence. The occasion concerns a trial between two local clans, the Olossitani and the Indicetani, plaintiffs and defendants, respectively. The issue seems to have been a border dispute. Those named include three prominent representatives of the Roman government in the region: (1) Marius Maturus, the *procurator Augusti,* who is also known from Tacitus's *Histories* (2.12.5–6, 3.42.2–4, and 3.43.2), where he is referred to as the procurator of the Maritime Alps. In the spring of 69 C.E., he failed to defend the region around Narbonne from an Othonian invasion. (2) The *legatus Augusti,* whom Pflaum has suggested must be the *legatus Augusti* for the province of Spain near Tarragon. (3) One of his assistants, a praetorian *legatus Augusti,* whom Pflaum suggests is the legate for the diocese of Tarragon.

---

90. The Greek verb here is *engraphein,* which does not occur in any other *defixio;* the closest parallel is "The Great Defixio," also from Selinous, which employs the cognate verb *katagraphein.*

91. The name means "the one from Selinous." Heurgon proposes to interpret the name as evidence that its bearer, while a citizen of the city, may not have been born into the status but more likely had acquired it, perhaps illegally, as an adult.

92. *Sundikoi* played an important role as prosecution witnesses in Greek courts and are thus named frequently in *defixiones* of a judicial character. But the *sundikoi* here are further called *xenoi,* "foreign." No such phrase appears elsewhere in connection with any Greek legal practice. Selinous was populated by residents of various national origins. Thus Heurgon (pp. 73–74) proposes the following scenario leading up to the deposition of the tablet: a dispute had arisen concerning the legality of the client's citizenship; certain witnesses were summoned to give testimony; those named on Side A include some noncitizens (*xenoi*) and, perhaps by extension, our Selinonntios; in any case, as Heurgon notes, we do appear to have a situation here where, in contrast to Athens, women were competent to provide legal testimony in court proceedings.

Lamboglia argues that the tablets were deposited by Sempronius Campanus Fidentius, on behalf of the Olossitani, and that the curses are directed against the highest authorities in the province and the *consilium legati*, the council with jurisdiction over the Indicetani. T. Aurelius Fulvus must have been functioning as advocate for the Indicetani. The text mentions no spirits or deities and uses no verb of cursing or binding. *Bibl.:* M. Almagro, "Plomos con inscripción del Museo de Ampurias," *Memorias de los Museos Arqueologicos Provinciales* 8 (1947): 123; *Année épigraphique* (1952), no. 122 (text); Nino Lamboglia, "Una nuova popolazione pirenaica: gli Olossitani," *Rivista di Studi Liguri* 25 (1959): 147–61; H.-G. Pflaum, *Les carrières procuratoriennes équestres*, vol. 1 (Paris, 1960), pp. 95–98; Solin, nos. 26–28.

1. (*Side A*) Marturus, Augustan Procurator; councillor of the legate, (that is) the councillor of the legate for the Indicetani; (the advocate?) for the Indicetani.

   (*Side B*) The Olossitani; Titus Aurelius Fulvus, Augustan legate; Rufus, Augustan legate.

2. (*Side A*) Councillor of M. Fulvus, the Olossitani, Campanus Fidentinus . . .

   (*Side B*) Fulvus, Augustan legate; Rufus, Augustan legate; Maturus, Augustan procurator; councillor of the legate; advocates of the Indicetani

3. (*Side A*) Sempronius Campanus Fidentinus of the Olossitani . . . oppose me unfairly . . .

   (*Side B*) Fulvus, Augustan legate; Rufus, Augustan legate; Maturus, Augustan procurator; councillor of the legate; advocates of the Indicetani.

53. Gaul, region of Aquitania; between modern Villepouge and Chagnon (France, Charente-Inférieure). Two tablets, measuring 8.5 × 10 cm.; one found near a crude monument, in a Gallo-Roman grave, with a coin from the reign of Marcus Aurelius, dated to 172 c.e. A dead puppy was part of the curse procedure, but it remains uncertain whether it was sacrificed specifically for the occasion. The tablets were pierced with a nail and, according to Audollent, joined in the manner of a diptych. Wünsch argued that the text was not written in the order in which it now appears. The translation is based on Wünsch's reconstruction of the text to its original order. The Latin points to a date in the late second century c.e. Pluto and Persephone are invoked. The occasion for the curse was a trial. *Bibl.:* DT 111–12; Wünsch (1900), no. 9.

(*First tablet*) I denounce the persons written below, Lentinus and Tasgillus, in order that they may depart from here for Pluto and Persephone. Just as this puppy harmed no one, so (may they harm no one) and may they not be able to win this suit; just as the mother of this puppy cannot defend it, so may their lawyers be unable to defend them, (and) so (may) those (legal) opponents

(*Second tablet*) be turned back from this suit; just as this puppy is (turned) on its back and is unable to rise, so neither (may) they; they are pierced through, just as this is; just as in this tomb animals/souls[93] have been transformed/silenced and cannot rise up, and they (can)not . . . (the rest is unreadable)

54. Original location unknown. Lead tablet measuring 8.3 × 11 cm., originally rolled up but with no sign of the customary nail holes. Eighteen short lines of Greek text on one side; one line on the other. The two lower corners are broken off, but even there the text can be reconstructed. The letters and the vocabulary of the tablet point to a date in the third or fourth centuries C.E. The figure invoked under a variety of well-attested names is the Egyptian god Seth. The text names two opponents in a dispute over property and is to be taken as a prelude to legal proceedings. The client is not named. The invocation occupies the first three lines of the text, while the petition takes up the remaining sixteen lines. The goal of the spell is to deprive the opponents of the mental faculties needed to make their case—thought, memory, and "hot" feelings. One unusual feature of the tablet is that it specifies the objects under dispute: slaves, personal property, and legal documents. *Bibl.:* P. Moraux, *Défixion judiciaire* (Brussels, 1960), pp. 3–61; SGD 179.

IAKOUB-IA IA AI[94] BOLCHÔSÊTH IÔRBÊTH NEUTHI[95] IAÔ IAÊ IÔ-SPHÊ[96] IÔ IÔ *ABRAÔTH. Make Akeilios Phausteinos[97] and Stephanos, my opponents in the

---

93. The term here is *animalia*.

94. The series of "words" IA, AI, IAO, IAE, and IO are to be taken as classic examples of the language/letter/sound games characteristic of invocations to higher beings in many different cultures. Such games are based on limited patterns of variations which are in turn based on associations and echoes of letters and sounds, for example, the combination in different order of the vowels, *i, a, o,* and *e.* These forms must have evolved long before they entered the handbooks of magicians as fixed formulas, although there is evidence to suggest that professional magicians themselves sometimes continued the tradition by adding variations of their own. In this instance, the presence of IAO is probably to be seen as an example of this game playing and not as an example of a conscious or "high level" combination of Egyptian and Jewish themes.

95. Various interpretations have been proposed for this term. If the diphthong *eu* is pronounced as *ef,* the result is a reference to the Egyptian goddess Nephthys, the wife of Seth-Typhon (so Plutarch, *Isis and Osiris* 356A, 375B). Another possibility, proposed by Moraux, is that the word may derive ultimately from the Coptic, *noute,* "god."

96. A variant of the term IASPHE, perhaps by association with Joseph, the biblical patriarch. IOSETH and IOSEPH appear side by side in the tablet from Beth Shean (no. 77, pp. 168–69).

97. The names are Roman, though written in Greek. There was a well-known Roman plebian family of the Acilii. M. Acilius Faustinus is attested as consul in 210 C.E. (*CIL* VI.1984).

matter concerning the slaves and concerning the personal property and concerning the papers and concerning the things of which they might accuse me[98]; and concerning these matters may they neither think (about them) nor remember (them); and cool off[99] their mind, their soul, and their passion, from today and from this very hour and for the entire time of (their) life.

55. Greece, Athens (Patissia). Wünsch dates the tablet to the fifth century B.C.E.; original location uncertain. Originally folded and struck with a nail. This curse belongs to a type where both the letters *and* the words are written backward, from right to left. It contains only the names of the cursed persons. *Bibl.:* DTA 26; Wilhelm, p. 105.

K[ro]nios
Sokratês[100]
K]ra[t]inos
Theo]dotos
Al]kaios
. . . . . . ês

56. Greece, Athens; original location not certain. Thick lead tablet measuring 7.5 × 8 cm., written on both sides, with the final fourteen lines on side B written sideways, running from bottom to top. Ziebarth dates the tablet to 350–325 B.C.E., based on the names of numerous well-known figures from that time. It should be noted, however, that two of the tablet's most famous names turn out to be "ghosts." Ziebarth identified

98. The verb *enkalein* was used of accusations by one party against another; more specifically, it was used of cases in which one party sought to repossess property wrongfully lost to the other. In our text, several pretrial items are mentioned that pertain to the case: the documents (*grammata*) that were a regular feature of such trials, through which the plaintiff sought to establish the legitimacy of debts, legacies, or property owed to the prosecuting party. The term translated here as "slaves" is *somata*. Since slave owners could sell, loan, rent, or donate slaves, legal disputes regarding ownership were common.

99. "Cooling off" one's opponents is a common petititon in curse tablets and binding spells. By contrast, anger is often associated with heat. Thus it is not surprising that numerous curse tablets contrast the present heated or emotional state (*thumos*) of their opponents with the desired future condition of cooling off.

100. Could this be the famous philosopher, whose death in 399 B.C.E. came as a result of a court action against him? Socrates was certainly a controversial figure in his time and the target of numerous accusations as one who invented new gods and denied the existence of the old ones (*Euthyphro* 3b). Elsewhere, he relates that his accusers (*katêgoroi*, a common term in curse tablets of a judiciary character) charge him with being "an evildoer (*adikei*) and a curious person who searches things under the earth and in heaven and makes the worse appear the better cause" and that he was a teacher who took pay for his instruction (*Apology* 19b–e). The name Socrates also appears in several other tablets (e.g., DTA 7, 10, 97, 106a, 170). The name was not uncommon at the time.

the noted philosopher, Aristotle, on side A, line 3, and the equally celebrated orator, Demosthenes of Paiania, on side B, line 1. D. R. Jordan informs us that on the basis of his own inspection of the tablet both names (and much else) must be read differently, as Aristogeiton and a different Demosthenes. The curse consists of three initial verbs, followed by two long lists of names, seventy-seven in all; the first thirty-three names appear in the accusative case, the rest, curiously, in the nominative. Many are accompanied by their demotic designation, the administrative unit to which all Athenian citizens were assigned on the basis of heredity. Thus the names belong primarily to citizens. Side B includes men and women, as well as a new binding formula in lines 11–12. The two sides may not refer to the same circumstances. No deities or spirits are invoked. No precise occasion is cited, but we must suppose that the targets constituted, at least in the eyes of the client, a party (political?) of some sort. The major political issue of the period was Macedonian rule over Greece which involved ever-changing alliances for and against that rule. It seems reasonable to suppose that our tablet, with its mention of numerous actors in pro- and anti-Macedonian affairs, may be a response to these events. The occurrence of "scribe" in line 52 of side A may point to an organized party. Other prominent figures of the period are also listed. We have not recorded all of the names from the tablet. *Bibl.:* Ziebarth (1934), no. 1A–B, pp. 1023–27; Robert, *Froehner,* 13–14; SGD 48; J. Ober, *Mass and Elite in Democratic Athens* (Princeton, 1989) p. 149.

> (*Side A*) I bind, I deeply bury, I cause to vanish from mankind: Eunomos of Pithos, Aristo(. . .),[101] Lusiklês of Acharnai,[102] Dêmocratês of Aixonê (line 5)[103] . . . , Archiadês of Thorikos (line 15),[104] Xenoklês of Sphettos (line 24),[105] Kallias (line 41), the resident alien Aristarchos the son of Aristarchos

101. Wessley (in Ziebarth, p. 1026) reads the name as Aristogeiton.

102. Probably the brother of Demophilus on the same tablet (side B, line 1).

103. Member of a distinguished Athenian family, noted for its wealth and active in horse breeding. Several of its members achieved victories in chariot races at important games. Our Democrates held the prestigious position of *choregos* (responsible for underwriting the performance of theater productions for a period of one year) for his deme in 326/325 B.C.E.; cf. Davies, *Families,* p. 360.

104. Possibly a member of a well-known family. The family of Archiades' father, Euthymachus, is known from a speech of the orator, Demosthenes, which mentions Euthymachus's son, Archiades, presumably also of Otryna (*Against Leochares* 44.9).

105. Possibly to be identified with the famous Athenian, Xenocles of Sphetta, the son of Xeinis. He held virtually all·of the major public offices open to or imposed upon wealthy citizens—gymnasiarch, trierarch, and superintendent of the mysteries. He was a friend of Lycurgus. His offices stretch from 346–306 B.C.E.

(lines 44–45),[106] Strombichos of Euonumon (line 53),[107] Strombichidês of Euonumon (line 54),[108] Polueuktos of Sphettos (line 56), Kalliphanês Kudanitês (a demotic—line 68), Nikoklês Kudanitês (line 70)[109] . . .

(*Side B*) Dêmophilos (line 1),[110] Dêmosthenês of P(. . .) (line 2), Menestratos (line 15),[111] Kleinis Laikastria[112] (line 16), Skulla Laikastr[ia] (line 17), Sôphronis Laikas[tria] (line 18), Achris Laikastri[a] (line 19), Onêsandros of Peiraeus (line 24)[113] . . .

57. Greece, Athens. The lead tablet was found in a well inside the Dipylon Gate in the Kerameikos, along with 574 pieces of lead, which record the annual evaluations of the Athenian cavalry. Jordan has argued that the well was used as a dump site and that all of its contents originated elsewhere. For this tablet he argues that it belonged originally to one of the fourth century B.C.E. graves near the Dipylon Gate. The sheet measures approximately 12 × 7 cm. and was folded into three sections. Like others of similar type, it mentions no spirits or deities and employs no verbs; the names are given in the accusative case. The inscription consists of two parts: one is written upside down—with respect to the other words—and consists of just one word; the other

106. Resident aliens or metics were numerous in Athens. They were required to have a citizen as sponsor. Many were quite wealthy and could assume various public duties, though never the highest.

107. Strombichus of Eunonymon was trierarch in 357 B.C.E. This office required great wealth in order to furnish and outfit a warship (trireme) for one year; cf. *PA* 13022 and Davies, *Families*, p. 163.

108. Quite possibly the son of Strombichos in the previous line; cf. Davies, *Families*, ibid.

109. This figure served as *parasitos* (a priest who served at public expense) in the late fourth century B.C.E.; cf. *PA* no. 10903 and Davies, *Families*, p. 409.

110. Demophilus held a religious office at Eleusis and has been identified with the Demophilus who lodged charges of impiety against Aristotle; cf. Diogenes Laertius, 5.5, and Athenaeus, *The Learned Banquet* 696a–b. There were, however, other prominent figures of the period who bore the same name; cf. Davies, *Families*, p. 498.

111. Listed in an inscription of the fourth century B.C.E. as a priest of Asclepius; cf. *PA* 10001.

112. Lines 16–19 contain names of women, each followed by the letters *laikast* (except in the first case, where the complete word *laikastria* is visible, the others are missing between one and four letters). These lines form part of two columns of names, at right angles to the main text running from bottom to top of the tablet. Ziebarth proposed that *laikas* was a mistaken abbreviation for the deme of the women, *lakiadai*. L. Robert (pp. 13–14) argued that there are problems with this reconstruction. First, women did not normally bear a demotic label. Second, a fourfold mistaken writing of *Lakias* seems unlikely. Instead, Robert proposes to read *laikas* as "prostitute," an intentional insult hurled at these women by the client. In line with this, Robert reconstructs the letters *tera* after the first name, Kleinis. If accepted, this would mean that the first-named of the women is thus labeled as "more of a prostitute (than the others)." Jordan informs us that the text clearly reads *laikastria* ("strumpet") and that the term was used merely to identify the women by their profession.

113. Mentioned in an inscription from the end of the fourth century B.C.E.; cf. *PA* 11449.

consists of five names and two more incomplete words. The names refer to famous figures in the troubled political life of Athens in the late fourth century B.C.E., the time of Cassander, the ruthless friend and successor of Alexander the Great.[114] From 319 B.C.E. until his death in 297, he controlled much of Greece under Macedonian power. *Bibl.:* K. Braun, "Der Dipylon-Brunnen B₁—Die Funde," *AM* 85 (1970): 129–290; Jordan, TILT; C. Habicht, *Pausanias' Guide to Ancient Greece* (Berkeley, 1985), pp. 81–82; SGD 14.

> (*First inscription, upside down*)
>   PLEI[S]TEA[115]
> (*Second inscription*)
>   Pleistarchos[116]
>   Eupolemos[117]
>   Kassa[n]dros[118]
>   Dêmêt[rios][119]
>   Ph[al]ê[rea]
>   . . knê. . . . Peir<a>iea[120]

58. Greece, Attica, specifically Halai. The lead strip measures 14 × 2 cm. and seems to have been folded. It is inscribed on both sides and shows traces of an earlier inscription. No deity or spirit is mentioned and no verb used. This curse belongs to the type in which the words are written backward, from right to left, though the letters are written "correctly," facing right. Wilhelm dates this tablet to the early fourth century B.C.E., based on his identification of the personal names with known historical figures from the period. The tablet reflects the troubled political circumstances at the end of the war between Athens and Sparta, the rise of Thebes as a new threat to Athenian power, and the formation of

---

114. Jordan (p. 234) comments that "the cause of the curse was political, that the tablet is the result of some Athenian's displeasure at the Makedonian domination of his city" and notes that the intense hostility toward the Macedonians is revealed in the "excessive honors that the people of the city bestowed on Demetrius Poliorketes and his father Antigonos when in 307 their forces freed Athens from Kassandros' hold."

115. Jordan suggests that the original engraver of the tablet began by misspelling the first name. His solution was simply to turn the sheet upside down and begin again.

116. Pleistarchus is the younger brother of Cassander, the Macedonian ruler and friend of Alexander the Great.

117. Eupolemon was Cassander's general in Greece.

118. A powerful figure who gained control over both Macedonia and Greece between 319 and 316 B.C.E.

119. The noted Peripatetic philosopher, from Phalerum, whom Cassander appointed governor of Athens.

120. Jordan interprets the word as a demotic, indicating someone from Peiraeus; he would be the fifth person named in the curse.

the Second Athenian League. A central figure in these events was Callistratus, a man frequently involved in lawsuits and a popular target in comedies of the period.[121] Our tablet mentions two brothers of Callistratus, Eupherus and Aristocrates. *Bibl.:* DTA 24; Wilhelm, pp. 115–22.

(*Side A*)
Phôkiôn[122] Ergokratês[123]
Eupheros[124] Aristocratês[125]

(*Side B*)
Mêdeia Pis[t]ocleês
Nikomenês[126] Euthêmôn[127] S[u]ra

59. Greece, Attica, again Halai. Four tablets, measuring roughly 9×2 cm., were inscribed by the same hand, written on both sides, and pierced by a single nail. These tablets belong to the type that uses a verb of cursing, in this case, *katadô*. Wilhelm argues that these tablets stem from the same time and circumstances as DTA 24 and 57, the fourth century B.C.E. *Bibl.:* DTA 47–50 (a–d); Wilhelm, pp. 114–15.

121. See the fragments of Theopompus (author of burlesques and other comedies), Antiphanes (author of many plays, mostly parodies), and Eubulus (author of some one hundred burlesques and parodies); *CAF* II.168: "There is this fellow, Callistratus . . . who had a big and lovely rear end."

122. Wilhelm tentatively identifies our Phokiôn with the famous Athenian statesman and general of the mid-fourth century B.C.E. Further, he proposes to connect the curse with disturbances surrounding a series of calendrical and administrative reforms concerning the offices of demarch and treasurer; cf. Wilhelm, pp. 117–18.

123. An Ergokratês appears on an inscription (*IG* 2.2, 1007), dating from the mid-fourth century B.C.E.

124. Eupheros is the brother of Callistratus of Aphidna, a powerful political figure in Athens; he was finally impeached and condemned to death. Callistratus himself is the target of two curse tablets (Ziebarth [1934], no. 2; cf. *DT* 63). The first reads simply, "I bind Callistratos, and all his associates/advocates (*sunêgorous*) I bind." The second, a fragment, mentions three Athenians (names missing) and "the accomplices (*sundikous*) [with Callist]ratos." Both reflect trials and, together with our tablet, are probably to be understood as judicial in nature.

125. A second brother of Callistratus; cf. Wilhelm, p. 117.

126. According to Wilhelm, p. 117, based on two inscriptions (*IG* 2.1.572 and 2.3.1208), Nikomenês, Euthêmon, and Astuphilos were associates in proposing a set of reforms that led to opposition from other parties. In both cases, Nikomenês' name appears along with Euthêmon and Astuphilos.

127. According to *IG* 2.1.571 and 572, Euthêmon initiated the administrative reforms, following difficulties with earlier officeholders. The precise time of these reforms is 368/367 B.C.E. Thus Wilhelm suggests that our curse stems from the same time and circumstances as *DT* 47–50 and 57. Thus, in Wilhelm's judgment, the curse must be attributed to the opponents of Euthêmon and his associates, the ruling elite of the time under the leadership of Callistratus. As Wilhelm argues, their opponents probably suffered legal punishments for their earlier misdeeds.

an abbreviated catalogue of those found among the curse tablets. More-over, he describes precisely the sorts of conditions under which we find *defixiones* in the ancient workplace—competition and rivalry. When we further consider that virtually all of the occupations mentioned on the tablets fall into the category of small-scale, marginal businesses, where survival and success were perennially uncertain, we should not be sur-prised to find individuals who were prepared to seek an advantage for themselves by cursing or binding the affairs of their nearby competitors.

In some of the tablets, where personal names appear together with the occupation, the underlying cause of tension may concern something other than professional issues; the person's occupation may have been added merely for the purpose of full identification. However, the fact that the occupation functions like a patronymic ("son of") or even a demotic ("belonging to such-and-such a deme") is in itself a telling indicator of the social importance of work. Thus there is an unavoidable element of uncertainty in our choice of tablets. For instance, a tablet from the Kerameikos area of Athens, dated to the late fourth century B.C.E., identifies several people by profession: two tavern keepers, a stallkeeper, a household slave, and a brothel keeper (SGD 11). But whether these occupations have anything to do with the occasion for the tablet is not certain. Other tablets of a similar kind mention a shield maker,[5] a painter, a flour seller, a scribe,[6] a seamstress[7] and a ship captain.[8]

In other cases, where in addition to the occupation, the tablet binds or curses the target's labor, products, income, and workplace, there can be no doubt that the root issue was competition between small businesses and their proprietors. Here it is worth noting that all of these tablets stem from Greece and the Greek colonies in Sicily and that their dates fall exclusively in the classical and Hellenistic periods.[9] Whether these limitations are significant, indicating that tablets of this nature occurred only in these times and places, or merely accidental, cannot be deter-mined. The earliest can be placed near 450 B.C.E.: one from Sicily ap-pears to condemn a list of names to a downturn in profits[10]; and one from Athens identifies its target as a bellows operator in the Athenian mint (no. 72). Once again, the majority of persons belong to the world of marginal laborers, some free and some slave—tavern keepers, carpen-ters, metalworkers, potters, prostitutes, and so on. In our selection, there are but two possible exceptions to this rule and both concern physicians (nos. 79, 81). But in both cases the physicians are the targets of the tablets, while the status of the client in at least one case (no. 79) seems to point in the direction of slaves or freedmen.

Among the occupations cited in the tablets, apart from chariot racers and gladiators, the most common is that of proprietors of taverns (*kapêleion* in Greek; *tabernae* or *popinae* in Latin). Of special note here is the fact that both women and men appear as proprietors and that women seem to have been particularly active as tavern keepers. From an aristocratic and literary perspective, the tavern and its proprietor were regularly treated with disdain, the very embodiment of disorderly and dishonest low-life in the city.[11] Rightly or not, they are often described as brothels. But for the working patrons of a local pub, the institution must have served—as has always been true—a whole range of social and personal functions that outside observers were simply incapable of discerning. In line with this, it would seem that the frequent use of *defixiones* in and around the tavern demonstrates that important issues were transacted there.

One final observation about the exigencies of certain crafts and their connection with cursing. If, as we have argued, curses were deployed in social and personal conditions of competition and uncertainty, we might expect to find an association of curse tablets with occupations characterized by a high potential for mistake and failure. Ceramics—the making of pottery through the use of high temperatures—was such an occupation.[12] A curious text from the classical period (sixth to fourth centuries B.C.E.) illustrates the hazards of firing pots as well as some of the customary ways of explaining the frequent disasters.[13] The so-called Potters' Hymn includes a passage that is almost certainly based on a once-extant formula for curse tablets against the kilns of rival potters:

> If you will pay me (Homer is the imagined speaker) for my song, O Potters, then come, Athena, and hold your hand above the kiln. May the cups and cans all turn a goodly black, may they be well fired and fetch the price asked. . . . But if you (potters) turn shameless and deceitful, then do I summon ravagers of kilns, Suntrips [Smasher] and Smaragos [Crusher] and Asbestos [Unquenchable] and Salaktês [Shake-to-pieces] and Omodamos [Conqueror of the unbaked] who cause much trouble for this craft.[14] Stamp on stoking tunnels and chambers, and may the whole kiln be thrown into confusion, while the potters loudly wail. As a horse's jaw grinds, so may the kiln grind to powder all the pots within it.[15] And if anyone bends over to look into the spyhole, may his whole face be scorched, so that all may learn to deal justly.

In light of this text, it comes as no surprise to find, some five hundred years later in Pliny's *Natural History* (28.4.19), immediately following his comment that no one was immune to the fear of curse tablets, his observation that "many people (presumably potters) believe that the

Philomêlos the son of Philomêlos from the deme of Melitê; and Phil . . . (?) from the deme of Melitê; and Eugeitôn the son of Eugeitôn from the deme of Acharnai.[3]

61. Greece, Attica. Inscribed in reverse, from bottom to top, from right to left. The tablet measures 11 × 5 cm. The invocation of the Praxidikai (female figures who deal out justice) is unusual among the surviving curse tablets. Here they occupy the place usually taken up by Persephone or Hermes. The immediate purpose of the spell is not clear, but since Manês' business is especially made the target of the curse, business rivalry between the commissioner of the tablet and Manês, the targeted victim, seems likely. The exchange here is based on the notion that "if you (gods) give now, I (person) will give later." In this case, *euangelia*, gifts in exchange for good tidings, were promised to the deities after the completion of the desired results. *Bibl.:* DTA 109.

> I bind and restrain Manês. And you, Dear Goddesses of Vengeance,[4] restrain him; Hermes the Restrainer, restrain Manês and the affairs of Manês and cause the entire business in which Manês is engaged to become entirely contrary and backward[5] for Manês. I will sacrifice thank offerings to you in exchange for the good news,[6] Goddesses of Vengeance and Hermes the Restrainer, if Manês fares badly.

62. Greece, Attica; original location not known. Wünsch dates this tablet to the fourth century B.C.E. It measures 41 × 4 cm. and is written in an elegant hand; originally folded and pierced with a nail. One line appears on the other side. Here the occasion is competition between small-scale merchants, mostly tavern keepers. In addition to the formula of binding and the mention of specific physical and mental features of the cursed persons, this tablet invokes Hermes as the agent of the binding action (see DTA 79–97). *Bibl.:* DTA 87; F. Bömer, *Untersuchungen*

---

3. These two names, given with both patronymic and demotic, belong to citizens. Thus the spell includes both slaves and citizens.

4. *Praxidikai* in the plural refers to the three goddesses of vengeance normally shown in images with heads only. According to the Suda (s.v. *Praxidikai*), they are the daughters of Ogyges, and are named Alalcomenia, Thelxinoea, and Aulis. The singular of *Praxidikai* was used by Orphic poets to refer to Persephone (see *Argon* 31 and *Hymn* 29.5). Jordan (SGD, p. 157) reports a tablet, found with six others in the Kerameikos, which begins with the phrase, *pros tas praxidikas* ("to the goddesses of vengeance").

5. The term *eparistera* here probably reflects the "backward" form of the writing on the tablet.

6. The Greek is *euangelia*, which can mean sacrifices in exchange for good news; cf. Xenophon, *Hellenica* 1.6.37.

über die Religion der Sklaven in Griechenland und Rom (Wiesbaden, 1963), pp. 984–85.

(*Side A*) I bind Kallias, the shop/tavern keeper who is one of my neighbors and his wife, Thraitta[7]; and the shop/tavern of the bald man[8] and the shop/tavern of Anthemiôn near (?) and Philôn the shop/tavern keeper. Of all of these I bind the soul, the work, the hands, and the feet; and their shops/taverns. I bind Sôsimenês, his (?) brother; and Karpos his servant, who is the fabric seller and also Glukanthis, who is called Malthakê, and also Agathôn the shop/tavern keeper the servant of Sôsimenês: of all of these I bind the soul, the work, the life, the hands, and the feet.

I bind Kittos my neighbor, the maker of wooden
frames[9]—Kittos's skill and work and soul and mind
and the tongue of Kittos.
I bind Mania (feminine) the shop/tavern keeper who is (located)
near the spring and the tavern of Aristandros of
Eleusis
and their work and mind.
The soul, hands, tongue, feet, and mind: all of these I
bind to Hermes the Restrainer in the unsealed[10]
graves[11]

(*Side B*) the servants of Aristandros.

63. Greece, Athens; original location not known. Thin lead tablet, with one-third of the original missing; pierced by a nail. Date uncertain, although probably no later than the second century B.C.E. and possibly as much as 150 years earlier. No deity is invoked and the simple formula "I bind!" is typical of early Greek *defixiones*. The curse is directed against a helmet maker, his wife who works with gold, their house, and their business. A mix of personal and business motives probably lies behind the tablet. *Bibl.:* H. Lechat, "Inscription imprécatoire trouvée à Athènes," *BCH* 13 (1889): 77–80; DTA 69.

7. *IG* 2.2.773 A, an inscription from the fourth century B.C.E. mentions a "Thraitta, the tavern keeper, who lives in Melita and who fled Menedemos (her husband?) who also lives in Melita; the Tavern of the Bald Man." Is it possible that our *defixio* was prepared and deposited by the same unfortunate Menedemos?
8. Probably better, "Baldy's Tavern," the common name for the place.
9. Wünsch gives the word as *kanabiourgos;* if we read *kannabiourgos* instead, the translation would be "rope maker."
10. Wünsch was unable to make sense of these letters and reads them as *asphragiai.* C. Faraone has suggested reading them as *asphragistois* ("unsealed").
11. We may safely assume that the tablet was originally deposited in a grave.

I bind Dionusios the helmet maker[12] and his wife Artemis the goldworker and their household and their work and their products and their life—and Kallip[pos . . .]

64. Greece, Attica; original location unknown. Wilhelm, on the basis of his identification of certain names with known figures, places it in the late fourth century B.C.E. The engraving is particularly elegant and resembles the lettering on public monuments of the period. The lead tablet measures 23 × 11 cm. and is written on both sides; probably folded. The curse seems to involve three separate issues and groups of persons—business competition, marital concerns, and hostility toward soldiers. Several of the names are written twice, once with the letters scrambled and once in normal fashion; the same technique shows up in a spell from Sicily in the fifth century B.C.E. (SGD 105). *Bibl.*: DTA 55; Wilhelm, pp. 107–8.

(*Side A*)

Dioklês the son of . . .[13]

I bind Ki[mônoklês] son of (?) the pipe maker and carpenter (?) and also his jar and the box where his pipes are carried, and also Athênagoras . . . (*Kimônoklês Oineis*)[14] Xenarch[os and Pa]taikion whom Epainetos claims to be his daughter . . . and whom he pledged (?) as wife to Exesthenês of Trozên (*Pataikion of Trozên?*)—a curse[15] (on them). Deinôn of Peiraeus son of Deisitheios, of Peiraeus (*Deinôn son of Deisitheios, of Peireus*); Oiniadês son of Apollodoros, of Eroiadai, the carpenter, who is serving with the soldiers in Peiraeus (*Oiniadês, son of Apollodorus, of Eroiadai*); Chaireleidês son of Chaireleidês, of Anaphlustos, the son of . . . , who is serving with the soldiers at Peiraeus (*Chaireleidês of Anaphlustos*) . . . Dêmostrat[os] son of Archamenês of Murrinous, son of Archamenês (*Dêmostratos of Murrinous*) . . . Hêrostratos (?) who is serving with the soldiers at Peiraeus . . . All of these I consign, (inscribed) in lead[16] and in wax[17] and in water (?) and to unemployment and to destruction and to bad reputation

---

12. The term *kranopoios* is not a common one. A character with this profession does show up in Aristophanes' play, *The Peace*, lines 1210ff., as part of a crowd of armor makers who complain bitterly that peace will drive them out of business.

13. The name appears centered above the rest of the text and in the nominative case. It seems likely that Diocles is the author or originator of the curse. If so, this is one of the rare instances in which the curser is named on a tablet.

14. The letters in parentheses are deliberately scrambled in the inscription; they represent a symbolic attempt to "scramble" the person whose name is thus miswritten.

15. The Greek term is *ara*, used for prayer and curse.

16. This undoubtedly refers to the tablet itself.

17. It is possible that the client also commissioned a wax figurine, representing (collectively?) his enemies.

and to (military?) defeat and in tombs[18]—both these and all the children and wives who belong to them.[19]

(*Side B*) Lu[sim]êdês . . . Philostratos Kei[. . . who is serving with the soldiers at Peiraeus. . . . I bind these in graves,[20] in distress, and in tombs.

65. Greece, Athens; near the railroad station. Lead tablet measuring 14 × 7 cm.; written on both sides and originally folded. No date is given, although certain formal aspects of the tablet suggest a date in the fourth century B.C.E. The editor describes both the character of the writing and the quality of the Greek as "quite bad," and considers the possibility that the writer might have been a non-Greek. As in other Attic *defixiones* of this period, there is no explicit mention of any spirit or deity, nor any detailed specification of the harmful effects of the spell on its targets. The setting appears to be competition and rivalry between small business owners, especially tavern-keepers. In addition to taverns, the spell mentions workshops and a store. Both men and women are named as targets. *Bibl.:* DTA 75; Faraone, "Context," p. 11.

(*Side A*) I bind[21] Anacharsis and I bind his workshop. I bind Artemis, the . . . and I bind the master of Artemis. I bind Humnis. I bind Rhodiôn the shop/ tavern keeper. May Rhod<i?>ôn perish along with his workshop . . . (?) who works (there?). I bind Rhodiôn the shop/tavern keeper,[22] I bind the shop/tavern, and I bind also the store.

(*Side B*) I bind Artemis and . . . and . . . may (?) gain power over Artemis . . . I bind the work . . . and the tongue. I bind Theodotos and the/this workshop. I bind Artamis and Philôn, his works . . . sister . . . friend . . .

66. Greece, Peiraeus; found in a tomb of uncertain location. Date uncertain. Written from right to left. The inscription consists of forty-one short lines. It belongs with DTA 96 to which it bears a close resemblance; the two were produced by the same hand and use the same formulas. The curse is directed successively at three targets: Mikion,

18. The Greek is *en mnêmasin*, which might designate either a public memorial or a grave. Either seems possible here. In the first case, the curse is designed to obliterate any memory of the person; in the second, the point would be that the person's death is wished or that the curse tablet itself was deposited in a grave.

19. As suggested by C. Faraone, the evil consequences seem to fall into pairs: lead and wax as the media of cursing; water and tombs as places for depositing the tablet(s); unemployment and bad reputation as social ills; and defeat and destruction as military disasters.

20. Here again the Greek is *en mnêmasin*.

21. The Greek verb is *katadênuein*, a variant of the more common *katadein*. In the curse tablets it occurs also in a *defixio* from Athens related to legal matters (DTA 94).

22. The writer's casual use of case endings and spellings makes it impossible to determine whether Rhodon and Rhodion are different persons.

then Hipponoides and Socrates together, and finally Aristo, a woman. The same phrases, slightly modified, are employed in each case. The precise occasion of the curse is not clear but business competition, a frequent source of tension, seems likely. No deity is mentioned. *Bibl.:* DTA 97.

> I have seized Mikiôn and bound his hands and feet and tongue and soul[23]; and if he is in any way about to utter a harsh word about Philôn[24] . . . may his tongue become lead.[25] And stab his tongue, and if he is in any way about to do business,[26] may it be unprofitable for him, and may everything be lost, stripped away, and destroyed.[27] I have seized Hipponôidês and Sôkratês and bound their hands and feet and tongues and souls; and if they are in any way about to utter a harsh or evil word about Philôn, or do something bad, may their tongues and souls become lead and may they be unable to speak or act; but rather stab their tongue; and if they have anything, or about to have anything, whether possessions or property or business, make it lost, stripped away and destroyed, and let them be destroyed for them. {I Aristô}[28] I have taken Aristô and bound the hands and the feet and the tongue and the soul; and may she be unable to speak any evil word about Philôn, but may her tongue become lead; and stab her tongue.

67. Greece, Attica; original location not known. Lead tablet measuring 19 × 6 cm.; originally folded and pierced by a nail. The editor offers no date but the tablet is certainly no later than the fourth century B.C.E. Several men are bound over to Hermes. Of the five lines, parts of the first and second lines are written so that the full line reads from left to right, while the individual words are spelled from right to left. This is another version of "garbled" writing used in Attic tablets of the classical period. Along with the names of the men, the spell specifies five items normally found in such texts (feet, hands, soul, tongue, products) and one unusual one (their profits or income). This final item is enough to suggest that the occasion was business competition of some sort. *Bibl.:* DTA 86.

---

23. The language here suggests that the tablet may have been accompanied by some sort of physical representation of the cursed parties.
24. Philon seems to be the most likely candidate for having commissioned the curse.
25. Here is a clear instance where the material of the tablet itself (lead) is used to symbolize the desired action: "make his tongue like lead—cold, heavy, and unable to move."
26. Similar phrases are used in trial curses, to incapacitate one's accusers in the courts, but the circumstances here need not point toward court actions. Rumors, backbiting, or competition between shopkeepers would do just as well.
27. The terms are *achôra, amoira,* and *aphanê.*
28. Here the writer inscribed two words but decided to begin again one line lower.

I bind over before Hermes the Restrainer Androsthenês[29] and (?) Iphemu-thanês and (?) Simias[30] (and **Dromôn**)—feet, hands, soul, tongue, products, and income.

68. Greece, Athens (Patissia). Lead tablet measuring 26 × 9 cm. Written in (often "misspelled") Greek, backward (in addition some letters are written facing left, or retrograde), and on both sides of the tablet. Highly repetitive formula, listing hands, feet, tongue, shop, and events (or contents) of shop. Only in the last line of the second side does the client add *katadô* ("I bind"). The client is evidently seeking to control an entire marketplace of shopkeepers, including a miller, a boxer, a madame, and at least one prostitute. Many of the "shopkeepers" listed on the first side are women, which might mean either that they are prostitutes (someone called *charitopôlis* on Side A, line 6) or that many shops in this market were owned by women. A number of those mentioned also bear nicknames indicating non-Athenian origins, such as Lukios from Lycia, Ludês from Lydia, and Lakaina from Lycaonia. The difference implied between those who own "shops" and those who are "dealers" is not clear. *Bibl.*: DTA 68.

> (*Side B*) (I bind) Diphilês: both the hands and feet and tongue and fee[t and shop and] everything in the shop[31]; Posis: hands and [feet and tongue] and the shop and everything in the shop [. . . hands] and feet and tongue and [the shop] and everything in [the shop]; Lusandros: hands and feet and sho[p and everything [in] the shop; Anutas the [de]al[er]: hands and [fe]e[t and the sh]o[p] and everything in the shop [. . .] hands and feet and tongue and shop and [everything] in the [shop]; Lukios: hands and feet and tongue and sho[p and] in the shop; Ludês: hands and feet and tongue and shop and everything in the shop; Killix [the . . . hands, feet] and shop and everything [in the] shop; Melas: hands and fe]et and feet and [ton]gue and shop and everything in [the shop]; Lakaina, the concubine[32] of Melanos: hands, feet and . . . I bind the slave (**masculine**) of Melas . . . hands and feet and sho[p . . .

69. Greece, Attica; original location not known. Found in a tomb; lead tablet inscribed on both sides. Metrical composition (poor elegaic couplets) with some Ionic forms. Third century B.C.E. Side B is "addressed" to the Furies; the spell itself involves Hekate and Tartarus. The occasion

---

29. A name attested several times in *PA* nos. 902–10.
30. A named attested several times in *PA* nos. 12664–667.
31. The term is *ergastêrion*, possibly signifying a brothel.
32. The Greek term is *pallaka*.

is apparently the jealousy of one woman (Bittos?) for another woman's assets and social position. The spell speaks of her marks of social and economic status, although it also mentions her mind (*nous*) as an asset to be bound. The content of the spell suggests that Bittos's motivation for commissioning was revenge. *Bibl.:* S. A. Koumanoudes, *Ephêmeris Archaiologikê* (1869): 333 (no. 405, fig. 49-gamma); G. Kaibel, *Epigrammata Graeca* (Berlin, 1878) no. 1136; DTA 108.

> (*Side A*) I will bind Sôsikleia an[d (her) p]roperty and great fame and fortune and mind. Let her become hateful to (her) friends. I will bind her under murky Tartarus[33]
>
> (*Side B*)[34] in troublesome bonds, with Hekate of the underworld.
>
>
>
> BITTÔ[35]   AIELKISÔS
>
> for the dizzying Furies.
>
>

70. Greece; purchased on the Athenian art market. The tablet is a thick sheet of lead measuring 11 × 7.5 cm., folded twice, with three nail holes. Peek dates it to ca. 350 B.C.E. The context for the curse is business competition among potters in the Kerameikos district of Athens, although the basis for this conjecture is limited to the references to the business of two of the secondary targets. In the background, there may well be the prospect of legal proceedings, for Nikias is somehow connected with the judicial council of the Areopagus. Litias may have been a witness, or even the defendant. *Bibl.:* Peek, no. 9, pp. 97–100; SGD 44.

> I bind Litias[36] before Hermes the Restrainer and Persephone, the tongue of Litias, the hands of Litias, the soul of Litias, the feet of Litias, the body of Litias, the head of Litias. I bind Nikias before Hermes the Restrainer, of the Areopagite,[37] the hands, the feet, the tongue, the body of Nikias. I bind Dêmetrios before Hermes the Restrainer, the body, the business of Dêmetrios the ceramic worker, the hands, the feet, the soul. I bind Epicharinos before Hermes the Restrainer. I bind Dêmadês the ceramic worker before Hermes the Restrainer, the body, the business, the soul. . . . I bind Daphnis

---

33. The same formula is used in Homer, *Iliad* 8.13, and Hesiod, *Theogony* 119.

34. The last metrical line of the spell proper continues onto the next side of the tablet.

35. The name is spelled backward. It is unclear whether Bittos is the name of the client or of the corpse in whose grave the tablet was deposited. Both names are written in large letters, from different directions and upside-down.

36. Peek proposes to identify our Litias with a shipbuilder mentioned in an inscription of 342/341 B.C.E. (*IG* 2.2. 1622).

37. Nikias is here connected with the Areopagus, the "Hill of Ares" on the outskirts of Athens northwest of the Acropolis. An Areopagite belonged to the ancient council, which met there until at least the fourth century C.E.

before Hermes the Restrainer. I bind Philônidês before Hermes the Restrainer. I bind, I bind Simalê Pistê before Hermes the Restrainer. I bind
Litias, the feet, the hands, the soul, the body of Litias, the tongue of Litias,
the will of Litias, which is carried out before Hermes the Restrainer and
Persephone and Hades.

71. Greece, Athens; discovered in "House D" in the industrial area of
the Agora. Other materials from the building, including a hearth and
fragments of iron and bronze, point to metalworking as the chief business. The editor suggests that the *defixio* may have been deposited
originally "into the foundations or under the floor." Lead tablet measuring 14.5 × 6.5 cm.; originally rolled up and pierced by a nail. The letters
and the archaeological context indicate a date in the fourth century
B.C.E. The Greek is quite simple and unsophisticated. The spirits invoked are called "those below." The binding formula itself is typical of
the simple forms in this period. The setting, as befits the place where the
defixio was found, involves some form of rivalry between smiths.
Whether the rivalry stemmed from business matters, as seems likely, or
personal factors, perhaps relating to the woman, is not clear. Several
men and at least one woman are mentioned; their names are badly
spelled, possibly intentionally. D. R. Jordan informs us that the word
translated by the editor as "bronze-worker" (*chalkea*) may instead be an
ethnic designation of origin, "of Chalkis." In this case, there would be
no reference to professional bronze-workers on the tablet. *Bibl.:* R. S.
Young, "An Industrial District of Ancient Athens," *Hesperia* 20 (1951):
222–23; Burford, *Craftsmen*, p. 163; SGD 20.

I bind Aristaichmos the bronze-worker to those below and also Purrias the
bronze-worker and his work and their souls and Sôsias of Lamia[38] and his
work and his soul and Alêgosi[39] (?) and strongly {and strongly} and Agêsion
the Boeotian woman.[40]

72. Greece, Athens; from the Kerameikos, but not in its original location. Lead tablet measuring 9 × 5 cm.; the upper right corner is partly
folded. Dated to the late fifth century B.C.E., this is probably the earliest
preserved *defixio* from the Greek mainland. Each line of the text is

---

38. Young (p. 223) identifies Sosias as a slave or metic, in any case not an Athenian since he
is from Lamia, some 150 kilometers northwest of Athens.
39. The editor is unable to make sense of these letters. Perhaps they are the scrambled letters
of another personal name.
40. The role of Agesion is not clear. The editor makes her the cause of the rivalry, but this
may be assuming too much.

(*Column B*) (Let) their work undergo a reversal and may their livelihood and life be of no profit.[61] May bad things destroy and harm (them?).[62] . . . Let them be foolish.[63] Let there be no more profit for them of any kind, but let them lose even their slaves/children.[64]

77. Palestine, Beth Shean in Galilee, near Jordan River. Found at Byzantine level of a well; measures from ca. 8.5 × 4 cm. to 10.8 × 8 cm. Lead tablet, in two fragments. Greek cursive writing; predominantly *voces mysticae* with one line of tau-rho and box symbols. Dated to the fourth century C.E. The major divine names are IÔ, a form of IAÔ, and EULAMÔN. But "lord angels" are also addressed, and the spell is full of Semitic and Egyptian names. The spell seeks to bind the psychological and physical capacities of one man and two women; the verbs are compounded for effect, extending from the simple *deô* to the more vivid *katadesmeuô*. The occasion appears to be business: the client seems to fear an audit of her accounts by three individuals. The spell is significant in demonstrating a woman's control over economic affairs. *Bibl.:* H. C. Youtie and Campbell Bonner, "Two Curse Tablets from Beisan," *TAPA* 68 (1937); 43–72; SGD 164.

(*Side 1, fragment I*) *CHUCH BACHUCH BAKACHUCH BAKAXICHUCH BAZABACHUCH BENNEBECHUCH BADÊTOPHÔTH *BAIN[CHÔ]ÔÔCH . . . ABRAZANOU SALBANA-CHAMBRÊ, Lord angels, bind, bind fast the tendons and the limbs and the thought and the mind and the intention of Sarmation, to whom Oursa gave birth, and Valent[ia], to whom Eva gave birth, and Saramannas, to whom Eusebis gave birth—muzzle them and blind them and silence them and make them dumb . . . blind in the presence of P[ancharia] to whom Thekla[65] gave birth. . . . IÔSETH *IÔSÊPH IÔPAKERB[ÊTH] IÔBOLCHOSÊTH IÔOS[E]SRO IÔ PA[TATH-NAX IÔ]APOMPS IÔTONTOLIPSKONTOLIPS IÔB[. .]LÔBRÔ IÔARISAXA IÔ . . . IOTRI . . . IÔDÔRUKUNXISITHIÔ IÔBOLCHOSÊTH MÔCHIÔ IÔALO ÔSORNOPHRIX* Come to me *E]ULAMÔN [Come] to me EULAMÔN ULAMÔN LAMÔN AMÔN MÔN ÔN N IÔ Come to me EUCHALÊ IÔLEU . . . SSKUPHIEU IÔLAKÔIUATH IÔMATHUTÔR IÔMANDOUÔR IÔCHACHACHOUÔ[R IÔ]DARDEUB IÔPHIBITAX IÔDEDOUXATH IÔSALATH   IÔSALILÊ   BAUI   IÔCHAM   IÔBACHEÔCH   IÔB . . .   CH   EÔOU

61. Our text reads *biou mê onainto*. The poetic collection, known as the *Greek Anthology*, includes an epigram (VII, no. 516) of Simonides (sixth or fifth century B.C.E.), written for the grave of a person murdered by robbers: "Oh, Zeus, protector of strangers, let those who killed me suffer the same fate! But let those who placed me in the ground enjoy a prosperous life (*onainto biou*)." The positive wish of the epigram has become negative in our tablet. Any direct borrowing seems unlikely; perhaps what we have is a common saying, reflected in both texts.
62. A possible alternative translation would be "May bad things come upon those who are destroying and harming (me)."
63. A similar phrase appears in DTA 65, line 8: *aphrones genointo*, "Let them become foolish."
64. The Greek *pais* may designate either children or slaves.
65. By the fourth century C.E. the name Thekla would probably have denoted a Christian woman, since the cult of her namesake was one of the largest in the eastern Mediterranean. On the names in this section, see Youtie and Bonner, pp. 58–59.

BAUZÔCHAIÔÔSDOUTH IÔ *MASKELLI MASKELLÔ PHNOUKENTABAÔTH OREOBAR-
ZAGRA [R]ÊXICHTHÔN HIPPOCHTHÔN PYRICHTHÔN PYRIPÊGANYX LEPETAN LEPETAN
IÔBEZEBUTH IÔTHOURAKRINI BRIA [BADÊ-] TOPHÔTH IÔDRAX IÔPHEDRA IÔARABAZA
Ô IÔIARBATHAGRA* MNÊPHI BLÔ CHN[ÊM]EO ARPO[N-] KNOUPHI BRINTATAÔPHRI
BRINSKULMA A . . . CHAR[.]TH MESONKRIPHI NIKTOU CHN[OUMAÔPHI OREO-]
[BA]RZAGRA KNÊMEÔPHI IÔARBATHA IÔCHTHECH . . . IA MUCHEÔ IÔPIP . . .
CH . . . ÔA . . . KANTOUNOBOÊTH DARDANÔ CHITHACHÔCHENCHÔCHEÔCHI
*ABRASAX IÔ . . . EUTHIN EUTHIN, I invoke you, SÊMEA KAN[TEU] KENTEU KONTEU
KÊRIDEU DAR[UNK]Ô KUKU[NX K]APCHUM[RÊ] *SEMESILAMPS, Lord angels, muz-
zle and subject and render subservient and bind and slave and restrain and tie
up[66] Sarmation, to whom Oursa gave birth, and Val[entia], to whom Eva gave
birth, and Saramanas, to whom Eusebis gave birth, in the presence of
Pancharia, to whom Thekla gave birth, choking them, tying up their thoughts,
their mind, their hearts, their intention, lest they inquire further after an account or a calculation[67] or anything else . . . from Pancharia, but (let) merciful fortune (come to) Pancharia throughout (her) life. [IÔ] *ABLANATHANALBA IÔ
*AKRAMACHAMARI IÔ *SESENGEN IÔ BARPHARN[GÊS . . . ÔTH IÔ
NEBOUTOSOUALÊTH AKTIÔPHI *ERESCHIGAL IÔ BERBITA IÔ THÔBAGRA BAUI . . .
A[BER]AMENTHÔOULERTHEXANAXETHRELUAÔTHNEMAREBA[68] the Great!
AEMINAEBARRÔTHER[RETHÔR] RABAEANI]MEA[69] IÔ SARCHACHATHARIA IÔ IAEÔBA-
PHRENEMOUNOTHEILARIKRIPHIAEYEAIPHIRKIRALITH[ONUO] MENERPHA] BÔEAI[70]
and the greatest name, PSI PSI PSI PSI PSI K K K CH CH CH PHI PHI PHI PHI K K K CH
CH . . . I K K K K [charaktêres: four tau-rho's,[71] two boxes, one partial box with
circle inside[72]] PPIIIYYYYDDDKKKKAKA . . . IÔSÊTH IÔ . . . BÊTH IÔBOLCHO[SÊTH]
IO[P]ATATHANAX B . . . EULAM[Ô ULAM]ÔE LAMÔEU AM[ÔE-] [UL MÔ]EULA
ÔEULA[M . . . AZAZA . . . [the name] of the great god IOU . . . IIIOUI . . .
g]reat[. . . I invo[ke you . . . (last two lines almost illegible)

78. Italy, Rome. Lead tablet, approximately 10 × 5 cm., originally
rolled up and pierced by a nail, with writing on both sides (Figure 19).
The content points to the fourth century C.E. The Greek is simple and
ordinary. The spirits invoked are addressed as "Lord Angels" and "Lord
Gods." The formula of binding is quite simple—"Restrain so-and-so."

66. The Greek verb is *katadesmeusate*.

67. These words, *epizêtêsôsin log[o]n êpsêphon*, would seem to indicate that the context of
the spell is a financial audit of Pancharias's business—that the *defixio* is meant to forestall any
investigation (and resulting prosecution?) of the client.

68. A lengthy palindrome, also attested in several recipes of *PGM* (e.g., I, line 294); see the
discussion in *GMP*, p. 331.

69. Palindrome turning on *theta*.

70. Another palindrome; cf. *PGM* III, line 60, with slight variations; this one begins and
turns upon vowel sets.

71. The use of this symbol may indicate Christian influence (cf. Papyrus Oslo 5.11). It should
be noted, however, that the same sign appears in a Jewish text from Aleppo (Syria); cf. Naveh
and Shaked, Amulet no. 4, line 8.

72. Preisendanz (*PGM*, vol. 2, p. 214) indexes similar signs from magical papyri that represent the Greek word *onoma* ("name").

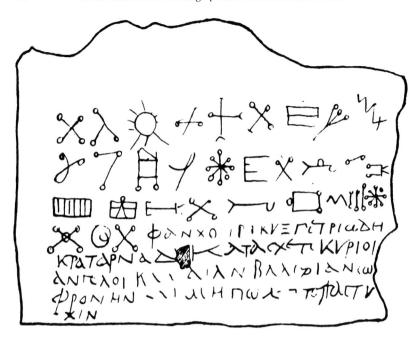

FIGURE 19. Lead *defixio* from Rome with perhaps thirty-two *charaktêres*. Several of the *charaktêres* are identical to those found on other tablets. As in other cases, the *charaktêres* appear here as the first lines of the spell.

The same target, a woman, is named on both sides of the tablet. The names of the persons involved, as well as the occasion, point in the direction of slaves or freedmen. The occasion seems reasonably clear. The client, a slave woman named Politoria, fearing that Clodia Valeria Sophronê was about to assign her to a workhouse, resorts to a binding spell in a desperate effort to avert this unhappy fate. She then deposited the tablet, as was customary, in a grave site. *Bibl.:* R. Wünsch, "Deisidaimoniaka," *ARW* 12 (1909): 37–41.

(*Side A*) PHANOIBIKUX PETRIADÊ KRATARNADÊ,[73] Lord Angels,[74] restrain Clôdia Valeria Sôphronê and may she not succeed in buying Pôlitoria.

73. Wünsch sees in these two words a garbled version of the phrase *petraios krateros Haides*, referring to the stone roof of Hades.
74. At a certain point, the traditional *daimones* of Greek culture came to be called *angeloi*, under Jewish influence. Other signs of "Jewish influence" in this tablet are the terms *semisilam* and *lailam*. Yet this influence is certainly not direct but rather part of the general contribution of Judaism to the culture of late antique spirituality.

(*Side B*) ARTHU*LAILAM *SEMISILAM *BACHUCH BACHAXICHUCH MENEBAICHUCH *ABRASAX,[75] Lord Gods, restrain the matron of the workhouse,[76] Clôdia Valeria Sôphronê and do not let her drag Pôletoria (as a workhouse laborer), to suffer (?) the fate of lifelessness[77] (there).

79. Italy, Rome; from a grave near the Porta Adreatina. Lead tablet measuring 10.4 × 3.5 cm.; the Greek text is written in two columns. Guarducci, the original editor, places it in the third century C.E., perhaps between 270–285 C.E., based on references in the text. The spirits addressed appear only at the end of the spell. They are addressed by their mysterious, secret, and holy names; some are familiar, while others are unprecedented. The curse involves the following elements: a Greek physician in the Roman army; two brothers who served as assistants to the physician, one of whom has died, while the other is probably the client who commissioned the tablet; and the curse, born of anger and frustration, against the land of Italy and the gates of Rome. *Bibl.:* M. Guarducci, "L'Italia e Roma in una *tabella defixionis* greca recentemente scoperta," *Bulletino della commissione archeologica comunale di Roma* 74 (1951–1952): 57–70; *BE* (1955): no. 292; *SEG* 14.615; Guarducci, *Epigrafia greca*, pp. 251–54; SGD 129 (text).

> Restrain Artemidôros the physician, of the Third Praetorian Cohort. The brother of the deceased Dêmêtrios, who has worked as his assistant, now wishes to depart to his own country.[78] Do not permit him,[79] but restrain the

75. A closely parallel set of *voces* appears on a tablet from North Africa (see no. 82).

76. The reconstructed Greek word is probably to be read as *ergastillarion*, reflecting the Latin word *ergastularius*, which in turn designated the person in charge of the workhouse (*ergastulum*). The *ergastulum* itself was a notorious institution, reserved for unruly slaves and noted for its harsh conditions. Whether or not it was an actual work site or simply the building where fettered slaves were quartered, it was used by Roman authors to express the lowest form of human living. Columella, the author of a treatise on agriculture (first century C.E.) writes as follows: "For those (slaves) who are in chains there should be an underground prison (*ergastulum*), receiving light through a number of narrow windows built so high from the ground that they cannot be reached by hand" (1.6.3).

77. The final phrase is uncertain. *Apsuchia*, meaning lifelessness, could be used as a term for death. The verb is lacking. Wünsch proposes *idein*, "to see."

78. The situation is thus as follows: the client has been serving as an assistant to the physician Artemidoros; following the death of his brother Demetrius, he now wishes to return to his home country; given his brother's name and the fact that Greek is their native tongue, their home lay somewhere east of Rome. The physician has refused to allow the departure, thus prompting the spell by his angry assistant. The brothers were probably slaves.

79. This phrase is not clear. The Greek is *mê easête oun auton*. It may mean "Do not allow him (the physician)" to prevent the departure. The text here follows Guarducci's reading, also adopted by Jordan.

land of Italy[80] and strike the gates of Rome[81] and also restrain Artemidôros
the physician, the son of Artemidôros. *EULAMÔN, LAMEILA . . ÔN, . REIOCHER-
SOPHRIX, OMÊLIEUS, AXÊIEUS, ARÊIEUS and, LATHOS and THAMBOS, restrain.

80. Italy; Nomentum (present-day Mentana, east of Rome). Lead tablet
with Latin writing on both sides; 9.2 × 5.2 cm. Date unknown. Two holes
in the tablet appear to have been made by nails. One of the points of
interest of these inscriptions is the fullness of their lists of body parts,
including repetition of a single part more than once. In both spells, the
targets are identified by their mother. The use of the term *quaestus* in the
first spell suggests that the occasion involved business competition of
some sort. *Bibl.: DT* 135; L. Borsari, "Mentana—Tombe Romane scop-
erte presso l'abitato," *Notizie degli scavi di antichità* (March 1901): 205–
10; R. Cagnat, "Revue des publications épigraphiques relatives à
l'antiquité romaine," *Revue archéologique* 39 (1901): 468–69.

> (*Side A*) Malcius (the son or servant) of Nicona: (his) eyes, hands, fingers,
> arms, nails, hair, head, feet, thigh, belly, buttocks, navel, chest, nipples,
> neck, mouth, cheeks, teeth, lips, chin, eyes, forehead, eyebrows, shoulder-
> blades, shoulders,[82] sinews, bones, *merilas*,[83] belly, penis, shin: in these
> tablets I bind[84] (his) business profits[85] and health.

> (*Side B*) Rufa (the daughter or servant) of Pu[b]lica: hands, teeth, eyes,
> arms, belly, nipples, chest, bones, *merilas*, belly . . . , shin, mouth, feet,
> forehead, nails, fingers, belly, naval, genitals, womb, groin: I bind [these
> parts] of Rufa (the daughter or servant) of Pu[b]lica in these tablets.

81. Metapontum in southeastern Italy. Excavated from a cemetery.
Lead tablet measuring 5 × 6.7 cm., originally folded, but no signs of nail
holes. The Greek of the tablet points to a date in the late fourth or early
third century B.C.E. No deities or spirits are invoked. The verb of binding
is also familiar in this period—*katadidêmi*. The client's name is not

---

80. The curse is addressed against Italy itself, where the disconsolate client must remain
against his will.

81. Here again the text is not absolutely clear. Guarducci has interpreted "the gates of
Rome" as a reference to the mouth of the Tiber; the verb she takes to mean "to cause to become
silted." The phrase thus expresses the wish that the lifeline of Rome, the Tiber, should become
useless. Robert offers a number of serious objections to this interpretation. Our translation
follows the text as reconstructed by Jordan. In general, the curse is directed against the gates of
Rome because they symbolize the client's inability to pass through them on his way home to his
native land.

82. *Scapulas* and *humerum*, respectively.

83. Not translatable: Audollent (*DT* 135) suggests an emendation to *meritas*.

84. The Latin verb is *defigo*.

85. The Latin term *quaestus* was used regularly of business matters. Here it occurs with
lucrum, which has much the same meaning.

given, but some seventeen targets, physicians in a medical clinic or group practice, are listed at the bottom. These names are well known from other sources, literary and epigraphic, especially from the nearby city of Tarentum. The editor of the tablet, Lo Porto, argues that a number of these figures may also have been associated with the medical branch of the Pythagoreans, established by Pythagoras at Croton in the sixth century B.C.E.[86] Pythagoras himself is known to have died in Metapontum. No specific occasion is cited, but the language points toward competition of some kind between the client and the targets. The spell itself seems to fall into two phases: the first binds the workplace, while the second focuses on the physicians who work there. *Bibl.:* F. G. Lo Porto, "Medici pitagorici in una *defixio* greca da Metaponto," *La Parola del Passato* 35 (1980): 282–88; M. Gigante, "Sulla *defixio* Metapontina," ibid., 381–82; *SEG* 34.1175; SGD 124 (text).

> Of these people I bind first the workplace.[87] I bind it so that it may not produce (anything) but rather be without work and suffer misfortune. Next, of the wicked people of this (workplace), I bind the (slaves?) of the physicians who are listed on this lead[88]—Philôn, Nearchos, Dikais, Theorôdus, Euklês (?), Simuliôn, Trê . . . , Leôn, Agias, Theodôridas, Bakalles, Philoklês, . . . , Zôilos . . .

82. North Africa; Carthage. Latin text on two lead tablets found in the "Fontaine aux milles amphores" area of Carthage beneath the Plateau de Sainte-Monique (Audollent's c and d respectively). Both measure roughly 6 × 5 cm. The tablets can be dated to the second or third century C.E. on the basis of the dating of Roman lamps found nearby.[89] The opening invocation consists entirely of familiar *voces mysticae* written in Greek letters; the remainder of the text is in Latin. The *voces* are inscribed in a continuous "box" fashion, the tablet being rotated one-quarter turn at the end of each line. After the tablet was rotated completely twice, the remaining area at the center was filled with the "normal" text. Audollent suggests that the *defixiones* were addressed to or

---

86. Caution is needed in assessing Lo Porto's claim about connections between the school of Pythagoras and the physicians of our tablet. The names are common ones and those identifiable as connected with Pythagorean circles derive from widely separated dates. On Pythagoreans and medicine in the period, see G. Sarton, *A History of Science*, vol. 1 (Cambridge, 1960), pp. 214–16, 333–34, and P. Wuilleumier, *Tarente des origines à la conquête romaine* (Paris, 1939), pp. 608ff.

87. The Greek term is *ergasterion*, which can designate a workplace of any sort.

88. One of the numerous instances where curse tablets refer to themselves in the course of the spell. The Greek term is *bolimos*, a form of *molibos*, itself an epic form of the more common *molubdos*—all meaning lead or, as here, the lead curse tablet itself.

89. Audollent, p. 120.

against the baths near the spring where the tablets were found, possibly by a competitor in business.[90] *Bibl.:* A. Audollent, "Les inscriptions de la 'Fontaine aux mille amphores' à Carthage," in *5ème Congrès Internationale d'Archéologie, Alger 1931* (Algiers 1933), pp. 129–38; J. Toutain, "L'histoire des religions au Congrès d'Alger," *Revue de l'histoire des religions* 101 (1931): 114–15; Solin, p. 31.

(c) ARTHU *LAILAM *SEMESE[I]LAM AEÊIOYÔ *BACHUCH BAKAXICHUCH MENE BAICHUCH *ABRASAX BAZABACHUCH MENE BAICHUCH ABRASAX Lord Gods, restrain and hinder the Falernian baths,[91] lest anyone should be able to approach that place; bind and bind up the Falernian baths[92] from this day, lest any person should approach that place.

(d) ARTHU LAILAM SEMESEILAM AE[ÊIOUÔ BACH]UCH BAKAXICHUCH MENEBAICHUCH ABRASAX BAZABACHUCH [ME]NEBAICHUCH ABRASAX Lord Gods restrain and hinder the Falernian baths, lest any person go there to bathe[93]; bind[94] the Falernian baths from this day, bind and bind up the Falernian baths,[95] lest anyone be able to go to that place from this day.

90. Ibid., pp. 132ff. C. Faraone has suggested by letter that "Falernian baths" might simply be the name of a tavern located near the baths.
91. The Latin reads *falernas*, probably for *balineas* ("baths").
92. Again *falernarum*, this time with *balineu*; probably to be read as *falernum balineum*.
93. The Latin is *lavarii*, the passive infinitive; thus probably "to be bathed."
94. *Nodiate = nodate*, "to knot."
95. Again *balineu* with *falernesi*.

5

Pleas for Justice and Revenge

Thus far we have employed a minimalist definition of curse tablets and binding spells as inscribed sheets of metal or other material that were used and generally commissioned, at least in the Roman period, by private individuals (clients) in order to influence—against their will and through the agency of spirits, *daimones,* and deities—the behavior and welfare of personal enemies and rivals (targets).[1] Beyond this, the forms and functions of *defixiones* vary widely with reference to such matters as where and how they were deposited, their specific uses (business rivalries, love spells), or their use of formulas and mysterious words.

Among the surviving *defixiones,* one distinctive category may be labeled as pleas for justice and revenge. H. Versnel, in a recent treatment, calls them judicial prayers for help from the gods and argues that they belong to a category quite different from curses on metal tablets.[2] In terms of their occasions, the most distinctive feature of this category lies in the explicit claim that the targets or enemies have somehow wronged the client. Sometimes this claim is quite vague (for example, "I have been mistreated!"), but more commonly the offense is spelled out in some detail, especially in the frequent cases concerning stolen property (see nos. 88, 94–100). In terms of their goals, these *defixiones* seek both justice and revenge—the recovery of the stolen goods as well as punishment of the alleged thief. But perhaps their most distinctive trait, as Versnel and others have shown, lies in the quasi-legal transaction that takes place between the client and the deity invoked[3]: in virtually every case concerned with stolen property, the client temporarily transfers ownership of the goods in question, sometimes even the culprits themselves, to the deity and thereby makes their recovery a matter of divine rather than merely human concern. For it is no longer just the human owner, but the gods themselves who have been deprived, offended, and dishonored. This transfer procedure seems generally to have taken place in a local temple, no doubt one associated with a deity known for special competence in hounding thieves. Ownership of the stolen property was ritually, if provisionally, transferred to the deity; the thief was required to return the goods to the temple, whence the owner presumably re-claimed them after paying a fee to the deity, that is, to the treasury of the temple. Behind this legal fiction, we glimpse a pressing need for redress against injustice, large or small, in what R. Tomlin aptly calls "an under-policed world."[4] Here the patron was no longer the emperor, far too remote from Roman Britain, or even a local magistrate, but the ever-vigilant and jealous gods themselves.

A striking confirmation of this unusual procedure has survived in a series of "confessional inscriptions" from Lydia and Phrygia (Asia Mi-nor) in the second and third centuries C.E.[5] In these remarkable texts, individuals who suspected that they had become the target of a plea for justice—those who have suffered the illnesses and misfortunes spelled out in numerous *defixiones*—set up tablets either proclaiming their inno-cence, since we may be certain that not all charged were guilty and that not all missing property had been stolen, or confessing their guilt, return-ing the property, and praising the power of the god who had tracked them down. One such text concerns a cloak stolen from the local bath. The victim had apparently lodged his complaint through the medium of a *defixio:* "The god was vexed with the man and after some time had him bring the cloak to the god, and he openly confessed his guilt. Then the god ordered him, through the agency of an angel, to sell the article of clothing and to publicize his (the god's) miracles on a stele.[6] Another text reveals a case of false accusations, which resulted eventually in the exoneration of the accused and the punishment of the slanderers by the god Men: "To Mên Axiottenos. Since Hermogenês, son of Glukon, and Nitônis, son of Philoxenos, have slandered Artemidôros with respect to (the theft/drinking of?) wine, Artemidôros has given a tablet. The god has punished Hermogenês, who has propitiated the god and from now on will extol (the god Mên)."[7]

These texts illustrate two important features of pleas for justice in particular and of *defixiones* in general: first, they indicate how deeply rooted was the belief in their effectiveness; and second, they tell us that the commissioning of a *defixio* was not, indeed could not be, an entirely private affair. In other words, the effectiveness of the process was depen-dent to a certain degree on public knowledge that "a fix" had been placed on a particular suspect. In turn, the suspect's response, whether claiming innocence or confessing guilt, makes sense only under these same assumptions—namely, that *defixiones* were believed to work and that one had been issued in a particular case.

The issue of balance between public and private aspects of pleas for justice also raises the question of how and where they were deposited or displayed. In some instances, public display in a temple area seems clear, whereas others have turned up in graves and wells where no one could read them. But as Versnel rightly observes, there were public and private aspects to the proceedings: "The most obvious—though by no means exclusive—procedure may be that the injured party first tries to draw a confession from the suspected culprits and then tells them explicitly that he is making a higher appeal to the god."[8]

Until recently, pleas for justice and revenge represented a modest share of the total corpus of *defixiones*. A rough count indicates nineteen in Greek and perhaps twelve in Latin. Since the 1970s, these numbers have increased dramatically, to the point where they now represent by far the largest single subcategory of all curse tablets and binding spells. The discovery of significant deposits at two British sites, Bath and Uley, has now raised the total in Britain alone to ca. three hundred, virtually all of them involving stolen property. Whether this means, as some have opined, that Britons in the period of late antiquity were especially preoccupied with recovering stolen property may be left as an open question. It seems much more likely that the current disproportion generated by the British tablets reflects nothing more than the hazards of preservation and discovery. What is certain, as Tomlin and Versnel have emphasized, is the remarkable similarity of atmosphere and formulaic language in pleas for justice from widely separated regions of the Greco-Roman world—Britain, Spain, Italy, and the eastern Aegean. Here again we encounter two recurrent features in the use of curses and spells: their unmistakably international character and the role of written handbooks or formularies, like the major collections of *PGM*, from which the spells were copied and by which they were transmitted from region to region.

Like other types of *defixiones*, pleas for justice and revenge look to the past as well as into the future. They arise from a perceived wrong already suffered—sometimes about to happen—while simultaneously anticipating future compensation. One subcategory of *defixio*, however, looks exclusively toward the future. These are curses on gravestones, invoking dire consequences for any who disturb or destroy the burial site. Here, of course, we encounter an arresting irony, for we now know that one of the most common reasons for disturbing a grave site was precisely to deposit a *defixio* inside. But the logic, or better the psychology, seems perfectly straightforward—the acknowledged power of one kind of *defixio* was deployed against another kind, based on the belief than "mine" is surely stronger than "theirs."

Such curses as those found on gravestones clearly cross the boundary between the private and the public sphere. The protection of graves and tombs was a matter not merely of private concern but of public policy as well.[9] There are other differences: they are inscribed on stone or wood rather than on metal sheets; and of necessity they name no specific culprit. They also represent a form of cursing with a remarkably well-attested history of use, reaching from Phoenicia in the eleventh century B.C.E. to early modern Europe.[10]

Altogether the few examples given here stand at one extreme of the spectrum covered by curse tablets and binding spells. Still, and the point is worth repeating, no subcategory is definitionally pure. Publicly displayed curses on ancient gravestones reveal not only common traits with private *defixiones*[11] but more generally they indicate the untenability of all traditional distinctions between "magic" and "religion." Thus we conclude with a few instructive examples from R. Lattimore's comprehensive treatment, *Themes in Greek and Latin Epitaphs;* their similarity to the language and the concerns of *defixiones* will be readily apparent.[12]

> I put an oath on, you who settle here, to treat this place with proper respect.[13]

> I, Idameneus, built this tomb to (my own) glory. May Zeus utterly destroy anyone who disturbs it.[14]

> . . . may he be guilty in the sight of all the gods and of Leto and her children.[15]

> . . . I invoke Selene.[16]

> . . . may he be guilty of impiety in the sight of the underground spirits.[17]

> If anyone does any harm to the statue, may he leave orphaned children, a bereaved estate and a desolate home behind him. May he lose all his goods by fire and die at the hands of evil men.[18]

> If anyone erases the dead image of this child, may he fall afoul of the curse of the untimely dead.[19]

> Whoever does anything counter to the injunctions set forth above, shall be held responsible to the authorities; and in addition, may he have no profit from children or goods, may he neither walk on land nor sail on sea, but may he die childless, penniless and ruined before death and all his seed perish with him, and after death may he find the underground gods to be angry avengers.[20]

Notes

1. The definition is essentially the one proposed by David Jordan in SGD, p. 151, with slight modifications.

2. H. Versnel, "Beyond Cursing," in *Magika,* pp. 60ff. Versnel insists on separating pleas for justice rather fully from *defixiones,* though he is prepared to recognize a border area where some pleas for justice (he enumerates some eighteen instances of this type) are difficult to distinguish from *defixiones.* Whether in this border area or in the "pure type" of the plea for justice (in this category Versnel is prepared to locate some twenty cases, apart from the British tablets), Versnel isolates the following distinctive characteristics: (1) the name of the author/client, although such is also the case in many spells concerning sex and love; (2) an argument defending the action; (3) a request that the act be excused; (4) the appearance of gods other than the usual chthonic deities, although "standard" *defixiones* also refer to a wide range of deities and spirits; (5) appeals to the gods through expressions of supplication rather than coercion, although the appeals in the *defixiones* are by no means always coercive; and (7) the occurrence of terms related to punishment and vindication, such as *ekdikô.* It should be noted that the use of *voces mysticae* is not characteristic of pleas for justice and revenge.

3. See the full discussion in Versnel, "Beyond Cursing," pp. 196ff. and in Tomlin, pp. 70–72 ("Quasi-legal language") with a full listing of the legal terms in the Bath tablets. Tomlin further notes (p. 71) that the petitioners at Bath follow prescribed procedures in cases of theft as they are spelled out in legal codes.

4. Ibid., p. 70.

5. Thorough discussions, with references to earlier literature, appear in Versnel, "Beyond Cursing," pp. 72–74; Tomlin, pp. 103–5; and E. N. Lane, *CMRDM,* vol. 3 (Leiden, 1976), pp. 17–38.

6. An inscription from Kavakh in Turkey, dated to 164/165 C.E. = Lane, *CMDRM,* vol. 1 (1971), no. 69.

7. An inscription from Kula in Turkey (undated) = Lane, *CMRDM,* vol. 1, no. 58.

8. Versnel, "Beyond Cursing," p. 81.

9. Perhaps the best-known instance is the so-called Nazareth decree of an early Roman emperor (usually identified as Claudius) proclaiming capital punishment for anyone caught disturbing graves or tombs; cf. *SEG* 8.13.

10. See now the survey essay by J. H. M. Strubbe, " 'Cursed be he that moves my bones,' " in *Magika,* pp. 33–59. Strubbe's essay, like the selection from tomb curses of Lattimore, suggests a particular emphasis on tomb protection in Asia Minor.

11. See the comments of R. Lattimore, *Themes in Greek and Latin Epitaphs* (Urbana, Ill., 1962), p. 122, n. 237. He observes there that "[s]epulchral curses,

both in Greek and Latin, are less fantastically specific than the curses by the living on the living, the *defixiones.*"

12. Although published in 1962, the material was first treated as a Ph.D. dissertation (University of Illinois, 1934). Lattimore gives numerous examples of such curses and lists many more in the footnotes.

13. Lattimore, *Epitaphs*, p. 106 (from the Greek island of Syros). The verb used is *enorkizomai*, the same one used frequently in *defixiones*.

14. Ibid., p. 109 (from the island of Rhodes; dated to the seventh century B.C.E.).

15. Ibid., p. 110 (from Pinara in Lycia, Asia Minor).

16. Ibid. (from the region of Cilicia in Asia Minor). The verb *enorkizô*.

17. Ibid., pp. 110–11. (from the region of Cilicia in Asia Minor). The spirits here are called *katachthonious daimonas*.

18. Ibid., p. 112 (from Iconium in Cilicia, Asia Minor).

19. Ibid. (from Aezani some 250 kilometers east of Pergamom in Asia Minor).

20. Ibid., pp. 115–16 (from Hierapolis in Phrygia, Asia Minor).

83. Greece, Athens (Patissia); original location not known. Lead tablet measuring 16 × 4 cm.; originally folded and pierced by a nail. Dated to the fourth or third century B.C.E. This is one of several tablets (cf. DTA 99–100; *DT* 72) addressed to Earth. The precise occasion is not clear, but the general intent is to exact revenge for a wrong suffered at the hands of two named individuals. *Bibl.:* DTA 98.

> Euruptolemos of Agrulê[1] I bind Euruptolemos and Xenophôn {Xenophôn} who is with Euruptolemos, and their tongues and words and deeds; and if they are planning or doing anything, let it be in vain. Beloved Earth, restrain Euruptolemos and Xenophôn and make them powerless and useless; and let Euruptolemos and Xenophôn waste away. Beloved Earth, help me; and since I have been wronged by Euruptolemos and Xenophôn I bind them.

84. Greece, Athens, in the Agora. Found in a deposit from a well dating to the first century C.E. Lead sheet measuring 23 × 11.7 cm., rolled up, with no sign of a nail. The script is described by the editor as "carefully written." The tablet includes several illustrations. Starting with line 16 and extending to the final line 29 is a crudely sketched "figure of a bat

1. The first names in the first two lines, Euruptolemos and Xenophon, are set apart from the rest of the text and should be taken as the heading or "title" (so Jordan) of the tablet.

FIGURE 20. Drawing of three-winged (armed?) Hekate and *charaktêres* from lead *defixio* found in the Agora at Athens. (*Gods and Heroes in the Ancient Mediterranean* [Princeton: American School of Classical Studies at Athens, 1980], p. 37 [Fig. 37]. Courtesy of the American School of Classical Studies.)

with outspread wings." Jordan has redescribed the figure as "a six-armed Hecate." In addition, there appear two or three "magical symbols" (Figure 20). Several deities are invoked, all well known from Greek myth, cult, and ritual and all associated with the "underworld": Pluto, the Fates, Persephone, the Furies, various unnamed gods (evil ones, underworld goddesses and gods), Hermes, and Hekate. It is clear, however, that Hekate is the central power for she alone reappears in subsequent lines. The verbs of binding—to register (*katagraphein*) and to consign, hand over or transfer (*katatithenai*)—are common in texts of this sort. The affairs of the target are placed under the temporary control of the deities invoked so that the desired result may follow. In this case, the issue is stolen property. What concerns the client is that the thieves are unknown to him and that no one has come forward to identify them. There is no mention of asking the gods to return the property. Nor is there a question, in a formal sense, of bringing them to public justice. In fact, the opposite would appear to be the case. Because there seemed no likelihood of public justice, the registering and transferring of the unknown thieves to infernal deities must mean that they, rather than human judges, will mete out the punishment. This is made explicit in lines 19–20, where Hekate is exhorted to "cut (out) the hearts of the thieves." The tone of the spell is most unusual. The client seems hesitant about commissioning the tablet. *Bibl.:* G. W. Elderkin, "Two Curse Inscriptions," *Hesperia* 6 (1937): 382–95; D. R. Jordan, "Hekatika," *Glotta* 83 (1980): 62–65; *SEG* 30.326; H. S. Versnel, "Religious Mentality in Ancient Prayer," in *Faith, Hope and Worship*, ed. H. S. Versnel (Leiden, 1981), pp. 22–23; SGD 21; H. Versnel, "Beyond Cursing," in *Magika*, p. 66.

I make an exception for the writer[2] and the destroyer, because he does this unwillingly, forced (into it) by the thieves. I register and hand over[3] to Pluto and to the Fates and to Persephone and to the Furies and to every harmful being[4]; I hand (them) over to Hekate,[5] eater of what has been demanded by the gods (?)[6]; I hand over to the goddesses and gods of the underworld, and to Hermes the helper; I transfer the thieves who stole from the little house in the quarter/street (?) called Acheloou[7]—(who stole) chain,[8] three spreads (one woolen, white, new), gum arabic . . . tools,[9] white piles of dirt,[10] linseed oil, and three white (objects): mastic, pepper, and bitter almonds. I hand over those who know about the theft and deny it. I hand over all of them who have received what is contained in this deposition. Lady Hekate of the heavens, Hekate of the underworld, Hekate of the crossroads, Hekate of the triple-face, Hekate of the single-face,[11] cut (out) the hearts of the

2. The verb here is *katagraphein* (cf. DTA 160 and the figurine from Karystos in Euboea, SGD 64). It has the sense of transferring something or someone by the act of recording the person's name under a new heading, in this case the heading of the gods being invoked. There may also be a play on the more literal meaning of the verb, "to engrave," that is, on the tablet itself.

3. The verb *katatithenai* conveys the sense of consigning or handing over, often in business matters.

4. The Greek is *kakos*, here meaning not "evil" but "able to cause harm."

5. In *DT* 38, a tablet from Alexandria (third century C.E.), Hekate appears with Pluto, Persephone, Hermes, and others. Such a full listing of gods and goddesses is unusual.

6. The Greek reads *theetophagos*.

7. Although the client claims not to know the identity of the thieves, he appears to know something about them, that they live in a particular part of Athens and are named after that section, described here as *litos* ("rundown" or "inexpensive"). At two places in the text, uncertainty about the theft expresses itself in the phrase "the thief or the thieves."

8. The Greek term is *katêna*, borrowed from Latin *catena*, which probably refers to a chain necklace.

9. The Greek term is *sunerga*, which can designate tools or other implements.

10. The Greek reads *leuka chômata*.

11. Three of the epithets of Hekate are traditional; they appear, for instance, in a fragment of the poet, Chariclides (*CAF* III, frag. 1): "Lady Hekate, of the crossroads, of the triple image, of the triple face . . ." The adjective *triprosôpos* is applied to Hekate in *PGM* IV, lines 2119 and 2880. But two of the epithets are decidedly not traditional: the first is *monoprosôpos*, "with one face or character"; and the second is *ouranios*, "heavenly."

thieves or the thief who took the items contained in this deposition. And let the earth not be walkable, the sea not sailable; let there be no enjoyment of life, no increase of children, but may utter destruction visit them or him. As inspector, you will wield upon them the bronze sickle, and you will cut them out (?). But I exempt the writer and the destroyer.

85. Greece, Megara, on the north side of the Corinthian isthmus, some twenty kilometers west of Eleusis; more exact place of origin not known. Lead tablet measuring 10 × 15 cm., in several fragments and written on both sides. Wünsch dates it to the first or second century C.E. Both sides of the tablet contain spells, with some similarities between them but also differences. The absence of names makes it difficult to determine whether the targets and the occasion are identical on the two sides. The figures invoked are addressed by mysterious and secret names. Among the recognizable deities on Side A are Althaia, Kore, Hekate, and Selene (the Moon); on Side B, Selene and Hekate. According to Wünsch, Hekate is the chief figure of the spell, the others being identified with her. In addition to the obvious Greek themes on Side A, there are indications of Jewish elements as well. *Bibl.:* Wünsch, DTA, pp. xiii–xiv; *DT* 41; Wünsch, *Antike Fluchtafeln*, no. 1, pp. 4–7; H. Versnel, "Beyond Cursing," in *Magika*, p. 65.

(*Side A*) ZÔAPHER TON THALLASSOSÊMON SEKNTÊAPAPHONOCHAI[12] the beloved child Panaitios inscribed (here?)[13] ECHAIPEN . . . We curse those EPAIPÊN . . . them and we anathematize[14] them. Althaia,[15] Kore, *OREOBAZAGRA Hekate Moon who devours its tail[16] . . . ITHIBI . . . we anathematize them—body, spirit, soul, mind, thought, feeling, life, heart—with Hekatean words and

12. This initial series of mysterious words contains elements that look like Greek; for example, *ton thalassemon* could be translated as "marked with the sea." Other words may also represent scribal mistranscriptions.

13. *Panaition* could be read as a proper name (well attested as such) or as an epithet of the spirit ("cause of all things").

14. Of the two binding words used here, *katagraphein* and *anathematizein*, the former appears frequently in Greek *defixiones*, whereas the latter is used exclusively in Jewish and Christian sources. In Deuteronomy 13:15 (LXX), it describes an oath taken by Israelites against those who introduce foreign rites. In his letters, Paul employs the form *anathema* as a curse formula several times (1 Corinthians 12:3; 16:22; Galatians 1:8; Romans 9:3). Each time it designates a powerful oath, consigning the target to death.

15. Althaia in mythological tradition was the mother of Meleager and wife of Oeneus. Pausanias reports the following story of her in describing a famous painting by Polygnotos: "As to the death of Meleager, Homer says that the Fury heard the curses of Althaia and that this was the cause of Meleager's death" (10.31.3). Althaia's success thus made her a likely candidate for selection by someone who wished to enlist powerful "cursers" at a later time.

16. A common Egyptian motif, both literary and artistic, in which a snake forms a circle and devours its own tail. It appears frequently in papyri and on gems. For literature see *GMP*, p. 337.

Hebrew oaths[17] . . . Earth Hekate . . . commanded by the holy names and oaths of the Hebrews—hair, head, brain, face, ears, eyebrows, nostrils . . . jaws, teeth . . . so that their soul may sigh, their health may . . . , their blood (and) flesh may burn[18] and (let) him/her sigh with what he/she suffers . . .

(*Side B*) I invoke[19] . . . also Moon, the triple-named, who (circulates?) in the middle of the night whenever the . . . walk about, who courses the heavens with a strong hand, the visible one with the dark-blue mantle . . . on land and sea, Einodia (?)[20] . . . , we anathematize (?) them . . . and enroll them for punishments, pain and retribution[21] . . . the body. Anathema.[22]

86. Greece, Chalcis. The author of this inscription[23] bears the name Amphicles and may have been a disciple of the famous orator, Herodes Atticus.[24] Its date falls in the second century C.E. Like the monuments of Herodes Atticus, which it resembles, this one is not a burial inscription but a warning designed to protect private property, in this case an installation of baths. But Amphicles has modified the text of his models in a most extraordinary way, by adding numerous allusions to the Septuagint and by deleting all traces of paganism.[25] Thus the author of our inscrip-

17. The Greek is *logois hekatikois horismasi abraikois*. The occurrence here of *abraikos* (*hebraikos*) confirms the earlier use of *anathematizein* as evidence of Jewish influence of some sort in the history of the spell's development.
18. This must mean fevers, a common "wish" in curses.
19. Side B is more fragmentary than A. It also lacks terms of Jewish origin. From what remains, Wünsch concludes that the lengthy series of epithets attached to Hekate must have derived from a Greek hymn.
20. A common epithet of Hekate in her capacity as guardian of crossroads; it should be noted that the text is not certain. On the goddess in general see Sarah Johnston, *Hekate Soteira: A Study of Hekate's Roles in the Chaldean Oracles and Related Literature* (Atlanta, 1990).
21. The Greek terms are *kolaseis, poinê*, and *timôria*. Versnel (p. 65, n. 26) proposes that *kolaseis* in particular refers to punishments in the afterlife and that it has the meaning here of "hell."
22. This is presumably the title or label for the tablet itself, "this is a/the curse."
23. This inscription and its twin are printed in *IG* XII, fasc. 9, 955 and 1179 (incorrectly listed as 1170 in Robert, p. 245, n. 30).
24. The view that Amphicles must be seen as a disciple of Herodes is developed by Robert, pp. 246–53. Philostratus, in his *Lives of the Sophists* 2.8 and 10, mentions an Amphicles of Chalcis as a favored pupil of Herodes. It must be noted, however, that the primary evidence for the connection between Herodes and the author of these two monuments is the "close parallels" in their texts. A close reading of the Herodes monuments, however, does not suggest such parallels. Those parallels that do exist are stock formulas, which might have come about from a number of sources, including collections of curse formulas available for purchase in the region.
25. Robert notes the following modifications: the opening appeal "to the gods and heroes," found in parallel versions has been dropped and been replaced by the singular "god"; the mention of the Erinyes, Grace (*Charis*), and Health (*Hygieia*), taken by some as proof that the inscription is pagan, is dismissed as mere moral personifications with no pagan implications; the use of the verb *eulogein* ("may he fare well") recalls the recurrent use of the root *eulog-* throughout Deuteronomy; and *epikataratos* ("There shall be a curse upon . . ."), though attested in purely pagan texts, is probably derived here from Deuteronomy 28:18–19.

tion was as an aristocrat of an important Greek city, a disciple of the most famous public figure in the Greece of his day, and in some sense a "Judaizing" Gentile. One lesson he had learned from his contacts with Judaism was the knowledge that the text of the Bible could be used as a source book for potent curses. The reference to the curses of Deuteronomy recalls the Jewish gravestones from Acmonia (see no. 91), which invoke "the curses written in Deuteronomy" upon anyone who should violate the burial site. *Bibl.:* R. Lattimore, *Themes in Greek and Latin Epitaphs* (Urbana, Ill., 1962), pp. 116–17; L. Robert, "Malédictions funéraires grecques," AIBL, *Comptes Rendus* (1978): 241–89.

> I declare to those who will possess this property: there shall be a curse upon the owner of this property who does not spare this place and the statue which has been erected but who instead dishonors or moves the boundaries or insults terribly or injures or breaks—partially or in whole—or overturns on the ground or scatters or obscures it. May god strike this person with trouble and fever and chills and itch and drought and insanity and blindness and mental fits[26]; and may his possessions disappear, may he not walk on land or sail at sea; may he (produce) no offspring. May his house not prosper; may he not enjoy crops, home, light, or the use and possession (of anything). May he have the Erinyes as watchers over him.[27] On the other hand, if anyone should look after and care for and protect (this property), may he fare well and enjoy a good reputation with everyone; may his house prosper by the birth of children and the enjoyment of his crops; and may Grace and Health watch over him.[28]

87. Greece, Rheneia (small island facing Delos); used primarily as the ancient cemetery of Delos. A marble tablet, measuring 31 × 42 cm.[29] The same text is carved on both sides; at the top, two hands reach upward in supplication toward heaven[30] (Figure 21). Unlike other curses in this volume, this tablet was a public document, meant to be seen and read by passersby. Such was the regular practice in relation to curses both for protecting graves from disturbance by robbers and for seeking justice for

26. This series of misfortunes is taken word for word from the Greek text (LXX) of Deuteronomy 28: 22 and 28.

27. The Greek term is *episkopous.*

28. The concluding blessing, unusual on tablets of this sort, is strongly reminiscent of the curses and blessings that accompany Moses' delivery of the commandments to the Israelites; see Deuteronomy 11:26.

29. A second marble tablet, with a virtually identical text, came from the same location and, presumably, the same time.

30. Another inscription from Delos (no. 2531), clearly not Jewish since it invokes "the holy goddess," shows a pair of raised hands, palms facing front as in our tablet, and begins "Theogenes raises his hands . . ."

FIGURE 21. Raised hands on a stone monument from the Greek island of Rheneia near Delos, inscribed with a plea for vengeance in the death of a Jewish woman. The cause of her death was not known, although the text of the inscription makes it clear that her relatives suspected that spells or poisons had been used.

those who had died in an untimely or violent manner. The letters indicate a date in the second century B.C.E.[31] Because of clear allusions to biblical passages from the Septuagint (LXX), interpreters of the tablet have generally taken it to be Jewish. While such a view is clearly possible, it must now be recognized that a second biblical community resided at Delos, namely, Samaritans,[32] and that a Samaritan origin must also be considered.[33] The text is an appeal for vengeance and justice in relation

31. See the discussion in Deissmann, pp. 432–34.

32. See the discussion in A. T. Kraabel, "New Evidence of the Samaritan Diaspora Has Been Found on Delos," *Biblical Archaeologist* 147 (March 1984): 44–46; Kraabel also discusses evidence for Jews on Delos.

33. Bergmann (pp. 507–6) has shown that prayers for vengeance in connection with the death of young persons was not uncommon. He cites a strikingly similar example from a burial inscription from Alexandria: "Arsinoê, untimely dead, raises her hands to the highest (*hupsistos*) god and to the one who oversees (*epoptês*) all things and to Helios and to Nemesis— whoever placed the spells (*pharmaka*) on her and whoever rejoiced or still rejoices at her death, attack them!"

to the untimely death of a young woman, Heraklea. The appeal is addressed to "the highest god, the lord of the winds and of all flesh and to the angels of God." No other figures are invoked. *Bibl.:* A. Wilhelm, "Zwei Fluchinschriften," *JOAI* 4 (1901): Beiblatt, cols. 9–18; A. Deissmann, *Light from the Ancient East* (London, 1911), pp. 423–35; J. Bergmann, "Die Rachgebete von Rheneia," *Philologus* 70 (1911): 503–7; P. Roussel and M. Launey, *Inscriptions de Délos* (Paris, 1937), no. 2532.

I[34] call upon[35] and beseech[36] the highest god, Lord of the spirits and of all flesh,[37] against those who by deceit murdered or cast a spell on/poisoned[38] miserable Hêraklea,[39] untimely dead, causing her to spill her innocent blood[40] in unjust fashion, so that the same happen[41] to those who murdered or cast a spell on/poisoned her and also to their children.[42] Lord who oversees all things[43] and angels of God, before whom on this day every soul humbles itself,[44] may you avenge this innocent blood and seek[45] (justice) speedily.

34. Bergmann argues (p. 509) that the speaker of the plea for justice is the dead woman herself. But the change from first to third person may indicate that the speaker is a surviving relative of the unfortunate Heraklea.

35. The verb, *epikalesthai*, appears together with the epithet, *hupsistos*, in several passages from the Septuagint—Sirach 46:5; 2 Maccabees 3:31.

36. The verb *axioun* is used commonly of prayer in Jewish and Christian texts and appears with *epikalesthai* in Jeremiah 11:14 (LXX).

37. This phrase is based on Numbers 16:22 (LXX); the spirits here indicate heavenly figures, in this case angels.

38. The Greek verb here is *pharmakeuein*. It has usually been taken as indicating death by poison. But the word was used just as frequently, if not more so, with reference to the casting of spells. Thus we must consider the possibility that the "cause of death" here was not poison but a binding spell. In view of this possibility, we must also reconsider the traditional interpretation, which simply takes for granted that the women were murdered. We know that they died, but the text, by its use of the phrase "murdered or poisoned/put under a spell," clearly indicates that the precise cause of their death was uncertain.

39. The woman's name on the second tablet is Marthina, a variant of Martha.

40. Again a clear allusion to biblical language as illustrated by the Septuagint. Deuteronomy 19:10 speaks specifically of not spilling innocent blood; the expression "innocent blood" occurs by itself several times.

41. The notion of exacting vengeance in conformity with the crime is again biblical, though not exclusively so.

42. As Deissmann notes, even the extension of punishment to the children of the criminal is biblical; cf. Exodus 20:5, "I visit payment for the sins of the fathers on their children."

43. A common expression in the Septuagint; cf. Job 34:23; 2 Maccabees 12:22.

44. The language here is thoroughly biblical. Leviticus 23:29 shows several precise verbal parallels; the text speaks of preparations for the Day of Atonement (Yom Kippur)—"Whoever (*pasa psuchê*, as in our text) does not humble himself on this day . . ." Thus Deissmann is led to the view that the stone tablets were set up in connection with the Day of Atonement. Bergmann (pp. 509–10) rejects this view and argues instead that along with its plea for justice, the tablet proudly proclaims in a public setting both the universality of the god of the Bible and the widespread appeal of Judaism among pagans of the Greco-Roman world.

45. The two verbs, *ekdikein* and *zêtein*, are used interchangeably in biblical texts; cf. Joel 3:21 (LXX) where one manuscript reads *ekdikêsô to haima* and another reads *ekzêtêsô to haima;* both mean "to avenge the blood."

88. Greece, the island of Delos; discovered in the pit of a private house. Lead tablet measuring 14 cm. on each side; written on both sides. The Greek contains numerous mistakes of spelling and grammar, suggesting that its author may have been a non-Greek, possibly Syrian. A date between the first century B.C.E. and the first century C.E. seems likely. The spell is addressed exclusively to Syrian deities. The occasion for the tablet is stolen property, here a necklace. The identity of the robbers was unknown to the owner. As Versnel has recognized, the tablet is a mix of a traditional *defixio* and a plea for justice. The same basic formula is followed on both sides; Side B adds the various parts of the body to be cursed. *Bibl.:* P. Bruneau, *Recherches sur les cultes de Délos à l'époque hellénistique et à l'époque impériale* (Paris, 1970), pp. 649–55; SGD 58; H. Versnel, "Beyond Cursing," in *Magika*, pp. 66–68.

> (*Side A*) Lord Gods Sukonaioi,[46] K . . , Lady Goddess Syria[47] Sukona, . . . , punish, show your power and direct your anger at whomever took (and) stole the necklace, at those who had any knowledge of it, at those who took part in it, whether man or woman.
>
> (*Side B*) Lord Gods Sukonaioi, . . . , Lady Goddess Syria . . . Sukona, punish, show your power. I register (with the gods) whomever took (and) stole the necklace. I register those who had any knowledge of it and those who took part in it. I register him, his head, his soul, the sinews of the one who stole the necklace/bracelet, and of those who know anything about it and who took part in it. I register the genitals and private parts of the one who stole (it); and of those who took (and) stole the necklace, the hands . . . from head to feet . . . toenails . . . of those who took the necklace . . . those who had any knowledge of it . . . whether man or woman.

89. Cnidus, in southwestern Asia Minor, near the island of Kos; discovered near some statue bases in the temple precinct of Demeter. Altogether, fourteen lead tablets made up the cache[48]; most were folded once. The editor claims that each shows holes in the corners (not always discernible from his drawings), perhaps to hang them on a wall in the temple. Some are inscribed on both sides. The language is an unpolished Doric Greek. A date in the first century B.C.E. seems likely. In addition to Demeter, the formulas invoke other deities normally associated with

---

46. This epithet is probably a hellenized semitic word, *sôkên,* meaning "governor," or "ruler."

47. The Syrian Goddess, known best from the treatise of pseudo-Lucian, *De Dea Syria,* is generally taken to be the Syrian goddess otherwise known at Atargatis; cf. *De Dea Syria,* ed. H. W. Attridge and R. A. Oden (Missoula, Mont., 1976). Atargatis is mentioned in other inscriptions from Delos where she is similarly addressed as *suria theos* (Bruneau, p. 655).

48. Due to the fragmentary character of several tablets, the enumeration differs from edition to edition. We have followed the original numbering of Newton.

the goddess: her daughter, Kore or Persephone; Pluto; and "the gods with them." In each case, the client dedicates (*anhieroi* and *anatithêmi*), through a ritual transfer, a personal enemy to the authority of the named gods for the purpose of subjecting the enemy to divine punishment (*kolasis* and *timôria*), torments (*basanoi*), and afflictions in the form of fever or illness (*peprêmenos*). Several of the spells specify provisional curses—that is, the punishments will expire if the target makes good the conditions that led to the curse in the first place. The tablets deal with four sets of circumstances: (1) false accusations against the client—nos. 81, 85; (2) stolen property (from a bathhouse?) and/or failure to return goods left on deposit—nos. 82, 83, 84, 86, 88, 89, 93a, 94; (3) a curse against a third party for seducing away the client's husband—no. 87; and (4) curses against personal enemies suspected of seeking to harm or kill the client—nos. 91, 95. It is noteworthy that all of the clients are women. The clients are usually named, whereas the targets remain anonymous. Thus most of the tablets from Cnidus fall under the rubric of pleas for justice and revenge of the sort familiar to us from the region of Asia Minor. *Bibl.:* C. T. Newton, *A History of Discoveries at Halicarnassus, Cnidus and Branchidae* (London, 1863), vol. 2, pp. 719–45 (with drawings, commentary, and partial translation of one tablet = no. 81); *DT* 1–13; E. Kagarow, *Griechische Fluchtafeln* (Leopoli, 1929), p. 52; Björck, *Der Fluch des Christen Sabinus* (Uppsala, 1938), pp. 121–25; and H. Versnel, "Beyond Cursing," in *Magika*, pp. 72–73 (with a translation of no. 82 = *DT* 2).

(*No. 81* [*DT 1*], *front*) I, Antigonê, make a dedication to Demeter, Kore, Pluto and all the gods and goddesses with Demeter. If I have given poison/spells[49] to Asclapiadas or contemplated in my soul doing anything evil to him; or if I have called a woman to the temple, offering her a *mina* and a half for her to remove him from among the living,[50] (if so) may Antigonê, having been struck by a fever,[51] go up to Demeter and make confession, and may she not find Demeter merciful but instead suffer great torments. If anyone has spoken to Asclapiadas against me or brought forward the woman, by offering her copper coins[52] . . .

49. The term is *pharmaka*, which might mean either poisons or spells.

50. In other words, Antigonê was accused of paying a woman specialist to kill her husband. Her accuser(s) had apparently persuaded the woman to testify against Antigonê.

51. Newton (pp. 726–29) interpreted the term *peprêmenos* to mean that the guilty party would be sold into a form of temple slavery, a reverse sort of manumission. Versnel (p. 73), based on parallels in similar tablets from other sites, argues that it must mean "burned" in the sense of suffering fevers and other afflictions.

52. Here Antigonê reveals her suspicion that the woman had been bribed to offer her testimony.

(*back*) Let it be permissible[53] for me (presumably in company with the target of the spell) to go to the same bath, under the same roof, or to the same table.[54]

(*No. 85* [*DT 4*]) I hand over to Demeter and Kore the person who has accused me of preparing poisons/spells against my husband. Having been struck by a fever, let him go up to Demeter with all of his family, and confess (his guilt). And let him not find Demeter, Kore, or the gods with Demeter (to be) merciful. As for me, let it be permissible and acceptable for me to be under the same roof or involved with him in any way. And I hand over also the person who has written (charges) against me or commanded others to do so. And let him not benefit from the mercy of Demeter, Kore, or the gods with Demeter, but instead suffer afflictions with all of his family.[55]

(*No. 95* [*DT 13*]) I hand over to Demeter and Kore and the gods with Demeter those who attacked and flogged me and put me in bonds and accused me . . . But as for me, let me be blameless . . .[56]

90. Asia Minor; original provenance not certain. Bronze tablet measuring 8.1 × 5.5 cm. with a hole at the top-middle, possibly for displaying the tablet in a public location.[57] The editor assigns a date between the first century B.C.E. and the second century C.E. The deity invoked is called "mother of the gods," most probably the goddess Cybele, whose cult center was Phrygia in Asia Minor. The tablet belongs to the common category of pleas for justice in connection with lost or stolen property. The owner-client temporarily transfers the property in question to the god so that its recovery and the punishment of the thief become a matter of divine justice and, in this case, of divine prestige as well. Presumably, at least a portion of this donation or transferred property remained the permanent possession of the god, that is, of the temple. *Bibl.:* Christiane Dunant, "Sus aux voleurs!" *Museum Helveticum* 35 (1978): 241–44; H. Versnel, "Beyond Cursing," in *Magika,* p. 74.

53. The Greek is *emoi d'hosia.* The phrase, which appears in several of the tablets from Cnidus (Newton's nos. 82, 84–87, 90–92), recalls the language of a plea for justice from Athens (no. 84). The language suggests some reluctance on the part of the client to undertake the action of commissioning the *defixio*, whether because of its illegality, its social unacceptability, or perhaps simply because of its great contagious power. Versnel (p. 73) comments that the formula also serves to protect clients in small communities where they might easily find themselves in the company of the unknown culprit at the moment when the gods chose to carry out the curse.

54. See the similar precautionary formula in a *defixio* from Italy (no. 92).

55. The phrase here is *meta tôn idiôn pantôn.*

56. The language of this fragmentary tablet suggests that the client belongs to the lower social classes, possibly even that he was a slave.

57. Bronze was not a common medium for *defixiones*, though Audollent includes two in his collection (*DT* 196 and 212).

I dedicate[58] to the mother of the gods all of the gold pieces which I have lost so that she will seek them out and bring all of them into the clear[59] and those who have (them) will be punished in a manner befitting her power,[60] so that she will not be made fun of.[61]

91. Acmonia, in the province of Phrygia, Asia Minor. A burial inscription, probably of Jewish origin. The date is clearly indicated on the stone itself as 248/249 C.E. This stone belongs with a number of others from the same city; together they curse anyone who disturbs the burial site. All are probably Jewish. On the opposite face of the stone, Aurelius lists his important civic offices: "stewardship of the marketplace and of corn purchasing, guardianship of public order; (I have) occupied all the municipal offices and held the post of stratêgos." This stone, and another one from the same place and time,[62] recall the curse bowl (see no. 109) from Mesopotamia which uses the formula "May the following verse (a biblical passage is cited) apply to him or her (the target) . . ." The region of Phrygia was noteworthy for the use of curses on epitaphs.[63] *Bibl.:* W. M. Ramsey, *The Bearing of Recent Discovery on the Trustworthiness of the New Testament* (London, 1915), pp. 358–61; *MAMA* VI.335a.

Aur(elios) Phrougianos, son of Mênokritos, and Aur(elia) Julianê (his) wife, to/ for Makaria (his) mother and Alexandria (their) sweetest daughter, constructed (this) as a tomb while still living. If anyone after their placement/burial {if} should bury another corpse or do harm on the pretext of (having made a) purchase, there shall be upon him the curses written in Deuteronomy.[64]

58. The verb *anatithenai* appears in similar tablets (*DT* 4, line 1, from Cnidus in Asia Minor; see no. 89); on other parallels between our tablet and those from Cnidus, see Dunant, pp. 243–44.

59. Following Egger's proposed scenario for the tablet from Innsbruck (no. 101), the client probably had the following sequence of events in mind: the tablet itself was nailed up in the local temple dedicated to the mother of the gods; the goddess would pursue the thieves and punish them until they decided to return the gold; the lost or stolen property would be returned to the temple where the owner would redeem it.

60. Versnel comments that the goddess's power (*dunamis*) corresponds to *aretê* (see no. 88) in other tablets where pleas for justice are made in the name of some deity. In all cases, these terms constitute a public appeal to the known powers of the god or goddess.

61. The notion that the goddess will be subject to humiliation if she fails to carry out the charge is unique in tablets of this kind but probably not to the psychology of curse tablets as a whole.

62. *MAMA* VI, no. 335, which similarly warns anyone who disturbs the tomb that "the curses (*arai*) written in Deuteronomy will come upon him." This stone no doubt belongs to the same period as the one cited previously.

63. See R. Lattimore, *Themes in Greek and Latin Epitaphs* (Urbana, Ill., 1962), p. 109 and L. Robert, "Malédictions funéraires grecques," AIBL, *Comptes Rendus* (1978): 253–54 and 267–69.

64. The Greek is *en tô deuteronomiô*. The reference is undoubtedly to the lengthy series of curses spelled out in Deuteronomy 28:15–68. The fact that the inscription of Amphicles of Chalcis cites some of these curses word for word and that Amphicles was himself probably not Jewish raises the possibility that our inscription from Acmonia, with its general reference to "the curses written in Deuteronomy," might not be Jewish.

92. Italy; from the region of ancient Bruttium. Bronze tablet measuring 10 × 14 cm. A date in the third century B.C.E. seems most likely. The letters were not carved with a stylus, as was customary, but hammered in with a chisel. This tablet appeals for the return of stolen property. The appeal is directed to an unnamed goddess; the names of both the thief and the client are given; the client promises to donate a portion of the stolen property (clothing and three gold pieces) in return for its return through the agency of the goddess; the thief is to return the stolen goods to the temple of the goddess; the goddess will bring suffering on the thief until he relents. Beyond these common features, this tablet shows one novel element—the thief must return not only what was taken but pay a penalty of twelve times its value and an offering of incense. *Bibl.: DT* 212; *SEG* 4.70; *IG* 14.644; V. Arangio-Ruiz and A. Olivieri, *Inscriptiones Graecae et Infimae Italiae ad Ius Pertinentes* (Milan, 1925), pp. 165–70 (with photograph and commentary); H. Versnel, "Beyond Cursing," in *Magika*, p. 73.

> Kollura dedicates to the priests of the goddess . . . the dark cloak (?) which Melitta received and has not returned; instead she is using it and knows where it is (?). Let her deposit with the goddess twelve times the amount with a measure of incense[65] according to the city ordinance.[66] Let the person who has the garment not recover until she deposits it with the goddess. Kollura dedicates to the priests of the goddess the three gold pieces that Melitta took and has not returned. Let her dedicate to the goddess twelve times the amount with a measure of incense according to the city ordinance. Let her not breath freely until she dedicates (it) to the goddess. And if she (Kollura) should unknowingly eat or drink with her (Melitta), let her remain unharmed, even if she comes under the same roof.[67]

93. Sicily; ancient Centuripae. Discovered in a tomb. Jordan proposes a date in the first century C.E. Although there is no indication of the specific cause behind the curse, the tablet clearly belongs to the category of pleas for justice and revenge. The deity is not named but rather addressed simply as "Lady." *Bibl.:* D. Comparetti, "Varietà epigraphiche siceliote," *Archivio Storico per la Sicilia Orientale* 16–17 (1919–1920): 197–200; *SEG* 4.61; SGD 115; H. Versnel, "Beyond Cursing," in *Magika*, pp. 64–65.

65. The word for incense here is *libanos,* widely used in religious rites and temples.
66. This requirement is without parallel in other tablets.
67. The precaution here presupposes not only that the client knows the suspected thief but that the social circumstances are those of a small, closed society where the two were likely to find themselves, in the normal course of daily living, under the same roof. Sensibly, Kollura asks that punishment not be visited on Melitta when they are together, lest it effect her too. Similar precautions are expressed in other tablets; the language is probably formulaic.

Lady,[68] destroy Eleutheros.[69] If you vindicate me, I will make a silver palm,[70] if you destroy him utterly from the human race.[71]

94. England, Bath; from the sacred spring whose matron deity was Sulis Minerva. Lead alloy tablet measuring 9.9 × 5.2 cm.; written in Latin capitals. The editor proposes a date in the second or third century C.E. for most of the tablets. Altogether some 130 curse tablets have been excavated and published; many more remain unexcavated.[72] Most deal with the loss or theft of personal property at the baths themselves, probably taken by bathhouse thieves.[73] The missing items included jewelry, coins, household items, gemstones, and, most of all, articles of clothing. Most of the clients appear to have come from the lower social classes (Tomlin, pp. 97–98). The texts are highly formulaic, obviously taken from available handbooks. The general procedures that underlie the purchase and deposition of these tablets are strikingly similar throughout the Greco-Roman world of late antiquity: the stolen property is ritually transferred to the appropriate deity, thus involving the god directly in the loss; normally the suspect is named and sometimes also the client (at Bath 21 tablets name the client); the client then urges the deity to visit various afflictions on the thief, including death, not so much as punishment but rather as inducement to return the property to the temple where, presumably, the owner would redeem it for a certain

68. The goddess invoked here might be any number of female deities, including Hekate, Demeter, or local goddesses addressed in other Sicilian *defixiones;* cf. the goddess of Selinus (no. 50).

69. The name, meaning "freedman," points to social circles of ex-slaves.

70. *Spadix* can designate a stringed musical instrument, like a lyre (so Comparetti) or the branch or frond of a palm tree. The latter might be more appropriate here if we see the promised offering as a decorative piece.

71. The double condition here ("If . . . if . . .") makes it clear that the donations and offerings in pleas for justice must be taken as provisional. The donation will be made *if* the god causes the property to be returned or otherwise vindicates the wronged party.

72. See the remarkable publication of the Bath tablets, a model of philological scholarship at its best and broadest, by Tomlin, pp. 59–278. Tomlin's publication includes a list of all curse tablets from England; a catalogue of all theft-related tablets from outside Britain; a complete study and catalogue of formulas, key words and phrases, language, and handwriting; and transcriptions and reconstructions of the 130 texts, with drawings and full commentary. Tomlin notes that as many as 500 more tablets remain to be uncovered, based on his assumption that no more than one-sixth of the spring's deposit has been excavated thus far.

73. Tomlin observes that the bathhouse thief was a well-known literary and presumably, social type: cf. Catullus 33, a poem addressed to an "outstanding" bathhouse thief; Seneca, *Letters* 56.2, which speaks of the noise created when one is apprehended in the act; Petronius, *Satyricon* 30, in a scene where a master beats his slave for having lost his clothing while in the bath; *Digest* I.15.3.5, which regulates those paid to watch over clothes in public baths. The curse tablets from Bath and elsewhere suggest that they were more than merely literary (pp. 80–81). In Greek law, such thieves could be executed; in the *Digest* of Justinian (XLVII.17), they occupy a full paragraph, where the penalty is forced labor in the mines.

fee. Thus they are pleas for both justice and vengeance. *Bibl.:* Tomlin, pp. 118–19 (no. 8).

> I have given to the goddess Sulis the six silver coins[74] which I have lost. It is for the goddess to exact (them) from the names written below: Senicianus and Saturninus and Anniola. The written page has been copied out.[75]
>
> An (n) iola
> Senicianus
> Saturninus.

95. England, Bath (see no. 94). Lead alloy tablet measuring 7.5 × 5.8 cm., written on both sides and folded twice. The editor proposes a date in the second century C.E. Like other tablets from Bath (nos. 4, 61, 62, 98, 99), this one employed a "mysterious" form of writing: the beginning and the end of each line is reversed so that it reads from left to right. *Bibl.:* Tomlin, pp. 164–65 (no. 44).

> (*Side A*) The person who lifted my bronze vessel is utterly accursed.[76] I give (him) to the temple of Sulis, whether woman or man, whether slave or free, whether boy or girl,[77] and let him who has done this spill his own blood into the vessel itself.[78]

74. In no. 54, a woman, Arminia, complains that Verecundius has stolen two silver coins (*argentiolos duos*). The editors comment that the coins are not likely to be the highly debased *antoniani* of the mid-third century, but the double *denarius* of Caracalla. By implication, then, the cost of a curse tablet would not have been more than two *argentioli/antoniani*, not a great sum.

75. This is the only reference in the Bath tablets to the actual process of copying the formulas from a master. In this case, the copyist made several errors, which were subsequently corrected by overwriting, presumably after rechecking the master.

76. The Latin term is [e]*xconic*[*tus*], read by the editor as *exconfixus*.

77. This formula, or variants of it, appears in numerous tablets from Bath and elsewhere, including a curse tablet from Delos dating from between 100 B.C.E. and 100 C.E. (see no. 88); cf. Tomlin, pp. 67–68 and 73. Similar phrases appear also in several other places: (1) the New Testament, in Paul's letter to the Galatians 3:28: "There is no Jew and Greek, slave and free, male and female . . ."; (2) Plutarch, *Life of C. Marius* 46: "Plato, when he was about to die, praised his companion spirit and Fortune that he was made a man and not an irrational beast, a Greek and not a barbarian, and beyond this that he had lived in the time of Socrates"; (3) Diogenes Laertius 1.33 (Thales): "Hermippus in his *Lives* refers to Thales the story which is told by some of Socrates, namely, that he used to say that there were three things for which he was grateful to Fortune, that he was made a man and not a beast, a male and not a female, a Greek and not a barbarian"; (4) Roman legal texts, for example, Justinian, *Novella* V.2: "as for the worship of God, there is neither masculine nor feminine, neither free nor slave," where Christian influence seems likely; and (5) various Jewish texts, of which the earliest is probably the *Tosefta* (Ber. 7:18): "Rabbi Judah says, 'One ought to say three blessings every day: blessed is he that he did not make me a Gentile; blessed is he that he did not make me a woman; blessed is he that he did not make me a boor.' " In its origins, the formula is probably Greek, whence it passed into Jewish circles and thence to Paul. There is no reason to posit Jewish or Christian influence on its use in the curse tablets.

78. The formula is unusual here, obviously modified from more standard formulas so as to suit the particular conditions of the theft. Other forms of punishment include loss of mind and

(*Side B*) I give, whether woman or man, whether slave or free, whether boy or girl, that thief who has stolen the property itself (that) the god may find (him).

96. England, Bath (see no. 94). Lead alloy tablet measuring 10.5 × 6.0 cm., written on both sides but not folded. This tablet is notable for several reasons. First, it is the only one with a text (Side A) completely written in reverse order: the first letter inscribed in the first line is actually the last letter of the text and vice versa. Second, the standard formula for identifying the suspected thief ("whether man or woman, . . .") is here supplemented in unique fashion by the addition of "whether pagan or Christian." The date is probably in the fourth century C.E. *Bibl.:* Tomlin, pp. 232–34 (no. 98).

(*Side A*) Whether pagan[79] or Christian, whether man or woman, whether boy or girl, whether slave or free, whoever has stolen from me, Annianus (son of) Matutina (?), six silver coins from my purse, you, Lady Goddess, are to exact (them) from him. If through some deceit he has given me . . . and do not give thus to him but reckon as (?) the blood of him who has invoked this upon me.[80]

(*Side B*)[81] Postumianus, Pisso, Locinna, Alauna, Materna, Gunsula, Candidina, Euticius, Peregrinus, Latinus, Senicianus, Avitianus, Victor, Scotius, Aessicunia, Paltucca, Calliopis, Celerianus.[82]

97. England; Kelvedon, Essex. Discovered folded up, in an oven dating from the third or fourth century C.E.; part of a Roman settlement. Lead tablet measuring 10.5 × 5 cm. Another in the series of tablets directed at recovering lost or stolen property by means of a tablet and by the customary donation to the god and his temple. The client's name is Varenus. The gods addressed are Mercury (the Roman equivalent of Hermes)[83]

eyes (no. 5); loss of blood and/or life (nos. 31, 65–66, 94, 99, 103); loss of sleep and/or children (nos. 10, 32, 52); loss of ability to eat, drink, defecate, and urinate (no. 41); blindness and childlessness (no. 45); cloud and smoke (no. 100).

79. The Latin term is *gen(tili)s*.

80. The tablet is corroded at this point and impossible to read. The editor proposes that the sentence is meant to be apotropaic, designed to turn aside any counterspell invoked by the thief (Tomlin, p. 234).

81. In this list, the names are written in the proper order from top to bottom, but the letters of each name are reversed.

82. The list contains eighteen names, ten of Roman or Greek form and eight of Celtic origin (Pisso, Locinna, Alauna, Gunsula, Senicianus, Scotius, Aessicunia, Paltucca). All are probably suspects in the theft.

83. Mercury is the god invoked in the ca. 140 tablets, most still unpublished, from Uley, also in England.

and personified Virtue. *Bibl.:* R. P. Wright in "Roman Britain in 1957," *JRS* 48 (1958): 150; R. Egger, *Nordtirols älteste Handschrift,* vol. 244 (Vienna, 1964), pp. 16–17; H. Versnel, "Beyond Cursing," in *Magika,* pp. 84–85.

> Whoever stole the property of Varenus, whether woman or man, let him pay with his own blood.[84] From the money which he will pay back, one half is donated to Mercury and Virtue.

98. England; Red Hill, Nottinghamshire. Ploughed up in a field along with other evidence for a Roman site. Lead tablet measuring 5.7 × 8.4 cm.; written on both sides and folded three times. The editor dates it to ca. 200 C.E. The occasion is stolen money, the considerable sum of 112 *denarii.* As with numerous other examples of this sort, the client donates a portion of the property—here one-tenth of the lost money—to the god. The deity invoked, Jupiter Maximus Optimus, is unusual, for such tablets are normally addressed to gods of the underworld, whereas Jupiter is the Roman counterpart to the Greek Zeus, a god of heaven and the world above. *Bibl.:* E. G. Turner, "A Curse Tablet from Nottinghamshire," *JRS* 53 (1963): 122–24; R. Egger, *Nordtirols älteste Handschrift,* vol. 244 (Vienna, 1964), pp. 17–19; H. Versnel, "Beyond Cursing," in *Magika,* p. 84, n. 104.

> Donated to Jupiter best and greatest, so that he may haunt (personal name missing) in his mind, in his memory, in his innards, in his intestines, in his heart, in his marrow, in his veins, in his . . . , whoever, whether man or woman, who stole the 112 *denarii* of Dignus (?) and that he (the thief) will personally[85] make a full settlement. To the god named above has been donated one-tenth of the sum when he repays it.[86]

99. England; Lydney Park, Gloucestershire. Discovered in the temple of the god Nodens. Tin tablet measuring 7.5 × 6 cm., with eleven lines of ordinary Latin. Once again the issue is lost property, in this case a gold ring, an item of considerable value. The unnamed owner dedicates half the value of the ring so that Nodens will search out its unlawful possessor, named Senicianus, and force him to return it to the

---

84. A common formula (*sanguine suo*) found in similar tablets from other sites in England; see Versnel, pp. 202–7. The sense is "with his life."

85. So Turner translates the phrase *in corpore suo.* From the parallel phrase, *sanguine suo,* in other tablets, a better translation here might be "with his body," that is, with death.

86. As usual in such tablets, the payment or donation to the deity is conditional. The payment will be made only when the property is recovered.

temple. What makes this tablet of more than usual interest is the fact that a gold ring was discovered at Silchester, some 50 kilometers to the southeast of Lydney Park. The seal or bezel of the ring bears the inscription, VENUS, a dedication to the pagan goddess. The hoop of the ring bears a secondary inscription, almost certainly later than the first, namely SENICIANE VIVAS IN DE [O], which is a common Christian exclamation ("Senicianus, may you live in/with God!"). The sequence of inscriptions suggests that the ring had two successive owners or possessors, the first a non-Christian and the second a Christian with the name of Senicianus. Both the ring and the tablet date from between 350 and 400 C.E. There is thus a real possibility, though nothing more, that we now possess not just the report of a lost or stolen ring, on the tablet, but also the ring itself, bearing the same name as the person cited in the tablet.[87] Here is a possible sequence of events, partially spelled out in the customary narrative recorded on the tablet: Silvianus, a non-Christian, lost his gold ring, inscribed with the name of Venus; Senicianus, a Christian, found it and had it inscribed with his own name and the Christian exclamation; Silvianus learned that a person named Senicianus had come into possession of the lost or stolen ring; Silvianus commissioned the tablet and deposited it in the local temple of Nodens; the ring was found some fourteen hundred years later by a farmer ploughing his field. *Bibl.: DT* 106 (*CIL* VII.140); R. G. Goodchild, "The Ring and the Curse," *Antiquity* 27 (1953): 100–102; H. Versnel, "Beyond Cursing," in *Magika,* p. 84; J. Toynbee, "Christianity in Roman Britain," *Journal of the British Archaeological Association* 27 (1953): 100–102.

> To the God Nodens. Silvianus has lost a ring. He has given half of it (its value) to Nodens. Among those whose name is Senicianus, do not permit health until he brings it to the temple of Nodens.[88]

100. Wales, Caerleon; discovered in the amphitheater, near the encampment of the Roman legion known as *Legio Secunda Augusta.* Lead

---

87. P. Corby Finney, who is working on a new study of the ring, kindly informs me that the name of Senicianus is not altogether rare in Christian circles of the period and could in this case refer to another person. The same name appears on one of the tablets from Bath, where its bearer is named as a suspect in the theft of silver coins; see no. 96.

88. It is clear that the stolen goods were to be returned to the temple of the gods in question. In a tablet from Uley in England (see Versnel, p. 88), a woman named Saturnina writes a letter (*commonitorium*) to Mercury concerning a lost piece of clothing. The thief, "whether man or woman, slave or free," is to return the item "to the temple mentioned above (i.e., of Mercury)."

tablet measuring roughly 7 × 7 cm. The tablet was not folded; it must have been deposited in the vicinity of the amphitheater or the nearby temple of Nemesis. A date in the first or second century C.E. seems likely. Its original form included two "handles" (*ansae*), one of which is now missing. The inscription covers eight lines in a cursive script of ordinary Latin. The deity addressed is Nemesis, saluted as "Lady."[89] The recurrence of the formulaic phrase "with his own blood and life" in several tablets from Bath and other sites in England suggests that the tablet was prompted by the loss of personal property to a suspected thief; the client invokes the power of Nemesis to avenge the loss. *Bibl.:* R. G. Collingwood, "Inscriptions on Stone and Lead," *Archaeologia* 78 (1928): 157–58; A. Oxé, "Ein römisches Fluchtäfelchen aus Caerleon (England)," *Germania* 15 (1931): 16–19; R. Egger, "Aus der Unterwelt der Festlandkelten," *Wiener Jahreshefte* 35 (1943): 108–10; Jordan, "Agora," p. 214; H. Versnel, "Beyond Cursing," in *Magika,* pp. 86–87.

> Lady Nemesis! I give you this cloak and these shoes. May the person who has worn (taken?) them not buy back/redeem (them) except with his own life and blood.[90]

101. Austria, near Innsbruck (ancient Veldidena, modern Wilten); discovered in a Roman burial site. Lead tablet measuring 5.7 × 2.6 cm. The text contains fifteen lines of unsophisticated Latin. Egger proposes a date near 100 C.E. The deities addressed are the familiar Mercurius; the uncommon Moltinus, a Celtic deity known only from one other inscription; and Cacus, an old Roman god or monster. The occasion is stolen property, valued by the owner at fourteen *denarii*. The owner-client is a woman, Secundina. The formulas find close parallels in Latin curse tablets from Britain (see nos. 94–100) and several Greek tablets. *Bibl.:* L. Franz, "Ein Fluchtäfelchen aus Veldidena," *JOAI* 44 (1959): Beiblatt, cols. 69–76; R. Egger, *Nordtirols älteste Handschrift,* vol. 244 (Vienna, 1964), pp. 3–23; H. Versnel, "Beyond Cursing," in *Magika,* pp. 83–84 (following his translation).

---

89. Nemesis does not appear frequently in *PGM.* She is cited in *PGM* VII, line 503, where she is identified with Isis and Adrasteia; and once in *PGM* XII, line 220, along with other gods of heaven and earth.
90. Egger's text thus reads: *domna nemesis, do tibi palleum et galliculas, qui tulit non redimat ni[si] vita sanguinei sui.*

Secundina commands of Mercurius[91] and Moltinus,[92] concerning whomever has stolen two necklaces[93] worth fourteen *denarii*,[94] that deceitful Cacus[95] remove him and his fortune, just as hers were taken,[96] the very things which she hands over to you so that you will track them down.[97] She hands them over to you so that you will track him down and separate him from his fortune, from his family and from his dear ones. She commands you on this; you must bring them to justice.[98]

91. Mercury, who was widely assimilated to the Greek god Hermes, was commonly associated with theft, which explains his appearance in spells against thieves. *PGM* V, lines 172ff., prescribes the following spell in order to catch a thief: "I summon you, Hermes, immortal god . . . I call upon Hermes, finder of thieves . . . to grab the thief's throat and bring him into the open today . . ." What the text in *PGM* makes clear in this case is that the tablet was to be deposited together with sacrificial offerings and spoken spells.

92. Moltinus is a Celtic deity, whose special concern seems to have been sheep and cattle; cf. Egger, pp. 11–12.

93. See Versnel, p. 83, n. 121.

94. Franz and Versnel translate "has stolen 14 denarii or two necklaces." They take *draucus* as a transliteration into Latin of the Greek *draukion*/necklace. Egger argues that it represents instead a Celtic loanword, with different meanings in Latin, among them "cow." The fact that Moltinus is a Celtic deity lends some plausibility to this argument.

95. Cacus and his sister, Caca, were known as fire deities who lived on the Palatine Hill in Rome; he is particularly known for having stolen cattle from Hercules. The appropriateness of such an act for the circumstances of our tablet, as interpreted by Egger, is readily apparent.

96. As frequently in *defixiones,* the punishment desired corresponds precisely to the crime; in this case, additional penalties are demanded—separation not only from the stolen property but from family and friends as well. What separation means here (death?) is not clear, although the endpoint is obvious, that social pressure should force him to return the stolen property.

97. The procedure, here as in similar texts (see no. 92), is for the owner to "dedicate" the lost property to the gods, that is, to transfer its ownership to them on a temporary basis, so that its theft becomes *their* loss, not just the owner's.

98. On punishments in Roman law for theft of cattle, see Egger, pp. 14–15; still open is the question whether Secundina appeals to the gods to bring the thief before the Roman court or to execute punishment themselves. Egger holds that in matters of theft involving relatively small amounts of property, it made better sense to turn to the gods and their powers than to the courts. Thus the ultimate purpose of the spell is to force the thief to return the stolen property to its rightful owner in the temple of the gods in question. Such a proceeding makes sense not in a Roman legal framework but rather in a small, rural community where Celtic customs prevailed.

6

Miscellaneous Tablets

Among the various types of *defixiones,* the miscellaneous examples collected in this chapter are surely the most arbitrary. Perhaps they should be labeled as "personals." Of course, all curse tablets and binding spells are personal *by definition,* in the sense that they represent a ritual medium whereby individual persons sought redress or advantage against other persons. In short, every *defixio,* of whatever sort, could be included among those brought together in this chapter.

Those actually located here meet one simple, if negative criterion: they fail to reveal enough information regarding their original causes or circumstances in order to qualify for one of the more precise categories. Put differently, the eclectic group of tablets in this chapter reminds us that our other categories (competition in public arenas; sex and love; legal and political affairs; business rivalries; and pleas for justice and revenge) do not completely exhaust the possibilities for strong emotional feelings, for personal conflict of various kinds that reveal themselves in ancient *defixiones.* For most of the tablets discussed here, just a little more information might enable us to relocate them to a different chapter. Still, there must have been some expressions of anger, jealousy, lust, and revenge that arose at less well-defined moments in the personal lives of our men and women. Thus some of the tablets in this chapter would probably remain here no matter what additional information we might learn about their original causes.

102. Greece, Attica; original location uncertain. Lead tablet dated to the first half of the fourth century B.C.E. Probably folded or rolled up;

18.5 × 6 cm. The occasion of the curse is not given. Several of the individuals mentioned are known from inscriptions of the fourth century. Of particular interest is the rare formula, "and I will not release." Numerous amulets existed for the express purpose of warding off the power of curses and binding spells. *Bibl.:* Wilhelm, pp. 120–22; A. Deissmann, *Light from the Ancient East* (London, 1911), p. 307; SGD 18.

> Gods. Good Fortune[1] I bind and will not release Antiklês,[2] the son of Antiphanês and Antiphanês, the son of Patroklês and Philoklês and Kleocharês and Philoklês and Smikrônidês and Timanthês and Timanthês[3] I bind all of these before Hermes—the underworldly, the treacherous, the restrainer, the roguish[4]; and I will not release (them).[5]

103. Greece, Attica; original location uncertain; probably from the fourth or third century B.C.E. Typical of the type that consists only of the name of the person(s) being cursed, with no reference to deities or spirits, the occasion, or any verb form. The lead strip measures 9 × 5 cm.; it was originally folded and pierced by a nail. *Bibl.:* DTA 4.

> Nikandros the freedman[6] of Euxitheos.

104. Greece, Athens; original location not known. Wilhelm dates it to the fourth century B.C.E. Lead tablet measuring 13 × 6 cm. Written on both sides (Figure 22). Side A comprises sixteen very short lines; Side B has nineteen somewhat longer lines. The order of several lines and of letters in individual words is deliberately scrambled. Like numerous others, this tablet is a catchall curse. It deals with the recovery of lost family members (or slaves); with professional fighters; and with love affairs involving several women, some of whom may have been courtesans or *hetairai*. *Bibl.:* DTA 102; Wilhelm, 112–13 (text); H. Versnel, "Beyond Cursing," in *Magika* p. 65.

> (*Side A*) I am sending a letter[7] to the *daimones* and to Persephone, and deliver (to them) Tribitis, (daughter of) Choirinê, who did me wrong, daugh-

---

1. The Greek is *agathê tuchê,* a popular phrase in Greece of the classical period and frequently used as a heading for public inscriptions; it later came to be personified as a popular deity. Here we may have an instance of a deliberate imitation of a public formula in a private document.
2. *IG* 2.2 1006 mentions an Antiklês the son of Antiphanês.
3. The names of Timanthês and Philoklês are repeated, probably by mistake.
4. The Greek is *eriounios,* a common epithet of Hermes; its meaning is uncertain.
5. The formula of not releasing is repeated here, as if to emphasize its unusual character.
6. The term *apeleutheros,* meaning freedman or ex-slave, locates the target of the curse among noncitizens and nonelite residents.
7. This form of communication with spirits and deities is not uncommon in spells; cf. DTA 103 and the comments of Wortmann, p. 81, on ancient Egyptian "letters to the dead."

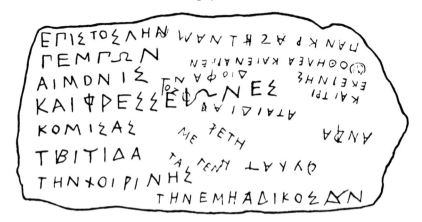

FIGURE 22. *Defixio* in letter form, with scrambled names and writing, from Attica (Greece) on the right side of the tablet. Such nonstandard forms of writing were intended to serve as a symbolic representation of the fate of the targets; that is, their lives were to be scrambled.

ter, husband, and three children, two female and one male. Pagkratês Mant[ias] Diophantos Metagenês

(*Side B*) Restrained are . . . the boxers, Aristomachos and Aristônumos.[8] Restrain all their power. Euandria, daughter of Charikleidês . . . ,[9] treacherous Doris, (daughter) of Aristokratês. May Persephone restrain all of her. Hermes and Hades,[10] may you restrain all of these. *Daimon*, (may you restrain) Galênê,[11] daughter of Polukleia,[12] by your side.

105. Greece, Athens, in the area of the Kerameikos. Found in a grave of the fifth century B.C.E. A long thin vertical strip, irregularly cut, originally rolled, now in eight pieces, and measuring 7 × 37.5 cm. Peek suggests that the Peithandros mentioned in the tablet is the father of the

8. Wilhelm notes that these two boxers may well be the source of the anguish that led to the commissioning of the curse. They may have won the affections of the women in question.

9. The female names that follow are the chief targets of the curse.

10. Hermes and Hades, the personifications of the underworld, are here invoked directly, using the vocative and the second person of the verb *katechô*.

11. A Galênê is also mentioned in DTA 107, a tablet from the fourth century B.C.E.

12. Galênê and Polukleia are two seemingly well-known courtesans (*hetairai*). The Greek writer, Athenaeus, who wrote his *The Learned Banquet* at the end of the third century C.E., states that Philetairos, a poet and author of comedies, wrote a comedy entitled *Kunagis/The Huntress* (587–88) in which he mentions a courtesan by the name of Galênê, and that Alexis, a comedian of the same period, wrote a piece entitled *Polykleia*, taken from the name of a courtesan (642c). Thus our curse may well take us directly into the complicated lives of these figures, of whom Wilhelm suggests that they were mother and daughter.

well-known politician, Blepuros Paionides.[13] The spell lists many other members of the Paionides family, as well as its retinue, including concubines and *hetairai.* The occasion for the spell seems more likely to have been a private dispute than a political trial. *Bibl.:* Peek, no. 3, pp. 91–93; Jeffery, p. 75; SGD 1.

> (*Side A*) . . . I bind . . . I bind Stephanos, son of Poluaratos at the side of Persephone and Hermes. I bind Theothemis. I bind Hêgemachos the son of Phanostratos. I bind Eukleia . . . the daughter of Dêmokratês. I bind the soul and tongue of Eukleia at the side of Persephone and Hermes. I bind Eukleia the daughter of Dêmokratês, soul and . . . words. I bind Philostratê. I bind Aristoboulê the concubine and the soul of Aristoboulê. I bind Charias, son of Pheidias, the soul of Charias and the tongue of Charias. I bind Ameinonikê the courtesan of Charias. I bind Ameinonikê at the side of Persephone. I bind also the soul and tongue of Ameinonikê and the words and deeds of Ameinonikê. I bind . . . I bind Charias; I bind Ameinonikê. I bind Timotheos, son of Paion and the tongue of Timotheos. I bind Mnesippos and the soul of Mnesippos. I bind Mnesias, son of Paion. I bind Demonikos (?) . . . I bind Plangôn, Mnesias's sister. I bind Kallippos, brother of Plangôn. I bind the soul and hands of Kallippos. I bind Ergasiôn and Puthios, his servant, and I bind the soul of Ergasiôn and the wife of Ergasiôn. I bind . . .

106. Palestine; near Hebron, south of Jerusalem; exact place of origin not known. Date uncertain, but probably third to fifth centuries c.e. Lead tablet measuring 7 × 4.5 cm.; originally rolled up. The powers invoked are the familiar *charaktêres,* represented by the figures drawn on four lines above the text. The occasion for the curse is not clear. *Bibl.:* B. Lifshitz, "Notes d'épigraphie grecque," *RB* 70 (1970): 81–83 (plate IX); SGD 163.

> I invoke you *charaktêres* to lay Eusebios low,[14] to whom the pious[15] mother Megalê gave birth, with suffering and injury; cast him into a fever. Lay him low with suffering and death and headaches. Quickly, quickly, now, now!

107. Palestine, Tell Sandahannah, probably to be identified with Marissa, approximately sixty kilometers southwest from Jerusalem. Discovered in a find of fifty-one limestone tablets, most quite small (4 × 6 cm.). Also discovered were sixteen lead figurines, most with their hands and feet bound (see no. 108). Many of the fragments contain nothing

---

13. See *IG* 2.2 1747, where Peithandros is described as the "secretary for the assembly and for the people."

14. The Greek is *kataklinai,* used here and elsewhere (*PGM* IV, line 2075) in the sense of putting someone to bed with illness.

15. The Greek adjective *hiera* ("pious") is rarely used of human beings. In this text it might, then, represent a Hebraism.

more than one or two personal names or scattered Greek letters. Only two contain full texts: the one translated here is a binding spell or, better, a counter binding spell. This tablet measures 14 × 16 cm. Some letters and perhaps one or two lines are missing. On the basis of letter forms and spelling, Wünsch assigns the tablets to the second century C.E. The unnamed client seeks retribution against Philonides, who has caused the client to lose his job, seemingly through a spell directed against the client. No deity or spirit is mentioned. The occasion is personal enmity, expressed here through financial concerns. The persons involved may well be Jewish, since the site itself was a Jewish town during the period in question. *Bibl.:* F. J. Bliss, "Report on the Excavations at Tell Sandahannah," *Palestine Exploration Fund. Quarterly Statement* (London, 1900), pp. 319–34; C. S. Clermont-Ganneau, "Royal Ptolemaic Inscriptions and Magic Figures from Tell Sandahannah," *Palestine Exploration Fund. Quarterly Statement* (1901): 54–58; R. Wünsch, "The Limestone Inscriptions of Tell Sandahannah," *Excavations in Palestine during the Years 1898–1900,* ed. F. J. Bliss and R. A. S. Macalister (London, 1902), pp. 173–76 (no. 34); R. Ganszyniec, "Sur deux tablettes de Tell Sandahannah," *BCH* 48 (1924): 516–21.

> I bind (?) Philônidês, son of Xenodikos. I demand that he be punished and that vengeance be exacted on the man who caused me to be expelled from the household of Dêmêtrios,[16] due to my headaches and other pains. Thus may oblivion seize the binding spell[17] that he pronounced against me. Let Philônidês, rendered harmless and incapable of harming others, be forever voiceless and destitute. Now, quickly.[18]

108. Same as no. 107. Sixteen lead figurines, ranging in height from 5 to 8 cm. (Figure 23). Most are bound at the hands or the feet, some at both. Some of the figurines show slight molding to indicate hair, head, face, navel, breasts, and limbs. Such figurines were a common feature of binding spells. These sixteen examples, especially given their varying shapes, may have constituted the inventory of a professional *magos* resident at Marissa. *Bibl.:* same as no. 107.

---

16. The language suggests that the client may have been an employee in the household of Demetrios and that he had been let go as a result of recurrent problems of health. Based on his analysis of the entire group of tablets, Wünsch concludes that most of the persons mentioned in them belonged to the category of slaves or freedmen.

17. The Greek term is *peridesmos,* not otherwise attested in the papyri or tablets. A form of the verb, *peridein,* does occur in *PGM* VII, line 453, as part of instructions for preparing a multipurpose binding spell (*katachos*) on lead.

18. This ending is reconstructed by Ganszyniec. The normal form would be "Now, now. Quickly, quickly."

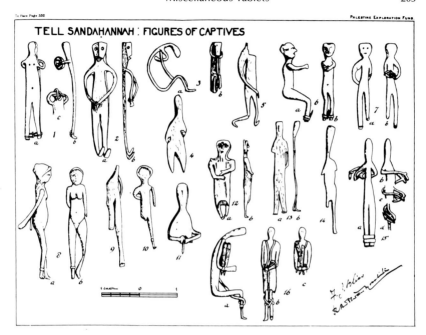

FIGURE 23. Bound lead figurines from Tell Sandahannah (Palestine). Some of the figurines show sexual features. The cache might represent the unused supply of a local *magos*. There is good reason to believe that professional *magoi* kept ready-made tablets and figurines on hand for potential customers.

109. Mesopotamia; original location unknown. From the invocation addressed to "the spirit who resides in the cemetery," it would appear that the bowl was originally deposited at a burial site. Like others of its kind, this bowl is made of earthenware and inscribed in ink. It measures approximately 12 cm. across the top. The language of the spell is Babylonian Jewish Aramaic; the numerous biblical quotations are in Hebrew, with occasional Aramaic "translations" and replacements. No date is given by the editors. At the bottom of the bowl a circle has been drawn; within the circle is a figure and a few words, of which only two make any sense. Outside the circle are thirteen lines of text in spiral form, from bottom to top. Judging by other bowls with similar figures (Figure 24), we should probably identify it with the angel or spirit invoked in the last line of the text. But it may also represent the target of the spell. The bowl is remarkable in several respects. It is unique in that unlike other bowls it is given over to cursing a personal enemy. The specific curses,

FIGURE 24. Inscribed bowl, probably from Mesopotamia, with figures at the bottom. The function of the bowl was to protect its owners from harmful spells and curses. The editors describe the central design as "a human face, perhaps with the body of an insect . . . Encircling this design is a number of figures, which include a serpent and two reptiles as well as a cross which ends its four corners with circles." (J. Naveh and S. Shaked, *Amulets and Magic Bowls: Aramaic Incantations of Late Antiquity* [Jerusalem: Magnes Press, Hebrew University, 1985], Plate 18 [Bowl 4, left side]. By permission.)

illnesses, diseases and other disasters, are drawn from biblical verses. Each verse, or cluster of verses, is introduced by the phrase, "May the following verse apply to him . . ." All of these features mark the bowl and its spell as distinctively Jewish. The professional who prepared the bowl, the unnamed client who commissioned it and the target are also certainly Jewish. The secret names of the angel(s) and the spirit(s) invoked are not attested elsewhere. *Bibl.:* Naveh and Shaked, pp. 174–79 (bowl no. 9).

(*Inside the circle*) 'HYD PKR TSMR W'QYM GBYH WSRYH and stars and planets.

(*Outside the circle*) and all the vomit and spit/saliva of Judah, son of Nanay, so that his tongue may dry up in his mouth, that his spit/saliva (?) may dissolve in his throat, that his legs may dry up, that sulphur and fire may burn in him, that his body may be struck by scalding, that he may be choked, estranged, and disturbed in the eyes of all those who see him, and that he may be banned, broken, lost, finished, vanquished, and that he may die, and that a flame may come upon him from heaven and shivers may seize him and a fracture catch him and a rebuke burn in him. May the following verse apply to him: they shall fall and not rise, and there will be no healing to their affliction. "Their eyes will darken, so that they see not and their loins will be made by you continually to shake" (Psalm 69:24). "Let their habitation be desolate, and let none dwell in their tents" (Psalm 69:24). May the following verse apply to him: "And my wrath shall wax hot and I will kill you with the sword and your wives shall be widows and your children fatherless" (Exodus 22:23). And the following may apply to Judah, son of Nanay: "The Lord shall smite you with consumption and with a fever and with an inflammation and with an extreme burning and with the sword and with the blasting and with [mildew] and they shall pursue you until you perish" (Deuteronomy 28:22). "The Lord shall smite you in the knees and in the legs with a festering eruption that cannot be healed, from the sole of your foot to the top of your head" (Deuteronomy 28:35). "The Lord shall smite you with madness and blindness and astonishment of heart" (Deuteronomy 28:28). "And you shall eat the flesh of your sons and the flesh of your daughters" (Leviticus 26:29). The throat of Judah, son of Nanay, shall not swallow and his gullet shall not eat; choking shall fall on his palate and paralysis shall fall on this mouth and tongue. . . . The following verse will apply to him: "The nations shall see and be confounded at all their might. They shall lay their hands on their mouth, their ears shall be deaf, they shall lick the dust like a serpent, they shall move out of their holes like worms of the earth" (Micah 7:16–17). "The Lord will not spare him but then the anger of the Lord and his jealousy shall smoke against that man and [all the curses that are written in this] book [shall lie upon him] and the Lord shall blot out his name from under heaven" (Deuteronomy 29:19). So shall the name of Judah, son of Nanay, be blotted out and [his memory] shall be uprooted from the world, just as the name of [Amalek] was blotted out . . . may his members be pressed down and may there be done to him (?), judgment will come, omen and misfortune swiftly, with an inflammation, a purulence, an itch, a vermin, a blackening, a shiver, a vermin . . . a pirate and a Satan. And in the name of SHSHŔB, the angel, and in the name of Mot and Yarod and Anahid and Istar Tura and . . . the spirit who resides in the cemetery, all should lean on Judah, son of Nanay . . .

110. Egypt, Alexandria. Lead tablet measuring 13 × 14 cm. Third century C.E. The client, Iônikos, seeks to bind another man, Annianos, through appeals to Hekate, Hermes, Pluto, Kore (here identified with the Babylonian goddess Ereschigal), and diverse *voces mysticae*. The spell uses several "binding" verbs. The precise purpose of the spell is unclear, although something to do with a love affair (homosexual) seems most likely. The spell repeats the invocational "refrain" four times. *Bibl.:* J. Zündel, "Aegyptische Glossen," *RM* 19 (1864): 483–96; C. Wessely, *Ephesia Grammata* (Vienna, 1886), pp. 23–24 (no. 244); E. Kuhnert, "Feuerzauber," *RM* 49 (1894): 37ff.; DTA, p. xv; *DT* 38; C. Harrauer, *Meliouchos. Studien zur Entwicklung religiöser Vorstellungen in griechischen synkretistischen zaubertexten* (Vienna, 1987), pp. 53–58.

TH[RÊ]KISITHPHÊ AMRACHARARA ÊPHOISKÊRE . . . Receive[19] Annianos! Hermes of the underworld ARCHEDAMA PHÔCHENSE PSEUSA RERTA THOUMISON and KT and Pluto *HUESEMMIGADÔN MAARCHAMA and Kore Ereschigal* ZA[BAR]BATHOUCH and Persephone [ZA]UDACHTHOUMAR, I invoke you by the name of the earth KEUÊMORI MÔRITHARCHÔTH and Hermes of the underworld ARCHEDAMA PHÔ-CHENSE PSEUSA RERTA THOUMISON and also Pluto HUESE[MM]IGADÔN MAARCHAMA and Kore Ereschigal ZABARBATHOUCH and Persephone ZAUDACHTHOUMAR. Let Annianos lose (his) memory and remember only Iônikos! I call upon you, mistress ruler of all mankind, all-dreadful one, bursting out of the earth,[20] who also gathers up the limbs of *MELIOUCHOS and MELIOUCHOS himself, Ereschigal NEBOUTOSOUALÊTH EREBENNÊ ARKUIA NEKUI Hekate, true Hekate, come and accomplish for me this very act![21] Hermes of the underworld ARCHEDAMA PHÔCHENSE PSEUSA RERTA THOUMISON and KT and Pluto HUESEMMIG[A]DÔN MAARCHAMA and Kore Ereschigal ZABARBATHOUCH and Persephone ZAUDACH-THOUMAR and *daimones* who are in this place. Restrain for me—Iônikos—the strength (and) the power of Annianos, so that you seize him and hand (him) over to (the) ones untimely dead, so that you melt away[22] (his) body,[23] (his) sinews, (his) limbs, (his) mind, so that he is unable to proceed against Iônikos, neither to hear or see anything evil about me, but rather (let him) be subject to me, under (my) feet, until he is subjected (to me)! For the mistress ruler of all has spun these things for him. Lady *MASKELLEI MASKELLÔ PHNOUKENTABAÔ

19. The Greek verb is *paralambanein.*

20. *Rhêxichthôn.* A word peculiar to invocations: for example, *PGM* IV, line 2727 (as an attribute of Hekate in spell instructions for a love binding ritual); and *PGM* VII, line 692 (as attribute of the celestial bear, in a spell of unclear purpose).

21. Behind this appeal, according to Harrauer (p. 57), lies an allusion to an important theme from Greek mythology; Zeus decreed that Persephone, the "queen" of the underworld, could not lay exclusive claim to the handsome Adonis but would have to share him equally with Aphrodite. Thus the idea at work in the spell runs as follows: "Just as Adonis/Meliouchos is forced by Persephone (a familiar figure in spells; here also called Ereschigal) to come to Aphrodite, so she must send Annianos to Ionikos." Behind the notion of gathering up the limbs of Meliouchos there must also lie an allusion to the story of Isis who, following the dissemination of Osiris's dismembered body, traveled around and reassembled every part but his genitals.

22. The same verb, *katatêkein,* is used in a love spell preserved in *PGM* XVI, line 3. The verb without the prefix *kata-* appears in *PGM* IV, lines 2931–32; XVIIa, lines 9–10; and XIXa, line 53; all are love spells.

23. *Sarkes* = *sarkas;* the plural of *sarx* is used commonly in Greek for the body as a whole.

*OREOBAZAGRA RÊXICHTHÔN HIPPOCHTHÔN PURIPÊGANUX, *Mistress Earth, of the underworld!* MEUÊRI MORITHARCHÔTH I invoke you by your name to effect this deed and watch over this binding spell for me and do it vigorously! Hermes ARCHEDAMA PHÔCHENSE PSEUSA RERTA THOUMISON and KT and Pluto [HU]ESEMMIGADÔN MAARCHAMA and Kore Ereschigal ZABARBATHOU[CH] and Persephone ZAUDACHTHOUMAR and d[aimon]es who are in this place, roaming about, acco[mplish] this deed and restrain . . .

111. Egypt, Eshmunein (Hermopolis Magna). Ostracon; date uncertain but likely no earlier than the third or fourth century C.E. The spell invokes the Greek god Kronos either to keep Hori from speaking *to* Hatros (perhaps for legal reasons), or to keep Hori from speaking *against* Hatros. The immediate milieu of the spell was probably Christian; the combination of (Jewish/)Christian and non-Christian names and invocations illuminates the syncretism of local Egyptian Christianity in this period. *Bibl.:* F. E. Brightman in *Coptic Ostraca,* ed. W. E. Crum (London, 1902), pp. 4–5 (no. 522); U. Wilcken, "Ostraka," *Archiv für Papyrusforschung* 2 (1903): 173; E. Preuschen, review of *Coptic Ostraca,* ed. W. E. Crum, *Byzantinische Zeitschrift* 15 (1906): 642; B. Couroyer, "Le 'doigt de Dieu,' " *RB* 63 (1956): 481–95; *PGM,* vol. 2 (Ostrakon no. 1).

> Kronos[24] who restrains the anger/passion of all mankind, restrain the anger/passion of Hôri, to whom Maria[25] gave birth, and let him not speak to/against Hatros, to whom Taêsês gave birth, for I adjure you by the finger of God[26] that he should not open (his) mouth (against/to) him, because he belongs to Kronos and is subject to Kronos. Let him not speak to him, neither by night nor day nor any hour.[27]

112. Egypt, Oxyrhynchus; original location not known. Like most tablets of its kind, this one must have been buried in a grave. Lead tablet measuring 15 × 19 cm. Dated to the third century C.E. Unusually for a tablet of such a late date, there is no mention of any deity by a tradi-

24. On Kronos's function in charms and spells, see S. Eitrem, "Kronos in der Magie," in *Mélanges Bidez,* vol. 2 (Brussels, 1934), pp. 351–60.

25. Maria: a Jewish name originally, Maria(m) gained great popularity as a Christian name in Egypt by the Byzantine period.

26. Evidently a reference to the angel Orphamiel, referred to as "the Great Finger of God" in Kropp, vol. 1, p. 48 and in a Coptic ostracon edited by W. Clarysse, "A Coptic Invocation of the Angel Orphamiel," *Enchoria* 14 (1986): 155. The roots of this invocation lie in biblical traditions surrounding references to the finger of god (Exodus 31:18; Deuteronomy 9:10). Luke 11:20 ("if it is by the finger of God that I cast out demons, then the kingdom of God has come upon you") shows the currency of the invocation in first-century Judaism.

27. A symbol is drawn here—a Greek *rho* through an *omega,* a common (e.g., *PGM* VII, line 537) abbreviation for *hôra,* "hour."

tional name. Only three mysterious names are inscribed in "wing" forma-tion[28]; the formations can also read as palindromes. The spell itself is written beneath and on the two sides of these figures; the text on the two sides is written at right angles, from top to bottom on the left and from bottom to top on the right. The immediate agent must have been the spirit of the dead person in whose grave the tablet was deposited. No occasion is mentioned. The client's expressed desire to silence his ene-mies might point to a judicial setting or more generally to a personal (erotic?) dispute of almost any sort. The mix of Greek and Egyptian names suggests a social setting in a relatively small community where Greeks and native Egyptians encountered one another regularly. All of the named parties are men. *Bibl.:* O. Guérard, "Deux textes magiques du musée du Caire," in *Mélanges Maspero*, vol. 2 (Paris, 1935–1937), pp. 206–12; SGD 154.

> Silence Chichoeis, to whom Tachoeis gave birth, in the presence of Hêraklios,[29] to whom Hêrakleia gave birth, and in the presence of Hermias, to whom Didumê gave birth. Let them hate Chichoeis. Let Hermias, to whom Didumê gave birth, hate Chichoeis, to whom Tachoeis gave birth. Silence Chichoeis himself in the presence of Hêraklios, to whom Hêrakleia gave birth. Let them hate him with a great hatred and let them not wish to see him. . . . Silence Chichoeis himself, today, this very hour, now, now, quickly, quickly.

113. Egypt; from the Cairo Geniza (for details, see p. 107). Two binding spells from scattered pages of a recipe book. There is no specific occa-sion given for their use, but they fall generally into the category of personal enmity. *Bibl.:* Naveh and Shaked, pp. 231–36 (Geniza 6).

> (*Page 3*) Another (spell). There should be written on a sheet of lead and buried in the house which you desire.[30] This is what should be written: "This writing is designated for X son/daughter of Y, so that he may melt and drip

---

28. The two outside figures form right-angle triangles, called *klimata* in *PGM* (cf., e.g., I, lines 10ff.). The central figure forms an isosceles triangle, called "heart" (*kardia*) and grape cluster (*botrus*) in *PGM* (cf., e.g., IV, line 12 and III, lines 69–70). Together the three triangles form a square.

29. Who were Heraklios and Hermias, such that the client of our tablet wanted Chichoeis to be unable to speak in their presence? Were they local governmental authorities before whom the client feared that Chichoeis might reveal damaging testimony? *DT* 139 (a lead tablet in Latin, from Rome, and dated to the time of Augustus) is quite similar to ours; it wishes that Rhodina, a woman, may be unable to speak and converse in the presence of Licinius Faustus, and it goes on to wish that she may always be hateful to him. Based on these rather close parallels, Guérard suggests that the proper translation for *murikoun* in our text might be "to make as mute as a dead person."

30. That is, the house belonging to the desired partner.

and groan and be cast down on a sickbed. In the name of 'W'W NWQ'K QDYTK 'PLWQ 'W'W KYT'WN WŠ'QSW ŠMW.[31]

(*Page 4*) You holy letters, cause there to fall on X son/daughter of Y, fire and fever and groaning and may he be cast down on a sickbed and may he have no healing for as long as I desire. A(men) A(men) S(elah) H(allelujah)

Another (spell). There should be written on an unbaked potsherd and buried in a furnace/bath: "The fire shall ever be burning against X son/daughter of Y; it shall never go out. For fire is kindled by my anger against X son/daughter of Y; it burns to the depths of Sheol; it devours earth and its harvest and it glows at X son/daughter of Y. Behold, the hand of the Lord is upon X son/daughter of Y. A(men) A(men) S(elah) H(allelujah)."

114. Egypt, from the Jewish collection of spells and recipes, *Sepher ha-Razim* (see p. 106 for details).[32]

If you wish to give your enemy trouble in sleeping,[33] take the head of a black dog that never saw light during its days and take a *lamella* from a strip of (lead) pipe from an aqueduct, and write upon it (the names of) these angels and say thus:

I hand over to you, angels of disquiet who stand upon the fourth step, the life and the soul and the spirit of N son of N so that you may tie him in chains of iron and bind him to a bronze yoke. Do not give him sleep, nor slumber, nor drowsiness to his eyelids; let him weep and cry like a woman at childbirth, and do not permit any (other) man to release him (from this spell).

Write thus and put (the inscribed lead) in the mouth of the dog's head[34] and put wax on its mouth and seal it with a ring which has a lion (engraved) on it. Then go and conceal it behind his house or in a place he frequents. If you wish to release him[35] (take the dog's head) away from where it is concealed and remove its seal and withdraw the text and throw it into a fire, and he will fall asleep at once. Do this with humility[36] and you will succeed.

115. Egypt; original location not known. Lead tablet measuring 19 × 23 cm.; originally folded and pierced by two nails. Dated to the fourth or

---

31. A typical set of mysterious words used to address the deity by his or her hidden name.

32. This passage is from the second firmament, lines 62ff.; *Sepher ha-Razim*, p. 49.

33. Similar spells to induce insomnia occur in *PGM* IV, lines 3255–74, and *PGM* LII, lines 20–26. A love spell designed to produce insomnia, also involving the use of a dog (the dog is to be made from dough or wax—it is a model), appears in *PGM* IV, lines 2943–66.

34. So also in *PGM* XXXVI, lines 370–71, where the love spell is to be placed in the mouth of a dead dog.

35. An example of a concern with the other uses of binding curses, namely, unbinding or releasing.

36. An interesting insight into the psychological preconditions prescribed for the effectiveness of these spells.

fifth century C.E. The tablet concerns an unspecified dispute between two men, Origen and Paomis. The spell is a *thumokatochon* ("restrains anger"), a type well attested among the recipes of *PGM* (e.g., IV, line 467). The deity is addressed as Brimô, a common epithet associated with various female figures, among them Artemis, Hekate, Selene, and Persephone. The layout of the tablet is unusually elaborate, even elegant (Figure 25). The text of the spell is laid out in three distinct sections: the initial invocation to the EULAMÔ rectangle—the command, "Restrain . . . ," appears once here; the lengthy palindrome in the form of a grape cluster—the command appears twice here, on either side of the triangle, near the bottom; and the final section, which reaches from BELIAS to the end and includes the only information regarding the nature of the dispute. The tablet was originally deposited in a grave. *Bibl.:* P. Collart, "Une nouvelle *tabella defixionis* d'Egypte," *Revue de philologie* 56 (1930): 249–56; SGD 162.

I invoke you, spirit of the dead person, whoever you are, by Lady Brimô, PROKUNÊTE NUKTODROMA BIASANDRA KALESANDRA KATANIKANDRA[37] LAKI LAKI-MOU *MASKELLI MASKELLÔ PHNOUKENTABAÔTH OREOBAZAGRA RÊXITHÔN HIP-POCHTHÔN PRUIPÊGANUX. Restrain the anger, the wrath of Paômios, to whom Tisatis gave birth. Now, now. Quickly, quickly.

(design)

EULAMÔ

ULAMÔE

LAMÔEU

AMÔEUL

MÔEULA

ÔEULAM

Yes, Lord. SISISRÔ SISIPHERMOU CHNOUÔR ABRASAX PHNOUNOBOÊL

OCHLOBAZARÔ

By the holy	ERÊKISITHPHÊRARACHARARAÊPHTHISIKÊRE
name IÔ BEZEBUTH	RÊKISITHPHÊRARACHARARAÊPHTHISIKÊR
BUTHIEZEU	ÊKISITHPHÊRARACHARARAÊPHTHISIKÊ
IÔ BARIAMBÔ	KISITHPHÊRARACHARARAÊPHTHISIK COPLOMURTILIPLÊX
MERMERIOU ABRASAX	ISITHPHÊRARACHARARAÊPHTHISI EXANAKERÔNITHA
EULAMÔ	SIKÊREARACHARARAÊPHTHIS LAMPSAMERÔ
EULAM ÔMALUE	IKÊREARARACHARA<RA>ÊPHTHI LAMPSAMAZÔN
EULA MALU EULAMÔ	KÊREARARACHARARAÊPHTH BASUMIAÔ
EUL AL ULAMÔ	PHÊRARACHARARAÊPH OPLOMURTILOPLÊX

---

37. This series of epithets or *voces mysticae*, with the name Brimo, appears in virtually identical form in *PGM* VII, line 696; it appears also in Wortmann, no. 1, line 52, as corrected by D. R. Jordan, "Love Charm," *ZPE* 72 (1988): 255.

FIGURE 25. *Defixio* from Egypt with elaborate layout, including *voces mysticae* in squares and wing formation; the spell appears only in the final lines in the bottom third of the tablet. Along the vertical line near the top of the tablet, several *charaktêres* have been inscribed.

EU	A	LAMÔ	ÊARARACHARARAÊ ANACHAZA
E		AMÔ	ARARACHARARA EXANAKERÔNITHA
		MÔ	RARACHARARA ANAXARNAXA
		Ô	ARACHARARR KERASPHAKERÔNAS

Restrain the anger, the ACHARA PHAMETATHASMAXARANA
wrath of Paômios, to RCHAR BASUMIAÔIAKINTHOU
whom
Tisatis gave birth.   Restrain the anger,
Now, now.   the wrath of Paômios,
Quickly, quickly.   to whom Tisatis gave
                                             birth. Now,
                                             now. Quickly, quickly.

BELIAS BELIÔAS AROUÊOU AROUÊL CHMOUCH CHMOUCH bind, bind up the anger, the wrath of Paômios, to whom Tisatis gave birth. Because I invoke you, the great bodylike bodiless one, who draws down the light, lord of the first creation, IAÔÊIÔIAIEOUIABOR. SABAOTH. LENTAMAOUTH ERÊKISITHPHÊARARA-CHARARAÊPHTHISIKÊRE[38] IÔ BEZEBUTH MERMERIOU ABRASAX IAÊIAÊE. Restrain the anger, the wrath of Paômios, to whom Tisatis gave birth, the mind, the wits, so that he may not speak against me (Origen) to whom Ioullê gave birth, who is also known as Theodôra. But let him be obedient to us. Now, now. Quickly, quickly.

116. Sicily; Messina. Discovered in a closed grave where it had been inserted through a pipe (Figure 26). Lead tablet measuring 16×2.6 cm.; written on both sides. Jordan suggests a date in the second century C.E. No deity is addressed and no verb of binding or cursing is used. The text consists exclusively of the name of the target, a woman, in the accusative case, that is, as the object of an implied verb, and several insulting epithets. No specific occasion is mentioned. At the very least, the tablet indicates strong feelings of animosity toward Arsinoê. *Bibl.:* P. Orsi, "Messana," *Monumenti Antichi* 24 (1916): 167–69 (with drawing and notes by D. Comparetti); *SEG* 4.47; SGD 114; H. Versel, "Beyond Cursing," in *Magika*, p. 65.

(*Side A*) (I bind?) Valeria Arsinoê, the bitch,[39] the dung worm, the criminal[40] and useless Arsinoê.

38. The same palindrome, with accompanying words, as previously presented.
39. The Greek word *skuza* is used to describe a dog in heat and thus came to be used as a term of abuse against women, just as with "bitch" in colloquial English.
40. The Greek word is *hamartôlos*, used widely for "sinner" in Jewish and Christian texts. Here it means something like "wrongdoer."

FIGURE 26. Drawing of tomb with pipe used to insert *defixio* into grave site; such offering pipes were common in Greek graves of the Roman period.

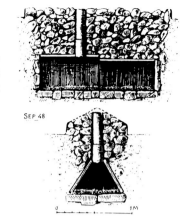

SEP 48

Fig. 33.

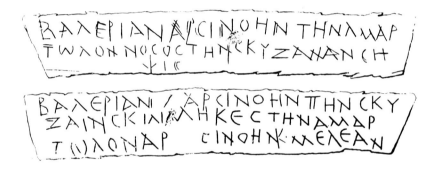

(*Side B*) (I bind?) Valeria Arsinoê, the criminal, sickness, the bitch, putrefaction.[41]

117. Morgantina, Sicily. Found in association with a well-altar of an early (pre-325 B.C.E.) sanctuary. Four of nine lead tablets, ranging from 4.6 cm. wide to 10.5 cm. high. The tablets date from the first century B.C.E. The key verb, *potidexesthe* ("admit, i.e., to Hades") allows the spells to be read either as death spells or as mortuary "aids," to ensure the subject's admission to a proper afterlife. The series of tablets is notable for several reasons: they were all deposited in a formal chthonic

41. "Sickness" (*nosos*) and "putrefaction" (*sêpsis*) appear in the nominative, that is, not as objects of a verb or as epithets of Valeria. They must be subjects of an implied verb; Comparetti suggests "May sickness and putrefaction lead you away!" The notion of rot, decay, or putrefaction appears also in *DT* 190.

sanctuary; the verbs in A, B, and D all belong to the traditional language of mortuary prayer, thus placing the spells on the border with cultic-liturgical modes of divine supplication; and all four tablets concern the same person, a slave girl named Venousta. *Bibl.:* N. Nabers, "Lead *Tabellae* from Morgantina," *AJA* 70 (1966): 67–68; *BE* 5 (1966): 381; N. Nabers, "Ten Lead *Tabellae* from Morgantina," *AJA* 83 (1979): 463–64, pl. 65; *SEG* 29.927–935; Jordan, TILT; SGD, pp. 179–80; Faraone, "Context," pp. 18–19; illustration of tablet A in William V. Harris, *Ancient Literacy* (Cambridge, Mass., 1989), fig. 2 (opp. p. 146).

(A) Gaia, Hermes, subchthonic gods, admit[42] Venousta the servant girl of Rouphos.

(B) Gaia and Hermes and subchthonic gods, admit Venousta the slave girl of Rouphos.

(C) Gaia, Hermes, subchthonic gods, take away Venousta the ser[vant girl[43]] of Rouphos.

(D) Gaia, [He]rmes, su[b]chthonic g[o]ds, admit [Ve]nous[t]a th[e] servant girl of Sex[tos].[44]

118. Italy, Puteoli (near Rome). Tablet measuring 11 × 13 cm.; dated to the second or third century C.E. On the top right hand side appear the name(s) of a deity; these names may refer to the figures on the right. Further below, on the left and next to the main inscription, are the words "the holy name" on two separate lines. Although the tablet is in Greek, the names are Latin. *Bibl.:* C. Hülsen, "Bleitafel mit Verwünschungsformeln," *Archäologische Zeitung* 39 (1881): 309–12; Wünsch, *Antike Fluchtafeln,* no. 2; *DT* 208.

SEÔTHE
SABAÔTH
SABAÔTH

IAO ÊL MICHAÊL NEPHTHÔ.[45]

the
holy
name   May Gaios Stalkios Leiberarios, to whom Philista gave birth, become

---

42. The verb, *potidexesthe,* a Doric form of *prosdexesthe,* is rare among *defixiones.* Verbs of "receiving" were normally employed in epitaphs to ensure the soul's easy admittance to Hades. Thus Nabers originally suggested that our tablets were not "curses directed against living persons, but pious prayers offered to the underworld gods on behalf of persons already dead at the time the tablets were written" ("Lead *Tabellae,*" p. 67).

43. On this reconstruction, see the suggestion of L. Koenen in Faraone, "Context," p. 31, n. 83.

44. Another owner, but presumably the same slave girl and the same "binding" formula.

45. The Egyptian goddess Nephthys.

an enemy of [or be hated by] Lollia Roupheina, may he become an enemy of Haplos, may he become an enemy of Eutuchos, may he become an enemy of Celer, may he become an enemy of Rouphos, may he become an enemy of the entire household of Rouphina, may he become an enemy of Polubios, may he become an enemy of Amômis (a woman), may he become an enemy of Thêbê . . .

7

Antidotes and
Counterspells

In his account of the Jewish uprising against the foreign dynasty of the Greek Seleucids, the pious author of 2 Maccabees relates the following episode (12:34–39): Judas, surnamed Maccabeus ("the hammerer"), lost a number of his men in battle; on the following day, when Judas went out to recover their bodies, he discovered that every fallen soldier had been wearing an amulet ("sacred tokens of the idols of Jamnia"), which, the author notes in a sanctimonious aside, "the Law forbids the Jews to wear." Whether Judas thought to check for similar amulets among the survivors, the author does not trouble to say, for the message is clear—the dead had fallen because of their forbidden compromise with heathen beliefs and practices. But the great likelihood is that the survivors, too, had fortified themselves against "anything harmful" by putting on their engraved stones or their inscribed sheets of metal and papyrus. To be sure, 2 Maccabees does not offer the sort of hard demographic data preferred by modern social scientists, but the fact remains that in this randomly chosen sample of ancient Jews, every one wore an amulet, as did virtually every sensible person of the time.

We would be committing a serious methodological error if we failed to balance our presentation of curse tablets and binding spells with a brief look at the repertoire of available countermeasures—phylacteries, antidotes, and amulets, as they are variously called. The result of such an omission would be an unacceptably, even unimaginably paranoid culture whose inhabitants would have been conscious not just of being under constant assault from *defixiones* commissioned by personal enemies known and unknown but, worse still, feeling utterly defenseless and without recourse.[1] At another level, such a mistake would lend unwelcome support to the widespread view that the use of spells and counterspells represented nothing more than a grab bag of traditional techniques, ineffective, limited to the "superstitious" lower classes, and without further significance for our understanding of the ancient world at large. Such has certainly been the predominant view. But once we complete our dynamic system by introducing amulets and counterspells, we not only undercut this view but at the same time open up the way to appreciating these techniques—in the words of Peter Brown—as elements "in the way in which men have frequently attempted to conceptualize their social relationships and to relate themselves to the problem of evil."[2] More particularly, we begin to see in them a strategy for dealing with social aggression and personal failure. In short, *defixiones* and amulets emerge, when seen together, as far more than simple coping mechanisms. They behave instead rather like traces in an X-ray, to borrow another phrase from Brown,[3] which point to powerful but invisible systems operating beneath the surface. For, as Hildred Geertz has argued, such practices make sense only "within the framework of a historically particular view of the nature of reality, a culturally unique image of the way in which the universe works . . . a hidden conceptual foundation for all of the specific diagnoses, prescriptions and recipes."[4] Thus, she continues, the historian must recognize "the fact that a particular notion is set within a general pattern of cultural concepts, a conventional cognitive map, in terms of which thinking and willing, being anxious and wishing, are carried out."[5]

Amulets were called *periapta* and *periammata*—"things tied around" parts of the body, usually the neck, an arm, or a leg.[6] These objects might be simple pieces of string; colorful embroidered bands; engraved stones and rings; or strips of metal, papyrus, and other materials inscribed with special formulas, then rolled up or folded and carried about on a string, in a pouch or in tubular containers. Thousands of such stones and strips have survived,[7] along with a considerable literature on their uses and preparation.[8] As the selected items below make abundantly clear, amulets were prepared and sold by specialists who produced them according to traditional recipes and consecrated them through ritual acts, thereby endowing them with effective power. C. Bonner cites a text of the Neoplatonic philosopher, Hermeias (sixth century C.E.) that illustrates the essential part of the process:

> We have explained how the soul is inspired. But how is an image inspired? The thing itself cannot respond to the divine, since it is lifeless; but the art of consecration purifies its matter and, by attracting certain marks and symbols to the image, first gives it a soul by these means and makes it capable of receiving a kind of life from the universe, thereby preparing it to receive illumination from the divine."[9]

Not every ordinary user would have been able to recite this theory, but something like it must have served as the cognitive map that gave amulets their power. Certainly, we find many of these "marks and symbols"—signs, figures, and *charaktêres*—on both *defixiones* and amulets. Together, the acts of consecration, engraving, or inscribing were taken as proof that amulets possessed special virtues of protection.

In practical terms, the purposes of amulets were rather simple. Most served to shield the bearer from all forms of harm and danger. By extension, that is by virtue of their ability to ward off unforseen disasters, some also guaranteed success and prosperity.[10] And not surprisingly, in a culture without aspirin or antibiotics, many aimed at protecting against common medical problems—digestive disorders, fevers, eye problems, scorpion bites, and various gynecological disorders.[11] In a few cases, we find amulets that functioned exactly like *defixiones:* love stones designed to attract or separate lovers[12]; a *thumokatochon* intended to subdue the anger of a personal enemy[13]; and several aggressive stones designed to harm or kill an opponent.[14] At least one stone appears to have been employed specifically to ward off the powers of *defixiones.*[15] Overall, it appears that amulets, like the bowls from Mesopotamia, originally served a single purpose—to protect the owner from a wide range of known and unknown evils; however, across time their uses expanded to cover other needs, so that the boundary lines between bowls, amulets, and *defixiones* gradually disappeared.

But who made use of amulets? In raising the question, we come up against a major issue regarding our general understanding of late antique culture. Do amulets—and their attendant beliefs and assumptions about how the world works—represent a basic and universal feature of that culture or just "an unswept corner of odd beliefs, surrounding unsavoury practices?"[16] For an answer we may turn to Pliny the Elder and recall his observation that "there is no one who is not afraid of curses and binding spells."[17] If this is so, we have our answer—everyone used amulets. What is more, given the conventional cognitive map of that world, it would have been foolish and unreasonable to behave otherwise. A list of oracular questions from Egypt (third or fourth century C.E.) may indicate just how widespread was the need for protection.[18] Along with a list of possible queries about the seeker's current circumstances ("Is my property to be sold at auction?" "Shall I become a city councillor?") appears the following question: "Am I under a spell (*pepharmakômai*)?"

To be sure, we hear occasional voices of protest and dissent. But they are just that, protests against universal assumptions and practices.

Among Christians, Gregory of Nazianzen insisted, in a baptismal sermon, that his flock had no need of amulets and spells, but most of his listeners probably remained unconvinced. Among Jews, the author of 2 Maccabees may have felt that amulets belonged to the forbidden practices of the wicked Amorites, but soldiers on the front line still took the necessary precautions, while the supposedly "rigorist" Rabbis not only allowed but prescribed amulets for a variety of purposes right down to the present day.[19] Among philosophers, Plotinus seems to presuppose that spells worked on all souls, except those of the wise, but he is a notable exception among his peers.[20] Far more representative of late antique philosophers was Iamblichus, whose theurgical treatise, *On the Mysteries of Egypt,* shows just how respectable such matters had become. Finally, among physicians, even the most empirical found themselves unable, or unwilling, to break entirely with a system that seemed to work. The noted physician Galen, for instance, prescribed the use of amulets, even while denying traditional explanations for their success.[21]

Galen's dilemma regarding amulets—they seem to be effective, even though his empirical medical theory allowed no room for them—prompts us to ask whether amulets *really* worked.[22] Before answering the question too hastily, and thereby falling once again into the Frazerian trap, we need to rephrase it slightly, for it is impossible to answer the question at all unless we know with greater precision what amulets were expected to do. What we need to ask is, Against whom or what did amulets offer protection? If we take our cue from Tambiah's observations that the audience of a ritual is identical with its performer[23]—in our case, the client—and if additionally we read the external threats and dangers as projections, at least in part, of the client's own internal condition, we find ourselves asking what it is in the *client's* life that the amulet represents and protects. Viewed in this setting, the answer to our reformulated question would have to include the following components: (1) the amulet itself, as a concrete physical object, shows forth the wearer's embeddedness in a concrete social system of exchanges between human actors; (2) as a protective device, the amulet points to an awareness that all social systems depend on an active yet invisible network of feelings, beliefs, and attitudes, whose particular feature here emerges as aggressiveness, hostility, and unpredictability; (3) from a dynamic perspective, these negative forces can add up to so many reasons not to act, to withdraw into the safety of solitude, to remain frozen; and (4) the protective function of amulets can thus be seen to embody a counterstrategy of individual action, undergirded by feelings of self-confidence, optimism, and the ability to formulate and achieve goals.

The Freudian analyst and anthropologist, Geza Roheim, has framed a similar explanation in his provocative reformulation of Freud's essentially negative view of "magic."[24] Beginning with a redefinition of magic (the term is Roheim's) as "the counterphobic attitude, the transition from passivity . . . (and) probably the basic element in thought and the initial phase of any activity,"[25] he concludes that we must postulate a third or *magical principle* that deals with the world as if it were governed by our wishes or desires or emotions."[26] Seen in this light, the amulet becomes the physical token of Roheim's "counterphobic attitude," the belief that we can achieve something despite all of the evidence to the contrary. In Freudian terms, the fearsome world of spirits against which the amulet offers protection represents the inhibiting, pessimistic, internal voice of the superego, urging caution and warning of failure, while the amulet itself manifests the countermove of the ego, the transition from passivity to activity. Perhaps this is what Malinowski had in mind when he spoke of "magic" as "the embodiment of the sublime folly of hope, which has yet been the best school of man's character."[27]

This brief excursion into Freudian theory leads us finally to the gods, spirits, and *daimones* invoked as protective agents in the amulets. Our excursion, and perhaps common sense as well, would lead us to expect a somewhat different set of agents from those invoked in the *defixiones,* inasmuch as the purpose of amulets was to overpower the force and, by implication, the agents of the *defixio.* As we shall see, this intuition proves largely correct, even taking account of differences due to different locales and changing tastes.[28] By and large, *defixiones* call upon chthonic figures and spirits of the dead, whereas amulets tend to invoke composite figures with solar connections (for example, the snake-legged god with cock's head), unmistakably solar deities, and, to a significantly greater degree, Egyptian gods. Some figures show up on both sides—for example, Hermes and Hekate appear on amulets—whereas the omnipresent Demeter on *defixiones* is virtually absent from the amulets. Elements common to both include the following: Jewish terms (for example, IAO and SABAOTH); *charaktêres* and vowel series; many of the common *voces mysticae;* and certain designs such as the headless *daimon* and trussed mummies.

Notes

1. Thus Preisendanz (1972), pp. 6–7, is wrong in asserting that it was generally not possible to counteract curses and binding spells. *PGM* IV, line 2177, offers

a spell to "break (the power of) curse tablets," while at a much earlier time (fifth century B.C.E.) a text of Magnes, a comic dramatist, indicates that there existed specialists (*analutai*) who offered to dissolve (*analuein*) spells directed against their clients; on Magnes, see *CAF*, vol. 1, p. 8, with additional references.

2. P. Brown, "Sorcery, Demons, and the Rise of Christianity," in *Religion and Society in the Age of Saint Augustine* (London, 1972), p. 120. Brown himself gives credit for the idea to E. E. Evans-Pritchard, in his book, *Witchcraft, Oracles and Magic among the Azande* (Oxford, 1976).

3. Brown, "Sorcery," p. 128.

4. "An Anthropology of Religion and Magic," *Journal of Interdisciplinary History* 6 (1975): 83. Geertz's remarks appear in her critical review of K. Thomas, *Religion and the Decline of Magic* (New York, 1971).

5. Geertz, "Religion and Magic," p. 84.

6. In general on these matters, see Bonner, *Amulets*, pp. 1–21, and more recently R. Kotansky, "Incantations and Prayers on Inscribed Greek Amulets," in *Magika*, pp. 107–37.

7. On stones, in addition to Bonner, see A. Delatte and P. Derchain, *Les intailles magiques gréco-égyptiennes* (Paris, 1964); and Hanna Philipp, *Mira et Magica* (Mainz, 1986).

8. On the numerous books on the preparation of amulets, using stones and plants, see A. Delatte, *Herbarius: Recherches sur le cérémonial usité chez les anciens pour la cuillette des simples et des plantes magiques*, 3d ed. (Paris, 1961); R. Halleux and J. Schamp, eds., *Les lapidaires grecs* (Paris, 1985); M. Waegeman, *Amulet and Alphabet: Magical Amulets in the First Book of Cyranides* (Amsterdam, 1987).

9. The comment appears in his commentary on Plato's *Phaedrus;* see the discussion in Bonner, *Amulets*, p. 16.

10. Bonner, *Amulets*, lists several examples, among them nos. 234–35 which read, "Be gracious to me and my children" and "Be gracious to me and my property."

11. See the numerous examples in Bonner, *Amulets*, pp. 51–94. In addition, there are full discussions of amulets and medicine in *Symposium on Byzantine Medicine*, ed. J. Scarborough (Washington, D.C., 1985).

12. Bonner, *Amulets*, nos. 150 ("Separate Hierakion . . . from Serenilla") and 156 ("Bring Achillas . . . to Dionysias").

13. Ibid., no. 149.

14. Ibid., no. 151, with commentary on pp. 108–10.

15. Ibid., no. 156, with commentary on pp. 116–17.

16. Brown, "Sorcery," p. 120.

17. See p. 253.

18. Papyrus Oxyrhynchus 1477.

19. See the discussions in J. Trachtenberg, *Jewish Magic and Superstition* (Cleveland, 1939), pp. 132–52; and S. Lieberman, *Greek in Jewish Palestine* (New York, 1942), pp. 100ff.

20. See p. 259.

21. See the discussion in L. Thorndike, *History of Magic and Experimental Science*, vol. 1: *The First Thirteen Centuries* (New York, 1923), pp. 172–74.

22. In the tradition that reaches from Sir James Frazer to Keith Thomas, the obvious answer has always been that such things do not work. Recently this assumption has come under attack and the old answer no longer seems self-evident (see pp. 21–24). The work of Claude Lévi-Strauss has been particularly influential in this shift of focus.

23. See Stanley J. Tambiah, "The Magical Power of Words," *Man* 3 (1968): 210–11.

24. G. Roheim, "The Origin and Function of Magic," in *Magic and Schizophrenia* (Bloomington, 1955), pp. 3–85.

25. Ibid., p. 3.

26. Ibid., p. 83.

27. B. Malinowski, "Magic, Science and Religion," in *Magic, Science and Religion* (Garden City, N.Y., 1954), p. 90.

28. Pliny, for instance, notes that in his time (first century C.E.) men were beginning to wear amulets decorated with images of Egyptian gods (*Natural History* 33.41).

119. Dalmatia, Tragurium (near modern Split, Yugoslavia). Lead tablet with two holes on the left edge, measuring 10 × 12 cm. The date is the sixth century C.E. Someone, perhaps the owner of the amulet, has added to the text in places. Greek formulas underlie some of the Latin text. The figure of Christ is invoked. The text begins with one cross and ends with three. Although the language of the text recalls formulas common to *defixiones*, this tablet was meant to be worn as a protective amulet. The apotropaic qualities of the amulet may have been sealed by dipping it in the waters of the Jordan, the river in which John had baptized Jesus.[1] In Christian art and literature, this event gave to the river and its waters a power to protect Christians against sin and evil. As was common in ancient spells, the power of the amulet is further guaranteed by associating it with three brief narratives (*historiolae*): the angel Gabriel's act of binding the "foul spirit of Tartarus"; Christ's successful banning of

---

1. On the apotropaic powers of Jordan water in Christian tradition, see Wünsch, *Antike Fluchtafeln*, pp. 29–30.

the spirit to unspecified hilly places where it could cause no harm to those who invoked the name of Christ; and finally, a confession from the spirit's own mouth[2] to the effect that it could not cross the Jordan, here seen as the boundary between this world and the fires of hell. *Bibl.:* Wünsch, *Antike Fluchtafeln*, no. 7 (with commentary); *CIL* III, p. 961.

(*Side A*) †In the name of the Lord, Jesus Christ, I denounce you, most foul spirit of Tartarus, whom the angel Gabriel bound with burning fetters,[3] (you) who hold ten thousand barbarians (?), came to Galilee after the resurrection. There (Christ) commanded that you be kept in the hilly, mountainous wild places[4] and only from that hour on to be invoked without difficulty. Therefore see, most foul spirit of Tartarus, that wherever you hear the name of the Lord or recognize the scripture, you are not able

(*Side B*) to harm when you wish. In vain you hold the Jordan River which you have not been able to cross. When asked how you are not able to cross, you said, because it runs there to the fire (which comes) from fiery hell, and for you everywhere and always, may it run to the fire from fiery hell. I denounce you through my Lord. Beware! †††

120. Amisus in Pontus, Asia Minor; found in a grave, with the silver tablet rolled-up inside a small bronze case. The tablet measures 4.5 × 6.7 cm. and is inscribed in Greek. Date uncertain; probably first century C.E. This is an amulet to ward off potential harm from enemies. One interesting feature of this tablet is the last line on Side A where the name of Moses is mentioned. The debate over the possible Jewish origin of Rufina, the owner of the amulet, remains inconclusive. *Bibl.:* S. Petrides, "Amulette judéo-grecque," *Echos d'Orient* 8 (1905): 88–90; R. Wünsch, "Deisidaimoniaka," *ARW* 12 (1909): 24–32, no. 4; J. G. Gager, *Moses in Greco-Roman Paganism* (Nashville, 1972), pp. 157–59.

I am[5] the great one who is sitting in heaven, the wandering hollow of the cosmos ARSENONEOPHRIS,[6] the safe name MIARSAU as the true *daimon* BARICHAA KMÊPHI[7] who is the ruler of the kingdoms of the gods *ABRIAÔTH ALARPHÔTHO *SÊTH. Never let evil appear. Drive away, drive away the curse[8]

2. Such confessions are a common feature in the *Testament of Solomon*.
3. On Gabriel's role as an angelic agent in spells, see no. 123.
4. Above the line, "where no man in[vades]" has been written.
5. "I am" or *anok* in Coptic, followed by the name of a deity, is a familiar formula in spells (e.g., *PGM* III, line 418). The formula asserts the great authority of the spell by claiming that it emanates directly from the named deity.
6. From the name *OSORONOPHRIS.
7. Kmeph is another name for Chnum from Elephantine, the Egyptian outpost on the Nile near Nubia. It is also an epithet of Osiris; see *PGM* III, line 142; VII, line 584.
8. The Greek is *hupothesis*. Here it means something laid down, perhaps in reference to the deposition of a curse tablet; see Wünsch, pp. 29–30.

from Rouphina; and if someone does me an injustice, revert (the curse) back to him. Nor let poison harm me. King of kings ABRIAÔN[9] TÔ ORTHIARÊ.[10] I am the one ruling the place[11] in Moses' name.

121. Greece, Beroea; thin silver sheet measuring 5.8 × 7.3 cm., fully preserved. The amulet was discovered rolled up and still within the bronze tube in which it was worn on the owner's body (Figure 27). Of the eleven lines, only four contain Greek; the others, as in most amulets of this kind, consist of mysterious words and names. The protective spell is addressed to "Lord Angels." *Bibl.:* D. M. Robinson, "A Magical Text from Beroea in Macedonia," in *Classical and Mediaeval Studies in Honor of Edward Kennard Rand,* ed. L. W. Jones (New York, 1938), pp. 243–53.

*ANOCH AI *AKRAMMACHAMARI *BARBATHIAÔTH LAMPSOUÊR LAMÊÊR LAMPHORÊ[12] *IAÔ *ABLANATHANALBA, Lord Angels, preserve Euphêlêtos, to whom Atalantê gave birth.

122. Mesopotamia; precise location not known. Of the many bowls like this one (more than seventy-two in Jewish Aramaic, thirty-three in Mandaic, and twenty-one in Syriac),[13] all seem to have been produced in Mesopotamia and Iran. Although the professionals who prepared them were probably Jews, it is not necessary to assume that all of the clients were Jews.[14] The texts reveal Jewish elements alongside themes reflecting native Syrian and Persian polytheism. Like the present bowl, most served as amulets, to ward off hostile spirits, to overturn binding spells, and to protect houses.[15] Often they include vivid illustrations on the

9. Probably a variant of *ABRIAOTH.

10. Wünsch, p. 31, suggests that this name comes from *Ares orthios*—"right-standing Ares," the god of war.

11. The Greek word *topos* is a common Jewish way of rendering the divine name (see Philo, *Dreams* 1.63).

12. The editor argues that the three preceding words are typical variants of the same name and provides a lengthy survey of its occurrences in a wide variety of spells (Robinson, pp. 249–50). He relates them to "some solar deity."

13. For a recent history of scholarship on these so-called magic bowls, see Naveh and Shaked, pp. 19–21. The most important publications are: J. A. Montgomery, *Aramaic Incantation Texts from Nippur* (Philadelphia, 1913); E. Yamauchi, *Mandaic Incantation Texts* (New Haven, 1967); V. P. Hamilton, "Syriac Incantation Bowls" (Ph.D. diss., Brandeis University, 1971); C. D. Isbell, *Corpus of Aramaic Incantation Bowls* (Missoula, Mont., 1975).

14. So Naveh and Shaked, pp. 17–18.

15. See, for example, Bowl 8 in Naveh and Shaked, pp. 173–75: "Bound are the demons, sealed are the devs, bound are the idol spirits, sealed are the liliths, male and female, bound is the evil eye . . ." For a thorough discussion of reversing and releasing spells and curses, see Levine, pp. 368–71. In the *Testament of Solomon* (7:5), a widely circulated compendium of spells and antidotes, an "evil" spirit (Lix Tetrax) says the following: "If I get the chance, I slither in under the corners of houses during night or day."

A.

B.

FIGURE 27. Protective amulet (above) on thin sheet of silver, unrolled and removed from bronze tube (below) where it was originally carried, probably around the neck of its owner.

FIGURE 28. Drawings of bound female figures from inscribed bowls found at Nippur (Mesopotamia). The figures represent the bound female spirits or *daimones,* often called Liliths in Aramaic texts, who were believed to cause troubles for both men and women.

bottom of the bowl (Figure 28). Many were unearthed in an inverted position; some were found as pairs, attached at the rim with bitumen to form a closed container. Theories regarding their use include the following: they served as demon-traps; they were to be filled with liquid, which the owner would then drink; the overturning of the bowl symbolized the overturning of the spirits.[16] The text (thirteen lines) begins at the bottom and spirals outward toward the rim. The language is Syriac. Naveh and Shaked cite four other bowls with virtually identical texts. From the numerous and extensive textual parallels, even on bowls in different languages, it is apparent that there existed written recipes which the professionals followed in preparing bowls for their clients. The date falls somewhere between the fourth and seventh centuries c.e. *Bibl.:* J. A. Montgomery, *Aramaic Incantation Texts from Nippur* (Philadelphia, 1913), pp. 242–43, no. 37; B. A. Levine, "The Language of the Magical Bowls," appendix to *A History of the Jews in Babylonia,* vol. 5, by J. Neusner (Leiden, 1970), pp. 343–75; T. Harviainen, *A Syriac Incanta-*

16. See the thorough treatment, with examples, in Montgomery, pp. 40–45, and the brief discussion in Naveh and Shaked, pp. 15–16.

tion Bowl in the Finnish National Museum (Helsinki, 1978); Naveh and Shaked, pp. 124–32 (Bowl 1).

> This bowl is designated for sealing and guarding the house, the dwelling and the body of Ḥuna son of Kupitay, so that tormentors, bad dreams, curses, vows, spells, magic practices, devils, demons, liliths,[17] encroachments, and terrors should leave him alone. The secret (amulet)[18] of heaven is buried in heaven, and the secret (amulet) of earth is buried in the earth. I speak the secret (amulet) of this house against all that is in it: against devils, demons, spells, magic practices, all the messengers of idolatry, all troops, charms, goddesses, all the mighty devils, all the mighty Satans, all the mighty liliths. I tell you this decree. He who accepts it, finds goodness; but for him who is bad (and) does not accept the mysterious words, angels of wrath will come against him, sabres and swords stand before him and kill him. Fire will surround him and flames come against him. But whoever listens to the decree will sit in the house, will eat and feed, drink and pour drink, rejoice and cause joy; he will be a brother to brethren and a friend to dwellers of the house; he will be a companion of children and is called one who fosters; he will be an associate of cattle and will be called (the source of) good fortune.[19] Accept peace from your father who is in heaven and sevenfold peace from male gods and from female goddesses. He who makes peace wins the suit. He who causes destruction is burnt in fire. (six *charaktêres*) Sealed and guarded shall be the house, the dwelling and the body of Ḥuna son of Kupitay, and the tormentors, bad dreams, curses, vows, spells, magic practices, devils, demons, liliths, encroachments, and terrors will leave him alone. His sickness shall be overcome and a wall of steel shall surround Ḥuna son of Kupitay. The sickness, devil, and demon of Ḥuna son of Kupitay shall be overcome, and he will be guarded night and day. Amen.

123. Mesopotamia; original location not known. This bowl belongs to a distinctive subcategory by virtue of two characteristics: its use of divorce

---

17. On liliths, their powers, and their reputations, see Neusner, pp. 235–36, and the fuller account in R. Patai, *The Hebrew Goddess* (New York, 1967), pp. 180–225. Liliths were sexually voracious creatures, mostly female though sometimes male, who joined with their gender opposites at times of special vulnerability—during menstruation, after childbirth, at night, and so on. The picture of these figures as sexually voracious finds its psychological complement in their murderous assaults on the children and mates of their human paramours. In all, the female lilith must be seen as the counterpart of the "sons of god" in Genesis 6. At some point, probably in the second century c.e., the belief arose that the only technique for dealing with them was to treat them as if they were "married" to their human lovers and to dismiss them by legally valid bills of divorce. On the legal background of divorce procedures in rabbinical circles, see Levine, pp. 348–50.

18. Here we follow the translation of Harviainen, who takes *raz* as referring to the bowl itself, together with its spell. The mythological notion here, according to Harviainen (p. 16), is that the bowl in the house is a miniature representation of the cosmic bowls (the cosmos itself) buried in heaven and on earth. As such it partakes of cosmic powers against the forces of evil.

19. The intended goal of the amulet is domestic tranquillity; the evils listed must be taken as manifestations of its opposite, domestic chaos.

and ban formulas as the means of separating harmful spirits, especially liliths, from the household; and the citation of Rabbi Yehoshua bar Perahia as the legal authority who is said to have issued valid decrees of divorce and banning. The use of these formulas and of Rabbi Yehoshua's name is certainly Jewish in origin, though they may have spread to non-Jewish bowls as well. As for Rabbi Yehoshua himself, who appears to have lived early in the first century B.C.E., he is cited in the Mishnah[20] and reappears—anachronistically—in later Rabbinic literature as one who taught Jesus.[21] The rudiments of his story are as follows: among the Jews of Babylonia in the sixth century C.E., long after his death, Rabbi Yehoshua developed a reputation for his knowledge and control of supernatural powers; his authority was put to use by the creators of bowls as a means of controlling harmful female spirits; they were subjected to authoritative bills of divorce and banned as though they were illicit wives; the same reputation probably explains the historically impossible link with Jesus, who was known independently in Jewish tradition as a sorcerer, that is, as one who exercised power over spirits. One important principle regarding interactions between humans and spirit-demons emerges with unusual clarity through this bowl and requires special attention here. By and large it has been assumed that these spirit-demons invaded humans on their own initiative. In the period of late antiquity, however, it would appear that this independent initiative and willfulness had been brought under human control in the sense that the behavior of spirit-demons came to be regarded almost exclusively as the result of curse tablets and binding spells. In other words, they were understood as a potential source of power, for good or evil, but they became actual only when summoned or invoked by widely known, "approved," and available techniques. *Bibl.:* Naveh and Shaked, pp. 158–63 (Bowl 5).

> By your name I make this amulet so that it may fortify this person and the threshold of the house and any possessions he may have. I bind the rocks of the earth and tie down the mysteries of heaven.[22] I overcome them . . . I

20. *Hagiga* 2.2 and *Avot* 1.6.

21. For a thorough discussion of these texts and their complicated history, see J. Neusner, *A History of the Jews in Babylonia,* vol. 5 (Leiden, 1970), pp. 235–43; cf. M. Smith, *Jesus the Magician* (New York, 1978), pp. 46–50, for a brief discussion of Jesus as magician in Jewish sources; and Jack N. Lightstone, "Magicians and Divine Men," in *The Commerce of the Sacred: Mediation of the Divine among Jews in the Graeco-Roman Diaspora* (Waterloo, Ontario, 1984), p. 51, for a discussion of Rabbi Joshue/Yehoshua. Among other complications, Rabbi Yehoshua lived some one hundred years before the time of Jesus, under King Alexander Jannaeus (ca. 80 B.C.E.; not 180 B.C.E. as given in an unfortunate misprint in Smith, p. 49).

22. The point here seems to be that the spell is more powerful than anything the author can think of in heaven or on earth.

rope, I tie, I overcome all demons and harmful spirits in the world, whether male or female, from their children to the old ones, whether I know their names or do not know them. And in case I do not know the names, they were already explained to me at the seven days of creation. What was not revealed to me at the seven days of creation was disclosed to me in the deed of divorce that came here from across the sea, written and sent to Rabbi Yehoshua' bar Peraḥya. Just as there was a lilith who strangled human beings—and Rabbi Yehoshua' bar Peraḥya sent a ban against her, though she did not accept it because he did not know her name[23]—her name was written in the deed of divorce and a proclamation was made against her in heaven by a deed of divorce that came here from across the sea; so you are roped, tied, and overcome, all of you, under the feet of this Marnaqa son of Qala.[24] In the name of Gabriel,[25] the mighty hero, who kills all heroes who are victorious in battle, and in the name of Yeho'el who shuts the mouth of all.[26] In the name of Yah, Yah, Yah,[27] Sabaoth. Amen, Amen, Selah.[28]

124. Mesopotamia; original location not known. Plain earthenware bowl measuring approximately 16 cm. across the top. Ten lines of text in spiral writing on the inside, beginning at a small circle near the bottom; two lines on the outside containing only personal names. The language is Babylonian Jewish Aramaic. The date is somewhere between the fourth and the sixth centuries C.E. The figures invoked are angels, with the usual range of secret and mysterious names. Naveh and Shaked cite five bowls which run parallel to portions of our bowl. The occasion for the commissioning of the bowl is clear; the clients have knowledge of a curse directed against them. No specific occasion for that curse is given, but the setting in what seems to be a small village points in the direction of

23. This interjection underlines the importance of names in ritual practices of all kinds. But at the same time, it underscores the continuity of that importance with the role of names in many other domains of human activity, in this case the legal sphere, where a bill of divorce was valid only if it could name the defendant. In this case, the first attempt to divorce the lilith had failed because her name had not been known. In the meantime, says the text, a new decree that included her name had appeared. Several bowls in Montgomery, *Incantation Texts* (nos. 8, 11, 17), identify the lilith by the names of her parents; none, to my knowledge, actually mentions the lilith's own name, though all claim to know it.

24. The name of the client.

25. There is a play on words here between the name of Gabriel and the Hebrew word for hero, *geber.*

26. Behind the authority of the bill of divorce stands the might of two great angels, Gabriel and Yeho'el (probably a variant of the more common Yo'el). The role of angels as guarantors of spells and curses is ubiquitous in late antiquity. Although they certainly originated in Jewish circles, they soon became universal figures. Their role in the world of spells and curses is fully illustrated by texts like the *Testament of Solomon* and *Sepher ha-Razim.*

27. A common variant in the papyri for the solemn name of the god of the Bible, normally written as IAO.

28. A solemn ending, found frequently in the biblical Psalms and later spells.

family disputes, perhaps over marriage, dowry, and inheritance. Although the text refers to a number of individuals in connection with the original curse, most of them women, one woman in particular seems to be the "target" of this anticurse. *Bibl.:* Naveh and Shaked, pp. 134–45 (Bowl 2).

(*Inside*) . . . Overturned . . . overturned, overturned, overturned, overturned is earth and heaven, overturned are the stars and the planets, overturned is the talk of all people, overturned is the curse of the mother and of the daughter, of men and women who stand in the open field and in the village, and on the mountain and the temple(s) and the synagogue(s).[29] Bound and sealed is the curse which she made. In the name of Betiel and Yequtiel and in the name of YYY the Great, the angel, who has eleven names—SSKB', KBB', KNBR', SDY', SWD'RY', MRYRY', 'NQP', 'NS, PSPS, KBYBY, BNWR'. Whoever transgresses against these names, these angels, bound and sealed are all demons and evil spirits.[30] All that is of the earth calls, and all that is of the heaven obeys.[31] I hear the voice of the earth and of heaven which receives all souls from this world. I heard the voice of the woman who cursed and I sent the angels against her (the intended target).[32] NKYR, NKYR, YY take vengeance; YY let us rejoice and rejoice; YY KYSS ṢṢṢ ṬYM', the woman who cursed. And they sent and injured her (away) from the eyes of the daughter, that she may not avenge or curse.

(*Outside*) Dakyâ, son of Qayyamtâ and Mahlepâ, son of (David?) and Šarkâ, daughter of Alištâ (?)

Miriam, daughter of Ḥoran.

125. Modern Lebanon, from a grave near Beirut. A narrow band of silver measuring 3 × 37.5 cm., originally rolled up and worn around the neck in a bronze cylinder. The text covers 121 brief lines (two to three words each) of text, some quite poorly written. The editor offers no exact date,

---

29. These two terms suggest a mixed society of Jews and pagans, not at all surprising in Mesopotamia, although the redundant character of the language of the spells may indicate that the terms do not refer to separate entities.

30. The sense here is that demons and evil spirits are warned not to transgress against the power of the angels whose power, through their names, has been invoked against them.

31. The precise meaning of the sentence is not clear. The general sense is that the incantation will be effective because it is based on the professional's knowledge of the proper links and sympathies between earth and heaven. Here these links are embodied in the secret names of the angels. Compare the formula in Amulet 1, lines 12–13: "I (the professional) have written; God will heal."

32. Again, these two sentences seem to provide reassurance for the client that the professional has the ability to detect curses (that is, to hear the voice of earth) and to bring about the desired response from heaven. In truth, we must suppose that in a small village society, the ability to discover curses depended more on a knowledge of village life than on an ability to listen in on celestial conversations.

but nothing before the third or fourth century is possible. Lines 15–33 of our text find an almost exact parallel in a papyrus spell from Oxyrhynchus (*PGM* XXXV), dated to the fifth century C.E. The function of our amulet was to protect its owner, named Alexandra, the daughter of Zoê, against a variety of evils, including bad demons and curse tablets (lines 8–13). At one point (lines 95ff., where the text is somewhat unclear), the amulet appeals for protection against the assaults of spells and curses. The invocations are addressed to a familiar array of angels and divine names, largely Jewish in character and origin, but conclude with an appeal for help from "the One God and his Christ." Alexandra would probably have called herself a Christian. There is nothing discernably "pagan" at any point. Bonner calls this object the "most remarkable petalon phylactery published up to this time (1950)." *Bibl.:* A. Héron de Villefosse, "Tablette magique de Beyrouth, conservée au musée du Louvre," in *Florilegium ou Receuil de travaux d'érudition dédiés à Monsieur Le Marquis Melchior de Vogüé* (Paris, 1909), pp. 287–95; Bonner, *Amulets*, pp. 101–2; D. R. Jordan, "A New Reading of a Phylactery from Beirut," *ZPE* 88 (1991): 61–69 (new transcription, with photograph and English translation).

I invoke you *SABAÔTH who are upon/above the heavens, who came (?) above ELAÔTH, who are above CHTHOTHAI. Protect Alexandra, to whom Zoê gave birth, from every *daimon* and from every power of *daimones* and from *daimonia* and from spells and curse tablets.[33]

I call in the name of the one who created all things[34];

I call upon the one who sits upon/over the first heaven *MARMARIÔTH;

I call upon the one who sits over second heaven OURIÊL;

I call upon the one who sits over the third heaven AÊL;

I call upon the one who sits over the fourth heaven GABRIÊL;

I call upon the one who sits over the fifth heaven CHAÊL;

I call upon the one who sits over the sixth heaven MORIATH;

I call upon the one who sits over the seventh heaven CACHTH;

I call upon the one (who is) over lightning RIOPHA;

I call upon the one (who is) over thunder ZONCHAR;

I call upon the one (who is) over rain TEBRIÊL(?);

I call upon the one (who is) over snow TOBRIÊL;

I call upon the one (who is) over the forests (?) THADAMA;

I call upon the one (who is) over earthquakes SIORACHA;

I call upon {I call upon} the one (who is) over the sea SOURIÊL;

I call upon the one (who is) over the serpents EITHABIRA;

I call upon the one (who is) seated over the rivers BÊLLIA;

33. A familiar pair of terms, *pharmaka* and *katadesmoi*.
34. The following series of invocations to the angels of the various heavens, using their secret names, is thoroughly typical of Jewish apocalypses and recipe books in late antiquity.

I call upon the one (who is) seated over the roads PHASOUSOUÊL;
I call upon the one (who is) seated over the cities EISTOCHAMA;
I call upon the one (who is) over the level ground NOUCHAÊL;
I call upon the one (who is) seated over every kind of wandering APRAPHÊS;
I call upon the one (who is) seated upon the mountains eternal (?) god EINATH
   ADÔNÊS[35] DECHOCHTHA, who are seated upon the serpents IATHENNOUIAN.
   The one (who is) seated over the firmament CHRARA; the one (who is)
   seated over seas (?) between the two CHÊROUBIN forever; the god of
   Abraam and the god of Isaach and the god of Iakôb. Protect Alexandra, to
   whom Zoê gave birth, from *daimonia* and spells and dizziness and from
   all suffering and from insanity. I invoke you, the living god in ZAARABEM
   NAMADÔN ZAMADÔN, who cause lightning and thunder EBIEMATHALZERÔ
   (with?) the new staff which tramples THESTA and EIBRADIBAS BARBLIOIS
   EIPSATHÔ ATHARIATH PHELCHAPHIAÔN, at whom all things male and all
   terrible binding spells shudder. Flee from Alexandra, to whom Zoê gave
   birth . . . under springs and the abyss of M . . . so that you[36] may not
   bring any stain on her—not by a kiss, a greeting, a meeting; not by drink
   or food, or through intercourse/conversation or by a look or through a
   piece of clothing; nor while she is praying or on the road or away from
   home, either in a river or in the baths. Holy, powerful and mighty names,
   protect Alexandra from every *daimonion,* whether male or female, and
   from every disturbance by *daimonia,* whether those of the night or those
   of the day. Remove them from Alexandra, to whom Zoê gave birth. Now,
   now. Quickly, quickly. One God and his Christ, help Alexandra.

126. An amulet of unknown date and origin; oval stone (carnelian) meant to be worn in a metal setting or carried in a pouch of some sort. 13 × 20 cm. The stone belongs to a subtype that consists exclusively, or largely, of text; this one bears no images (Figure 29). Both the front and the back are covered with inscriptions; in addition, a third inscription appears on the edge or rim. Apart from the usual number of *voces mysticae,* the names on the stone are exclusively Jewish. Thus we call this object Jewish, although similar formulas also appear on Christian amulets. The purpose of the stone was to protect its owner from unspecified misfortunes, which are commanded "not to approach (the owner)" and "not to disobey the name of god." The name of god is no doubt identical with the *voces mysticae. Bibl.:* A. Delatte and P. Derchain, *Les intailles magiques gréco-égyptiennes* (Paris, 1964), no. 460 (with photograph); L. Robert, "Amulettes grecques," *Journal des savants* (1981): 6–27 (with photograph and extensive commentary); *SEG* 31.1594.

35. No doubt a variant of Adonai.
36. Jordan's reading of the text is significantly different from the original edition at this point.

FIGURE 29. Impression of a protective amulet on stone, inscribed with *charaktêres* and spell on both sides. As with many spells, the *charaktêres* occupy the opening lines. (L. Robert, "Amulettes grecques," *Journal des Savants* [Paris, 1981], p. 7 [Fig. 1]. By permission.)

(*Front*—2 lines of *charaktêres* followed by 12 lines of text) I invoke you, god, great *BARBATHIÊAÔTH, *SABAÔTH, god seated[37] upon the mountain of violence,[38] god seated above the bramble,[39] god

(*Back*—continuation of text from front; 13 lines) seated upon the Cherubi(m).[40] He is the all-powerful one. He addresses you . . . every unfortunate encounter, *MARMARAUÔTH IÊAÔTH.

37. The phrase *kathêmenos*, in the sense of "seated" or "enthroned," appears in numerous amulets and spells, but is especially common in Jewish settings; cf. Robert, pp. 9–12.
38. Robert (pp. 14–18) interprets the phrase *epanô tou orous palamnaiou* as a reference to the sacrifice of Isaac by Abraham and lists numerous occurrences of the motif on amulets, lamps, and frescoes of Jewish and Christian origin.
39. Robert takes *batos* as a reference to the burning bush of Exodus 3 (the Septuagint uses *batos* for the bush).
40. A similar expression appears in *DT* 241 (see no. 12); in another amulet (no. 125); and in several texts of *PGM* (e.g., VII, line 634); cf. Robert, pp. 8–12. In the Septuagint, the phrase appears in the Psalms (79:1, 98:1) and Daniel 3:55.

(*Rim*) This spell is from SABAÔTH *ADÔNAI. Do not come near, because (the owner) belongs to the lord god of Israel. *AKRAMMACHAMAREI BRASAOU ABRABLAIN. I invoke you god, ENATHIAÔ PHABATHALLON BABLAIAIAÔ THALA-CHEROURÔSARBÔS *THÔUTH. Do not disobey the name of god.

127. Palestine, Nirim, in the Negev desert; the ancient name was Ma'on/Menois. Bronze tablet measuring 3.5 × 8.8 cm.; found rolled up, with traces of the original wrapping (woven material) still visible. Nineteen amulets were discovered in the small apse of the ancient synagogue. A date in the fifth or sixth century c.e. is most likely. Esther herself was undoubtedly Jewish, as the quotation from Exodus 15:26 confirms, but the misfortunes against which the amulet was designed to protect her were universal. *Bibl.:* Naveh and Shaked, pp. 98–101 (amulet no. 13).

> A good amulet for Esther, daughter of Tettius (?), to save her from evil tormentors, from the evil eye, from spirits, from demons, from night ghosts, from all evil tormentors, from the evil eye, from . . . , from impure spirits. "If only you will obey the Lord your God, if you will do what is right in his eyes, if you will listen to his commandments and keep all his statutes, I will not bring upon you any of the sufferings which I brought on the Egyptians; for I am the Lord, your healer."[41] . . .

128. Egypt, original location not known. Fragmentary papyrus sheet measuring 8.5 × 14 cm., with portions missing from the top, bottom, and left side. A date in the third or the fourth century c.e. seems most likely. What remains visible are the following: (1) two columns of nouns representing various kinds of misfortune; the columns are separated by a vertical line and labeled, respectively, with supralined *alpha* and *beta;* (2) remains of text above the columns, probably containing the original request or spell; (3) traces of a double-lined circle surrounding the spell and columns, identified by the editor as an ouroboros/serpent; and (4) traces of a design under column a. Like other amulets on papyri, this one was undoubtedly meant to be folded and carried about on the owner's body. *Bibl.:* G. Geraci, "Un *actio* magica contro afflizioni fisiche e morali," *Aegyptus* 59 (1979): 63–72; *PGM* CXXI.

> . . . for release from (?)
>
> (*Column A*) death, darkness, diversion, suffering, fear, feebleness, poverty, disturbance.[42]

41. As in other Jewish spells, a biblical verse is cited, here Exodus 15:26 (the same verse is cited in Naveh and Shaked, Geniza Amulet no. 8, lines 22–27.
42. The Greek terms are, in order, (column a) *telos, skotos, ektropê, lupê, phobos, astheneia, penia, thorubos;* (column b) *apotomia, ponêria, baskanos, asôtia, douleia, aschêmosunê, odurmos, loimos, kenôsis, melania, pikron, hubris.*

(*Column B*) dire straits, evil, the evil eye, wastefulness, slavery, indecency, grief, plague, emptiness, blackness, bitterness, pride.

129. *PGM* XXXVI, lines 178–87; from a typical collection (294 lines) with varied contents. This and the following spell indicate the preoccupation not with imposing curses on others but instead on averting or breaking their curses on oneself. The drawings (Figure 30) do not follow the accompanying instructions.

> A spell to break spells (*lusipharmakon*). Take a piece of lead and inscribe on it the unique figure, holding a torch in its right hand, a knife in its left hand (on the left), on its head three falcons, under its legs a scarab and under the scarab an ouroboros serpent. These are the things to be written around the figure: (*charaktêres* and figures).

130. *PGM* XXXVI, lines 256–64 (see no. 129).

> Take a triangular ostracon (broken piece of pottery) from the intersection of a road—pick it up with your left hand—write on it with myrrh-ink and then hide it—"ASSTRAÊLOS[43] CHRAÊLOS,[44] destroy every spell (*pharmakon*) prepared against me (so-and-so), for I invoke you according to the great and frightful names which the winds fear and which make the rocks split apart at its sound: (*charaktêres*)."

131. Pseudo-Apuleius, *Herbarius* 7.1; a collection of traditional lore regarding plants and their powers, falsely attributed to Apuleius and probably composed in the fifth century C.E. The plant described here (*pedeleonis,* "lion's foot") possesses the power of undoing the effects of curse tablets. What we read here is a unique instance of an herbal antidote to *defixiones.* The recipe itself is to be prepared for the client by a trained professional. *Bibl.: DT,* p. cxx; *Pseudo-Apulei Platonici Herbarius,* ed. E. Howald and H. E. Sigerist (Leipzig, 1927), pp. 37–38; cf. A. Delatte, *Herbarius: Recherches sur le cérémonial usité chez les anciens pour la cuillette des simples et des plantes magiques,* 3d ed. (Paris, 1961).

> If some one should be charmed and cursed,[45] this is how you can release him: Cook seven *pedeleonis* plants, without roots, when the moon is de-

---

43. G. Scholem, *Jewish Gnosticism, Merkabah Mysticism, and Talmudic Tradition* (New York, 1965), pp. 95–96, comments on the Jewish character of this name.

44. See E. Peterson, "Engel- und Dämonennamen. Nomina Barbara," *RM* 75 (1926): 421, for an occurrence of the same name in a Mandean bowl.

45. The terms are *devotus* and *defixus.* As commonly in Latin texts, they probably designate a single action of being put under a spell by a curse tablet.

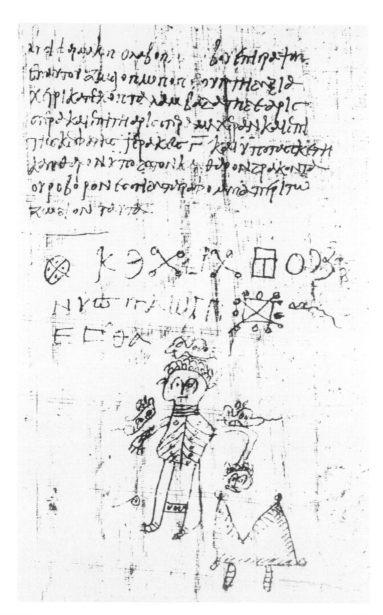

FIGURE 30. Designs on a recipe for preparing a lead *defixio* (*PGM* XXXVI, lines 186ff.). The last line of instructions, just above the designs, reads "These are the things to be written around the figure." The lines and circles on the body and the tie between the legs represent typical symbols of binding; the three snakelike figures surrounding the head symbolize *daimones*. The animal-like figure (bottom right) and the head in the right hand remain obscure.

creasing and without using water; cleanse it as well as yourself, as you do this before the threshold outside the house on the first night[46]; burn and fumigate the birthwort plant; then return to the house without looking behind you and you will release him (from it).

132. Pseudo-Orpheus, *Lithika Kerugmata* 20.14–18; a treatise on the special properties of various stones. Another pseudepigraphic work preserved under the name of Orpheus, a poem entitled *Lithika,* similarly describes the powerful virtues of various stones. In lines 410–11 of the *Lithika,* the author assures his readers that "there is great power in herbs, but far greater in stones." In this particular section of the *Kerugmata,* the stone under discussion is the coral (*koralios*). *Bibl.:* E. Abel, ed., *Orphei Lithica* (Berlin, 1881); L. Thorndike, *History of Magic & Experimental Science,* vol. 1 (New York, 1923), pp 293–96; R. Halleux and J. Schamp, eds., *Les lapidaires grecs* (Paris, 1985).

> For it is the stone of Hermes. It works even on dreams and it drives away apparitions by virtue of its repellent power. And it is a powerful phylactery against the anger of one's master once the image of the figure of Hekate or of the Gorgon is carved into it. Anyone who wears it will never succumb to spells, thunder, or lightning, nor be wounded by evil demons. It makes its wearer invulnerable to suffering and it also releases from all forms of pollution[47] and curses[48] . . . it works to ward off all life-threatening spells and to release (the wearer) from all forms of pollution and curses, like an antidote.

133. Place of origin unknown; a date in the fourth century C.E. seems likely. The thin gold sheet measures 2.7 × 8 cm. and is almost complete. Of the thirty-two lines, only six contain legible Greek; the rest consist of vowel patterns and mysterious names of a familiar sort. The writing is not elegant; at about line 16, the author began to trace underlining in an apparent effort to create a more orderly appearance. The writing is particularly cramped at lines 14–16 and 26–29, where the client's name and matronymic appear. The editors suggest that this crowding is the result of the normal procedure for preparing such amulets, whose message was prepared well in advance with the personal names being added only at the moment of sale. In this case, the names were too long to fit conveniently into the available space. The deities addressed are "Lord Gods," invoked by an elaborate series of mysterious names, vowels, and

---

46. The first night might refer to the wedding ceremony mentioned in at least one manuscript; otherwise it probably means the first night of the month.

47. Here and below, the term is *miasmata.*

48. Here and below the term is *katadesmoi,* the standard term for curse tablets.

charaktêres. Bibl.: C. Faraone and R. Kotansky, "An Inscribed Gold Phylactery in Stamford, Connecticut," *ZPE* 75 (1988): 257–66.

(*charaktêres*) *BAROUCH *IAÔ (*charaktêres*) ANOCH U Ô O E A EANTHOUKÔIA ARTAEMMIEM.THAR BAROUCH MARITHA[49] OO OO AA OU OU II OO UU O . . OUTHIÔSU (sign) O A''E Ê Ô I U Ø' IAÔ ARBARBAPHRARAPHRAX RATHRATHAX[50] *BAINCHÔÔÔCH AEÊIO[UÔ] AEMMIEINATHÔRA A AP,[51] lord gods, cure and pre-serve Eugenia, to whom Galêneia gave birth, AÔ AÊ AÊ ÔAAÔ OOO IIII IAA AEÊIOUÔ EA EA AAA EAÔ IÊÊUE IÊ IÊ IÔA ÔAI UÊ OU ÊU ÔÔ AE IA ÔA EÔ AEÔ IÔ IÔAEOÊU UU OOO ÔÔÔ EÊA OUÔ ÊIIÊ ÊA ÊEO ÊEOUÔAI ÊÊÊ OOOOOOO . . . *DAMNAMENEUS IÔ CHURBURETHBERÔCH, lord gods, protect Eugenia, to whom Galêneia gave birth, from every evil and from all wickedness. A EE ÊÊÊ IIII OOOOO UUUUU ÊÊÊÊÊÊÊÊ,[52] lord gods[53] . . .

134. Italy, Rome; exact location not known. Together with four other lead tablets, all in fragmentary condition, this one was found rolled up and pierced by a single nail. The original size was approximately 11.3 × 31.6 cm. The original editor, W. S. Fox, assigns them to the mid-first century B.C.E.[54] The Latin texts of the five tablets are almost identical and were inscribed by the same scribe, clearly using a common model. No clear occasion is indicated, though personal enmity of some sort seems most likely. The deities invoked are the familiar figure of Persephone, here in her Latin version as Proserpina; next, her husband Pluto, the traditional lord of the underworld; and finally, the much less common figure of Cerberus.[55] Cerberus is the otherwise well-known "hound of

49. Perhaps derived from Aramaic *mry t'* ("my lord, come!").

50. The editors note that the numerical value of the letters in the three preceding words is 2662, that is, a palindrome. Similarly, the value of the following word is 3663, also a palindrome.

51. The editors argue that these two letters are probably a false start for the verb at the end of the line, *apallaxete* ("cure").

52. Frequently this series of vowels was written out in pyramid or ladder fashion; so *PGM* I, lines 13–19.

53. At this point, as in the two earlier occurrences of the phrase, a new invocation was apparently meant to be copied. Space for a few additional lines remains at the bottom of the sheet.

54. Fox also concludes from various characteristic spellings that the persons involved in the spell represent "the lowest classes of the population" (Fox, *Tabellae*, p. 35). If true, however, this would apply to the professional who prepared the tablets and not necessarily to the clients.

55. Cerberus appears in no other Latin *defixio*. Fox, *Tabellae*, notes, however, that the figure of Cerberus appears in two Greek tablets from Attica (*DT* 74, line 5, and *DT* 75A, line 9). In both cases, it is called *phulax* or "guardian." Other figures mentioned in both texts include Hermes, Hekate, Pluto, and Persephone, all standard denizens of the underworld. Cerberus also appears in *PGM* IV, lines 2264 and 2294, as part of a lengthy invocation to the Moon, written in iambic trimeter; it appears again in line 2861, this time in an invocation to the Moon in dactylic hexameters. Finally, Cerberus is invoked in lines 1911ff. of the same text as part of a love spell designed to attract a woman to a man.

hell" or canine guardian of the entrance to Hades in Greek and Roman mythology. Two features of this tablet deserve special mention: first, the client promises a set of sacrificial offerings to Cerberus, provided that the curse is carried out by a certain date; and second, the tablet contains one of the most complete listings of anatomical parts known to us. They are clustered by groups with a view to incapacitating and afflicting certain anatomically "appropriate" actions of the target, named Plotius. *Bibl.:* W. S. Fox, *The Johns Hopkins Tabellae Defixionum* (Baltimore, 1912); Fox, "An Infernal Postal Service," *Art and Archaeology* 1 (1914): 205–7; E. H. Warmington, ed., *Remains of Old Latin*, vol. 4, Loeb Classical Library (Cambridge, Mass., 1953), pp. 280–85 (no. 33); N. Lewis and M. Reinhold, *Roman Civilization. Sourcebook I: The Republic* (New York, 1966) pp. 479–80.

> Good and beautiful Proserpina or Salvia, if you prefer that I call you so, wife of Pluto, snatch away the health, the body, the complexion, the strength, and the faculties[56] of Plotius. Hand him over to Pluto, your husband. May he not be able to escape this (curse) by his wits. Hand him over to fevers—quartan, tertian and daily[57]—so that they wrestle and struggle with him. Let them overcome him to the point where they snatch away his soul. Thus I give over to you this victim,[58] O Proserpina or Acherusia[59] if you prefer that I call you so. Summon for me the triple-headed hound[60] to snatch away the heart of Plotius.[61] Promise that you will give him three victims (gifts)[62]—dates, figs, and a black pig—if he completes this before the month of March. These I will offer you, Proserpina Salvia, when you complete this in an orderly fashion.[63] I give over to you the head of Plotius, the slave/son of Avonia.
>
> Proserpina Salvia, I give over to you the head of Plotius.
> Proserpina Salvia, I give over to you the forehead of Plotius.
> Proserpina Salvia, I give over to you the eyebrows of Plotius.

56. The Latin terms here are *vires* and *virtutes*, which might be translated any number of ways.

57. Cf. *DT* 74, line 6, with which our tablet shows other similarities.

58. The term *victima* is normally used of animal sacrifices but seems to be applied here to the human target, Plotius. Alternatively, the victim may designate an offering made at the time of depositing the tablet.

59. Acheron was the name of the river located at the mythical entrance to Hades; thus it might also be translated as "of the underworld."

60. Here as elsewhere, Cerberus's name is not mentioned; yet its description as "triple-headed" was quite common.

61. Among other attributes, Cerberus was known as a devourer of human flesh; cf. Hesiod, *Theogony* 311: "Cerberus who devours raw flesh, the brazen-voiced hound of Hades."

62. Three gifts, no doubt, for the three heads; customarily, underground deities preferred pigs, especially black ones.

63. The idea seems to be that the client will offer the gifts to Proserpina who would in turn convey them to Cerberus.

Proserpina Salvia, I give over to you the eyelids of Plotius.
Proserpina Salvia, I give over to you the pupils of Plotius.[64]
Proserpina Salvia, I give over to you the nostrils, lips, ears,
nose, tongue, and teeth of Plotius, so that he may not be able to say what
is causing him pain; the neck, shoulders, arms, and fingers, so that he
may not be able to aid himself in any way; his breast, liver, heart, and
lungs, so that he may not be able to discover the source of his pain; his
intestines, stomach, navel, and sides, so that he may not be able to sleep;
his shoulder blades, so that he may not be able to sleep soundly[65]; his
"sacred organ" so that he may not be able to urinate; his rump, anus,
thighs, knees, shanks, shins, feet, ankles, heels, toes, and toenails, so that
he may not be able to stand by his own strength. No matter what he may
have written, great or small,[66] just as he has written a proper spell and
commissioned[67] it (against me),[68] so I hand over and consign Plotius to
you, so that you may take care of him by the month of February. Let him
perish miserably. Let him leave life miserably. Let him be destroyed miserably. Take care of him so that he may not see[69] another month.

64. This extraordinary list of body parts moves more or less systematically from head to toe.
Following the initial listing of separate parts, the succeeding groups of parts culminate in an
affliction to which they are immediately relevant—for example, the speech organs are affected
so that the target cannot express his suffering and so on.

65. Here the person copying made a mistake, as is clear from the parallel passages in the
other tablets. Having omitted "shoulder blades" from the preceding list, the copyist made up a
new phrase ("not sleep soundly") rather than leave them out altogether.

66. The formula s(e)ive . . . s(e)ive, reminiscent of similar phrases from the tablets of Bath,
indicates once again the legal flavor and atmosphere found in many *defixiones*.

67. The verb *mandavit* might also mean that the tablet was deposited in the proper location
or handed over to some other person.

68. As this line indicates, the client knew that Plotius had previously commissioned his own
spell against the client. Thus our tablet is a counterspell. As such it constitutes important
evidence for the public character of spells and counterspells. They were clearly not completely
private or secret actions. Indeed, we must suppose that some public knowledge was essential to
their effectiveness. Fox, *Tabellae* (p. 46) translates these difficult lines as follows: "In what
manner he has according to the laws of magic composed any curse (i.e., against me) and
entrusted it to writing, in like manner do I consign him to thee." The underlying Latin reads:
*seive plus, seive paruum scriptum fuerit, quomodo quicquid legitime scripsit, mandauit, seic ego
Ploti tibi trado, mando.*

69. As is typical of spells and prayers, the text uses three different verbs (*aspicere, videre,
contemplare*) to express the same idea.

8

Testimonies

In this chapter, we endeavour to collect all references to *defixiones* from various types of ancient literature—histories, fiction, drama, philosophical and theological treatises, encyclopedias, Christian saints' lives, legal codes, poetry, manuals of various kinds (including those for producing *defixiones*), and even public inscriptions. In addition to the tablets themselves, these references testify to the broad and deep dissemination of *defixiones* in a variety of ancient societies.

By and large, the materials in the present collection may be said to speak for themselves. And yet, ancient texts never speak *by* themselves, which explains in large part why so little has been heard from *defixiones* in modern scholarship. Ancient voices come to life only through the active intervention of the modern interlocutor; few students of ancient Mediterranean cultures have known of these curse tablets and binding spells or thought them of sufficient interest to engage them in conversation. They belong truly to the category of the silent majority. And that is *the* argument of this book: *defixiones,* though widely ignored and disregarded, introduce us to the cultural *koinê,* the universal religious discourse of the ancient Mediterranean world. The tablets presented in the preceding chapters—each one representing at least ten others, scattered across international journals, museums, and storage rooms—give us access to the concerns of individuals whose voices would otherwise remain silent.

Once engaged in conversation, these tablets make a powerful impression through their immediacy and directness. They add depth and texture to the—one is tempted to say "real"—life of ancient men and women. Since this is so, we must ask why they have received so little attention. Obvious answers might be found in the obscurity of the journals in which they have been published; a tendency to classify them as "magical," "superstitious," or "popular" and thus not the stuff of serious history; and the claim that ancient authors themselves, our best guides to what "really" mattered in antiquity, paid little attention to them. It is this final claim that we seek to address in this chapter on ancient testimonies. Although it is certainly true that discussions of curse tablets would not rank high on any statistical listing of topics in ancient literature generally, or in any individual author, the claim that they played no significant role in ancient life remains flawed in two respects: first, the aristocratic bias of ancient authors *and* modern scholars has produced an enormous distortion in our conception of ancient life and culture, a distortion now under attack from various quarters; and second, the list of authors who do mention *katadesmoi* and *defixiones* is not unimpressive, including Homer, Plato, Ovid, Tacitus, Apuleius, Plotinus, and Eusebius of Caeserea. In addition to such traditional literary testimonies, curse tablets and binding spells show up in other sorts of ancient texts: public inscriptions (nos. 135–36); legal codes (no. 157); and handbooks used by professionals in preparing spells and charms for their clients. In short, the literary evidence has been present all along; it has simply been ignored.

In most respects, these testimonies confirm the picture painted by the tablets themselves. First and foremost, it is now beyond dispute that nearly everyone—99 percent of the population is not too high an estimate—believed in their power. That great fact gatherer, Pliny the Elder, was certainly correct when he observed (no. 146) that "there is no one who does not fear to be spellbound by curse tablets." More precisely, this belief seems to have transgressed every significant ancient boundary—social, cultural, linguistic, geographical, and religious. When added to the great reach of time embraced by the earliest literary testimony (Homer in the eighth century B.C.E.) and the latest (Eustathius in the twelfth century C.E.), the apparent ease with which the use of *defixiones* managed not merely to survive but to cross these boundaries suggests that we are dealing with a truly universal feature of ancient civilization.

But if the testimonies confirm some impressions left by the tablets, they also point to certain persistent disjunctions. Most notable among these is the matter of gender. In his remarkable study of love charms and tablets, the late J. Winkler observed that in literary texts, the clients are usually female, while the experts to whom they resort for help are male.[1] But the tablets show a preponderance of men in pursuit of women—just the opposite of the literary portrait. Beyond the sphere of eros, literary texts also consistently portray the ominous purveyor of horrific spells as a woman, like the bloodthirsty Thessalian "hag" in Lucan's *Pharsalia* (6.413ff.). In fact, the majority of practicing professionals appears to have been men. In this instance, the unique evidence provided by *defixiones* makes it possible both to recognize and to isolate this gender disjunction and thus to make appropriate corrections in our perception of the "real facts." The truth is that it made little difference who you were—man or woman; Greek, Roman, Jew, or Christian; commoner or aristocrat; unlettered peasant or wise philosopher. In matters of the heart, as in many other affairs of daily life, anyone could play the role of client or target. For there was no one who did not fear the power of *defixiones*.

Note

1. "Constraints," pp. 71–98.

135. Italy, Tuder (central Italy); dated to the first century C.E. The inscription appears on the base for a public statue erected by a local freedman to thank the god Jupiter for having miraculously saved members of the city council (*decuriones*). According to the inscription, the councillors' names had been attached (*defixa*) to monuments/tombs (*monumentis*). If *defixa* simply means "attached" and if *monumentis* refers to a public monument, the result would be a public act rather than an instance of a *defixio*. But if *defixa* is used in a technical sense, that is, in preparing a curse tablet, and if *monumentis* refers to tombs, we would have a clear reference to a *defixio*. The circumstances behind the drama are not given, but the general scene is familiar enough. As a result of some grievance against members of the city council, the aggrieved party sought redress. The inscription identifies the culprit as an unnamed public slave, someone employed by the city. If the identity of the slave ever became known, execution certainly followed. *Bibl.: DT,* p. cxxi; *CIL* 11.2.4639; T. Wiedeman, *Greek and Roman Slavery* (London, 1981), p. 189; G. Luck, *Arcana Mundi* (Baltimore, 1985), pp. 90–91.

FOR SALVATION
of the colony of Tuder, both of its city council and of its
people
TO JUPITER OPTIMUS MAXIMUS, GUARDIAN, KEEPER
because he by his own divine power has removed and vindicated

the councillor's names attached to monuments
through the unutterable crime of the public slave
and has freed colony and citizens from fear of perils
L. CANCRIUS, FREEDMAN OF CLEMENS PRIMIGENIUS
member of the six-priest colleges both of Augustus and of the Flavians first
of all to be given these honors by the council
HAS FULFILLED HIS VOW

136. North Africa; Lambaesis in ancient Numidia (modern Tunisia), the site of a Roman army colony; *CIL* 8.2756. Burial inscription on marble. No date is given though a time after 212 C.E., when Roman soldiers were given official permission to take wives, seems most likely. This poignant monument attests once again the belief that personal misfortune came about as the result of curses invoked by one's enemies. *Bibl.:* L. Renier, *Recherches sur la ville de Lambèse (Province de Constantine), accompagné d'un receuil d'inscriptions Romaines par M. le commandant De La Mare, avec commentaire des inscriptions* (Paris, 1852), p. 124; review of this work by J. Baehr, *Heidelberger Jahrbücher* 46 (1853): 716–18; *DT*, p. cxxi.

> . . . Here lies Ennia Fructuosa, most beloved wife, of unmistakable modesty, a matron to be praised for her unusual loyalty. She took the name of wife at age fifteen, but was unable to live with it for more than thirteen years. She did not receive the kind of death she deserved—cursed by spells,[1] she long lay mute so that her life was rather torn from her by violence than given back to nature. Either the infernal gods or the heavenly deities will punish this wicked crime which has been perpetrated. Aelius Proculinus, her husband, a tribune in the great Third Legion, the Augusta, erected this monument.

137. Asia Minor; the region of Lydia, near the ancient city of Maeonia. A marble stone measuring 44.5 × 103 cm. The date is 156/157 C.E. The inscription belongs to the category of "confession texts," of which many examples have survived.[2] Its interest for the study of curse tablets lies in the fact that this text narrates a sacral-legal proceeding against a woman, Tatia, who was suspected of having placed a curse on her son-in-law. The curse is likely to have been a *defixio*. As part of her ultimately unsuccessful efforts to exonerate herself, Tatia posted in the temple what Versnel calls "conditional self-curses," curses that the gods would carry out only

1. The key phrase here is *carminibus defixa*, which must mean that her husband believed her to have been the target of a curse spell.
2. For a discussion of these inscriptions see Versnel, pp. 75ff., and the extensive literature cited there.

if the individual was shown to be guilty. Since Tatia and her son both suffered misfortunes that some had interpreted as divine retribution, her offspring undertook efforts to avert further damage by removing them from the temple and propitiating the gods. They set up *stelae,* or honorific inscriptions, which described the gods' great deeds. The proceedings take place under the aegis of three deities: Great Artemis, Anaeitis, and the moon-god Men. The inscription sheds light on several problems in the study of *defixiones:* (1) the fact that Tatia was believed by others to have engaged in suspicious practices suggests that such things were not entirely private, secretive acts—indeed, she appears to have been moved to vindicate herself due to public rumors about her; (2) it illustrates a tendency in the Greco-Roman world to attribute personal misfortune to spells and charms; and (3) the fact that these events take place in a local temple points to the difficulty in maintaining traditional distinctions between the spheres, the beliefs, the gods, and the actions of magic and religion. *Bibl:* J. Zingerle, "Heiliges Recht," *JOAI* 23 (1926), Beiblatt, cols. 16–23; Björck, *Sabinus,* pp. 127f.; *Tituli Asiae Minoris,* vol. 5.1 (1981), no. 318; E. N. Lane, *CMRDM,* vol. 1, pp. 27–29 (=no. 44) and vol. 3, pp. 27–30; *SEG* 64.648; Versnel, "Beyond Cursing," pp. 75ff.

> The 241st year, the month of Panemos, the 2nd day. Great Artemis, Anaeitis[3] and Men of Tiamos.[4] Because Ioukoundos fell into a condition of insanity and it was noised abroad by all that he had been put under a spell[5] by his mother-in-law Tatia, she set up a scepter[6] and placed curses in the temple in order to defend herself against what was being said about her, having suffered such a state of conscience.[7] The gods sent punishment on her which she did not escape. Likewise also her son Sôkratês was passing the

3. Anaeitis was a goddess of Babylonian origins (the Semitic name was Anat) who traveled to Asia Minor, where she became associated with Artemis and the moon-god, Men; cf. Lane, *CMRDM,* p. 83. vol. 3, p. 83.

4. On Men, the moon-god widely worshipped in Asia Minor, see Lane, *CMRDM,* vol. 3, passim.

5. The text uses the word *pharmakon.* Versnel translates it as "poison," which is misleading; *pharmakon* was used most commonly for spells, in this case almost certainly a curse. The effect of the curse was a condition of insanity (*mania*). In DTA 65, a judicial curse tablet takes aim at a number of potential prosecution witnesses and expresses the desire that they may all loose their senses (*aphrones genointo*).

6. This was a standard procedure for opening legal proceedings under the aegis of the gods; see the discussion in Lane, vol. 3, pp. 28–29. The placing of the scepter meant that accusations were lodged against known or unknown persons and that the prosecution of the case was thereby turned over to the gods.

7. Lane's comment here is apropos: "One may well wonder . . . whether Tatia was not in fact innocent. . . . Certainly, it would require a great deal of temerity to forswear oneself by a god known for such efficacious punishments" (vol. 3, p. 30).

entrance that leads down to the sacred grove and carrying a vine-dressing sickle and it dropped on his foot and thus destruction came on him in a single day's punishment. Therefore great are the gods of Axiottenos! They set about to have removed/canceled the scepter and the curses that were in the temple, the ones the estate of Ioukoundos and Moschios had sought to undo. The descendants of Tatia, Sôkrateia and Moschas along with Ioukoundos and Menekratês, constantly propitiate the gods and praise[8] them from now on, having inscribed on (this) stele the powers/deeds of the gods.[9]

138. Homer, *Iliad* 6.168. The *Iliad* probably reached its present form in the eighth century B.C.E. In book 6, Diomedes the Achaean challenges Glaucus, a Lycian fighting with the Trojans, to identify himself. In reply, Glaucus tells the story of his family, including an incident concerning his grandfather, Bellerophon, who had been accused and exiled by King Proteus. The question here is what is meant by the folded wooden tablet on which are written *sêmata*, described as *lugra* and *thumophthora*. *Bibl.:* W. Leaf, ed., *The Iliad* (Amsterdam, 1971), p. 270.

> The king stopped short of killing him (Bellerophon), for he feared in his heart to do so, and sent him off to Lukia and gave him sinister signs, numerous and life-threatening,[10] written on a folded tablet. He told him to show them to his (Bellerophon's) father-in-law and thereby be put to death.

139. Sophocles, *Ajax* 839–42; the author probably produced his play before 441 B.C.E. In this scene, Ajax is about to kill himself. In anticipation, he calls for vengeance against those who had dishonored him by awarding the arms of the dead Achilles to Odysseus rather than to him. His earlier attempt to restore his honor by killing Agamemnon and Menelaos, had been turned aside by an attack of madness brought on by Athena. Instead, he assaulted a flock of sheep. There is no question of a curse tablet in this passage, but rather a reference to gods who show up

---

8. The verb *eulogein* indicates not a private act of personal piety but rather the public action of setting up an offering to the gods.

9. The actions of the family consisted in offering sacrifices and setting up a *stele* (probably this very stone), or inscription, in which they "confessed" publicly the great feats of the gods.

10. How to read these words? Do they simply refer to the contents of the message, which presumably contained the command to arrange for Bellerophon's death? In this case, the tablet would simply be a letter. Against this view, it can be argued that *sêmata* does not designate words but some other kind of writing, and that in the *Odyssey* (2.329) the phrase *pharmaka thumophthora* clearly points to deadly spells. Thus there is at least a possibility that the folded tablet (*pinax ptuktos*) was a prototype of later curse tablets (also folded and also making use of signs), with the signs intended somehow to bring about Bellerophon's death. In his note on this passage (p. 270), Leaf comments that "writing was regarded as a form of magic." It is worth noting that among the earliest Greek *defixiones*, some take the form of letters to the underworld.

commonly in such tablets as well as a phrase which may well contain a curse formula. *Bibl.:* M. Delcourt, *Héphaistos ou la légende du Magicien* (Paris, 1957), p. 160; Anne-Marie Tupet, *La magie dans la poésie latine* (Paris, 1976), pp. 261–62; J. C. Kamerbeek, *The Plays of Sophocles. Commentaries,* part I: *The Ajax* (Leiden, 1953), pp. 169–177.

> This much I ask of you, O Zeus. And so also I call on Hermes of the underworld,[11] the guide of souls, to lay me out with one quick and easy blow, as I plunge this sword into my side. I call for help, too, from those propitious maidens, who with eternal vision look upon the many woes of humans, you dread, swift-footed Erinyes/Furies,[12] that you may learn how I have been done in by the sons of Atreus. Just as they see me falling by my own hand, so may they be undone by beloved kin.[13]

140. Plato, *Republic,* book 2 (364C); its date probably falls near 375 B.C.E. In a preliminary discussion concerning the nature of true justice, Adeimantos reports a number of widely held views that deprived justice of any serious meaning, citing as examples popular religious figures who claim to possess the power to manipulate the gods for any purpose, good or evil, by spells and charms. Two points are worth noting: first, the speech assumes that curse tablets were a common feature of Athenian life; and second, the wandering professionals who sold them directed their attention at the wealthy. In short, the use of curse tablets was not limited to the "ignorant" and "superstitious" lower classes as often claimed.

> Begging priests and soothsayers go to the doors of the wealthy and convince them that if you want to harm an enemy, at very little expense, whether he deserves it or not, they will persuade the gods through charms and binding spells[14] to do your bidding.

141. Plato, *Laws,* XI (933A); probably the last of Plato's dialogues, written near 355 B.C.E. Book XI deals with a variety of actions that the

---

11. Hermes *chthonios,* a familiar figure in curse tablets of just this period, the fifth century B.C.E.

12. The Erinyes also figure in a number of curse tablets (see E. Kagarow, *Griechische Fluchtafeln* [Leopoli, 1929], p. 62).

13. In her study of the god Hephaistos, Marie Delcourt (p. 160) argues that Ajax's words must be interpreted as a formulaic expression taken from a cursing ritual; cf. also Tupet, pp. 261–62. Certainly the expression (lines 840–41) "just as (*hôsper*) . . . so (*tôs*)" is reminiscent of similar formulas in numerous curse tablets.

14. The Greek terms are *epôdai* and *katadesmoi.* The latter term is the standard word for curse tablets inscribed on metal plates. Many of the Greek *defixiones* date from Plato's time and before.

state must either encourage or sanction. The issue here involves two kinds of *pharmaka*, poisons and spells. This passage confirms several important notions about the use of binding spells in classical Greek culture: they were widespread; they were feared by most people; they were made available by professionals; they were commonly deposited in cemeteries; and they used human figurines. As in the previous passage, Plato finds himself in a dilemma. On the one hand, he claims not to believe that such things really work; on the other hand, he must ban them because of their noxious effects on the broad public. *Bibl.:* J. de Romilly, *Magic and Rhetoric in Ancient Greece* (Cambridge, Mass., 1975), pp. 23–43 ("Plato and Conjurers").

> There is also another kind that persuades reckless persons that they can do injury by sorceries, and incantations and binding spells,[15] as they are called, and makes others believe that they are likely to be injured by the powers of the magician. And when men are disturbed in their minds at the sight of waxen images fixed at their doors, or in a place where three roads meet,[16] or on the graves of parents,[17] there is no use in trying to persuade them that they should despise all such things, since they have no certain knowledge about them . . . and we must entreat, exhort, and advise men not to have recourse to such practices, by which they scare the multitude out of their wits, as if they were children. . . . Let the law about poisoning or witchcraft run as follows. He whoever is suspected of injuring others by binding spells, enchantments or incantations, or other similar practices, if they are soothsayers or diviners, let them die. But if he is not a soothsayer and is convicted of witchcraft, let the court fix what he ought to pay or suffer.

142. Ovid, *Amores/Loves* 3.7.27–30; three books of elegiac love poetry written and revised over a period of time, from around 20 B.C.E. to perhaps 1 B.C.E. The poet laments his inability to "perform" in the arms of his lover. His first explanation for this humiliating failure takes the form of imagining that he has been the target of a love spell, in this case the kind exemplified by numerous separation tablets. The ease with which Ovid considers this explanation suggests that the practice was well known at the time.

---

15. The Greek terms are *manganeia, epôdai,* and *katadesis.* The last two appear in *Republic* 364; *katadesis* here must be identical with *katadesmos.*

16. In *PGM* IV, line 2955, a love spell is to be deposited "at a crossroad." It is interesting to note that the same spell requires the use of wax to be shaped as a small dog. The papyrus is, of course, much later than the time of Plato, but many of the practices involved in these rituals appear to have survived over long stretches of time.

17. Curse tablets, as we know, were regularly deposited in graves and tombs.

Has some Thessalian[18] poison[19] bewitched[20] my body, is it some spell or drug[21] that has brought this misery upon me. Has some sorceress written[22] my name on crimson wax, and stuck a pin in my liver.[23]

143. Ovid, *Heroides;* a series of poetic love letters from legendary women to their absent lovers. Letter VI is written by Hypsile, queen of Lemnos, to Jason, following his arrival in Thessaly. Their brief love affair had taken place on Jason's expedition to Argos. In the course of her letter, Hypsile gives vent to her anger at reports that Jason had subsequently taken up with Medea, herself notorious in literary traditions for skill in deploying powerful herbs and spells. In lines 82–94, an angry Hypsile asserts that Medea could only have won Jason's affections through love spells. In this instance, the accusation of magic turns out to be a face-saving device for failure. *Bibl.:* A. Palmer, ed., *Heroides* (Oxford, 1898), pp. 332–33; H. Jacobson, *Ovid's Heroides* (Princeton, 1974), pp. 94–108.

> (Hypsile complains . . .) Her charm for you is neither in her beauty nor in her merit; but you are made hers by the incantations she knows. . . . Among tombs she stalks, ungirded, with hair flowing loose and gathers from the yet warm funeral pyre the appointed bones. She curses those who are absent, she fashions waxen images[24] and into their wretched liver drives the slender needle.[25]

144. Ovid, *Fasti* II, 571–82; Ovid's poetic explanation of the Roman religious calendar. At this point Ovid is discussing the Feralia, or festival

18. In the literary tradition of Rome, Thessaly was known as the (Greek) home of charms and spells; it is the setting for Lucan's gruesome account of the famous "Thessalian witch"; the play, *Heracles on Mount Oeta,* attributed to Seneca, identifies Thessaly as the place where powerful herbs grow; and in Apuleius's *Golden Ass,* the hero, Lucius, travels to Thessaly in order to study "the magical arts." In these stories, the practitioners are always women.

19. The Latin term is *venenum,* probably derived from Venus, the goddess of love. It could be used narrowly of a love spell or more broadly of spells of any kind.

20. The Latin term is *devota,* a technical term for spells on curse tablets; such tablets were called both *devotiones* and *defixiones.*

21. Like Plato, Ovid considers two possible forms of enchantment, drugs taken from plants (*herba*) or recited spells (*carmina*).

22. The Latin verb is *defixit;* again this is a technical term for curse tablets.

23. The practice of sticking pins in small figurines, especially in curse tablets associated with love, whether for attraction or separation, is well attested in literary texts, in the papyri, and in surviving figurines. The liver was the proverbial seat of the emotions.

24. The two verbs here are *devovet* and *figit;* both are technical terms for love spells on metal tablets.

25. It should be noted that some modern editors regard the details spelled out here ("Among . . . needle") as a later interpolation. But in either case, the details are consistent with what we know from curse tablets themselves: they were used in love affairs; they were frequently deposited in tombs; and they made use of material from or associated with the target of the spell.

of the dead, celebrated from February 17–21. It was a time when spirits were around and about. Perhaps inspired by the thought of spirits and cemeteries, Ovid describes certain ritual activities of an old woman. *Bibl.:* J. G. Frazer, *Publii Ovidii Nasonis Fastorum Libri Sex: The Fasti of Ovid,* vol. 2 (London, 1929), pp. 446–52; F. Bömer, *Die Fasten,* vol. 2 (Heidelberg, 1958), pp. 126–27; Anne-Marie Tupet, *La magie dans la poésie latine* (Paris, 1976), pp. 408–14.

> An old woman sits among girls and performs rites in honor of Tacita ("Silent One"), though she herself does not remain silent.[26] With three fingers she puts three lumps of incense under the threshold where the little mouse has made a secret path for her. Then she binds enchanted threads together with dark lead[27] and mutters with seven black beans in her mouth. She roasts in the fire the head of a small fish which she has sewn up, sealed with pitch and pierced with a bronze needle. She also drops wine on it. . . . Then, as she leaves, she says, "We have bound up hostile tongues and unfriendly mouths."[28] . . . At once you will ask me, "Who is this goddess Muta (Mute One)?"[29]

145. Seneca, *On Benefits* 6.35.4; this essay deals with the theme of how one can be the source of benefit toward others. At this point in the essay, Seneca is discussing those who seek to escape the responsibility of repaying a benefit received from others, in this case the repayment of a debt.

> If you decided to repay a debt to him with money from his own pocket, you would appear to be very far from being grateful. But what you desire is even

26. Tupet, pp. 409–10, suggests the following scenario: the young women come to the old woman for the preparation and consecration of lead *defixiones;* the association of the festival of all souls with the dead is appropriate in that the young women would then deposit their tablets in the cemeteries where the holiday would take them in any case.

27. The connection between threads and lead tablets appears in several love spells in *PGM* as well as in a number of preserved tablets. The connection is probably a physical one, that is, the recipes call for a figurine to be attached by threads to the tablet on which the spell is inscribed. Alternatively, the threads might be pieces taken from the victim's clothing. Bömer (p. 126) and Tupet (pp. 409–10) correctly see the passage as a clear reference to spells on metal tablets, namely, *defixiones.*

28. In her parting words, the old woman indicates that the spell belongs to the particular category of "silencing charms," which are amply illustrated among both surviving *defixiones* ("I bind the tongue of so-and-so") and in *PGM* (e.g., IX, lines 1–14).

29. The underworld goddess (so named by Ovid himself in line 610, *infernae nympha paludis* or "goddess of the infernal marsh") is attested not just in Ovid but in a lead *defixio,* with Latin text; cf. R. Egger, "Zu einem Fluchtäfelchen aus Blei," *Römische Antike und frühes Christentum,* vol. 2 (Klagenfurt, 1963), pp. 247–53. Frazer (p. 446) and others had previously argued that Ovid had simply invented the name Muta in order to explain certain features of the Feralia. The tablet itself is an appeal to the deity Muta Tacita that she should silence a certain Quartus and that he should run around like a scurrying mouse ("O Muta Tacita, let Quartus be silent, let him run around disturbed like a scurrying mouse").

more unjust. For you invoke curses upon him and call down[30] terrible imprecations upon someone who ought instead to be sacred to you.[31]

146. Pliny, *Natural History* 28.4.19; probably published in the 70s of the first century C.E. In this enormous compendium of fact and fancy, Pliny includes several informative excurses on *magia,* its origins and its expansion; he also indulges in a consistent polemic against an imprecisely defined group whom he calls *magici* (28.10–31; 30.1–13). Pliny certainly attests the wide public belief in "this most fraudulent of arts (which) has held complete sway throughout the world for many ages. Nobody should be surprised at the greatness of its influence" (30.1). As for his own views, Pliny is probably representative of Roman writers and intellectuals in that "while appearing to condemn magic most severely, (he) really believed in the detested art much more than he thought."[32] *Bibl.:* E. Tavenner, *Studies in Magic from Latin Literature* (New York, 1916).

There is no one who does not fear to be spellbound by curse tablets.[33]

147. Tacitus, *Annals* 2.30; a year-by-year account of Roman history from the death of Augustus in 14 to 68 C.E., the year of Nero's death. Their date of publication lies close to 115 C.E. Here Libo Drusus is described by Tacitus as a prominent young Roman, whose friend Firmius Catus had persuaded him to consult with "the forecasts of astrologers, the rituals of magicians, and the interpreters of dreams" (2.27). Eventually Libo's activities were revealed to the emperor Tiberius and to the senate. His prosecutors produced some of his personal papers. Before the trial could be completed, Libo committed suicide. *Bibl.:* E. Tavenner, *Studies in Magic from Latin Literature* (New York, 1916), pp. 50–51; E. Massoneau, *La magie dans l'antiquité romaine* (Paris, 1934), pp. 177–78; E. Koestermann, *Cornelius Tacitus. Annalen,* vol. 1 (Heidelberg, 1963), p. 304; F. R. D. Goodyear, *The Annals of Tacitus. Books 1–6,* vol. 2 (Cambridge, 1981), p. 276.

---

30. The verb here is *defigis,* a technical term for invoking curses via tablets.

31. A later Christian text, *The Miracles of Saints Cyrus and John,* reports an instance in which a certain Stephanos became the target of curses invoked by his relatives who had borrowed money from him and later resorted to spells in order to avoid repayment; see the discussion in H. J. Magoulias, "The Lives of Byzantine Saints as Sources of Data for the History of Magic in the Sixth and Seventh Centuries A.D.: Sorcery, Relics and Icons," *Byzantion* 37 (1967): 234.

32. Tavenner, p. 56.

33. The Latin verb here is *defigi,* once again a technical term for curse tablets.

In one paper (of Libo) the accuser argued that a set of marks, sinister or at least mysterious, had been appended in Libo's own hand to the names of the imperial family and a number of senators.[34]

148. Tacitus, *Annals* 2.69. As the adopted son of Tiberius and the adopted grandson of Augustus, Germanicus was the intended heir to imperial power at Rome. In the year 19 C.E., he fell seriously ill, believing that he had been put under a spell[35] by his rival, Piso. He died later in the same year. Others clearly suspected that Piso had commissioned and deposited a curse tablet; Piso was later put on trial for various wrongdoings, among them the use of curse tablets and spells.[36] Tacitus conveys a full and accurate description of the implements and beliefs associated with the use of *defixiones*. Much the same account, including details regarding lead tablets, appears in the Greek historian of Rome, Dio Cassius (ca. 230 C.E.).[37]

It is a fact that explorations of the floor and walls brought to light[38] the remains of human bodies, spells, curse tablets, leaden tablets engraved with the name Germanicus,[39] charred and blood-smeared ashes and others of the implements of witchcraft by which it is believed that the living soul can be devoted to the powers of the underworld.[40]

149. Tacitus, *Annals* 4.52. Several years after Germanicus's death, a cousin of his widow, named Claudia Pulchra, was put on trial and accused of various crimes, among them that she had commissioned spells and curse tablets against the emperor Tiberius.

34. The charge, then, was that Libo had prepared curses against a number of personal enemies, perhaps using mysterious names and *charaktêres*.

35. The word used, *venenum,* can mean either poison or spell.

36. *Annals* 3.13, where the same terms, *venenum* and *devotio,* are used. The Roman historian Suetonius, a slightly younger contemporary of Tacitus, reports the same events (4.3), speaking of Germanicus's reluctance to break with Piso "until he received information that he had been attacked by spells (*veneficiis*) and curse tablets (*devotionibus*)."

37. Dio Cassius, in his history of Rome, based on earlier sources (57.18): "For human bones were found buried in the house where he (Germanicus) lived, along with lead tablets (*elasmoi molibdinoi*) on which were curses and his own name."

38. What brought about the excavations is not indicated. This much seems clear from the story: misfortune, in this case a serious illness, was attributable to curse tablets; and the deposition of the tablets was not an entirely private affair.

39. The Latin reads as follows: *carmina et devotiones et nomen Germanici plumbeis tabulis insculptum.* F. R. D. Goodyear, *The Annals of Tacitus. Books 1–6,* vol. 2 (Cambridge, 1981), pp. 409–10, observes that the three terms designate but a single thing, lead curse tablets.

40. The Latin here is *numinibus infernis.* Once again, Tacitus accurately describes what appears in the tablets themselves.

He (Domitius Afer) accused her of (invoking) spells/poisons[41] and curse tablets against the emperor.

150. Tacitus, *Annals* 12.65. Under the reign of Claudius, his wife Agrippina sought to eliminate one of her female rivals, Domitia Lepida. In typical fashion, Agrippina attacked her rival by accusing Lepida of directing curse tablets against her.

Charges were made that she had assaulted the emperor's wife by the use of curse tablets.[42]

151. Tacitus, *Annals* 16.31. In yet another trial, during the reign of the notorious Nero, Soranus and his daughter Servilia were put on trial. Servilia in particular was accused of consulting magicians against the emperor.

When the accuser asked if she (Servilia) had sold her bridal ornaments . . . in order to earn money for performing magical rites, she . . . exclaimed, "I have resorted to no impious gods, to no curse tablets."[43]

152. Lucian of Samosata, *Dialogues of the Courtesans* 4.4. These dialogues reflect Lucian's lifelong production of satirical literature, often directed against contemporary styles of religious practices and philosophical beliefs. This particular set of dialogues appeared near the middle of the first century C.E. The situation here is that Melitta fears the loss of her lover, Charinus, and turns to another woman, Bacchis, for assistance. What she seeks is the name of a professional who will be able to provide her with a love spell. *Bibl.:* C. P. Jones, *Culture and Society in Lucian* (Cambridge, 1986).

MELITTA: Bacchis, do you know any old woman of the kind called Thessalian.[44] There are said to be a lot of them around. They use incantations and can make a woman loved, no matter how much she is hated before. . . . (Melitta relates how her troubles with Charinus began) . . . I ran over to him in my usual way, but he pushed me away when I tried to embrace him and said, "Go off with you to Hermotimus the shipowner or read what's written on the walls in the Kerameikos, where your names are scribbled on a tombstone."[45]

41. Again, the word is *venenum*.
42. *Devotionibus*.
43. *Devotiones*.
44. Once again, the official practitioners of the magical arts are identified as women and with the region of Thessaly in Greece.
45. The connection between spells of various kinds, the Kerameikos, and the writing of names on or in tombs is well attested in the *defixiones* themselves. Of course, the names may simply refer to graffiti of the sort, "A loves B!"

BACCHIS: Well, my dear, there is a most useful vendor of spells, a Syrian, who is still very fresh and firm. Once, when Phanias was angry with me without reason, just as Charinus is with you, she reconciled him with me after four months, when I had already despaired of him; she brought him back with incantations. . . . She doesn't charge a big fee—just a drachma and a loaf of bread. Besides that you must put out seven obols, some sulphur, and a torch along with salt. These are taken by the old woman. She must also have a bowl of wine mixed and drink it by herself. You'll also need something belonging to the man himself, such as clothing or boots or a few of his hairs or anything of that sort. . . . She hangs these (boots) on a peg and fumigates[46] them with sulphur, sprinkling salt over the fire, and mumbles both your names. Then she plucks out a wheel[47] from her bosom and whirls it around, rattling off an incantation full of horrible outlandish names.[48] That's what she did on that occasion, and shortly afterward, though at one and the same time his friends warned him off, and Phoebis, the lady whose company he was keeping, pleaded desperately with him, he returned to me.

153. Apuleius, *Metamorphoses* 3.17. In this episode of his picaresque novel, the young Lucius tries unsuccessfully to imitate the spells produced by a professional woman, Pamphile. In this scene, Pamphile's servant girl, Fotis, describes her mistress's workshop for Lucius. Here and elsewhere in his writings, Apuleius reveals his extensive knowledge and experience in the world of spells. *Bibl.:* A. Abt, *Die Apologie des Apuleius von Madaura und die antike Zauberei* (Giessen, 1908); R. T. van der Paardt, *L. Apuleius Madaurensis. The Metamorphoses. A Commentary on Book III with Text and Introduction* (Amsterdam, 1971); J. Winkler, *Auctor & Actor: A Narratological Reading of Apuleius' "Golden Ass"* (Berkeley, 1985).

There she (**Pamphile**) began by arranging in her infernal workshop all the customary implements of her art—aromatic herbs of all kinds, metal strips

46. *PGM* IV, lines 295ff., speaks of the use of smoke (*thumiatêrion*) in the preparation of a metal tablet.

47. The Greek term is *rhombos,* probably to be taken as identical with the Greek *iunx.* The rhombus is mentioned in *PGM* IV, line 2336, a (love?) spell addressed variously to Hermes, Selene, Kore, and Hekate. The instrument is described as belonging to "her who rules Tartaros." It appears also in the same text, line 2296, as an instrument connected with the invocation of the deity. A. S. F. Gow, ed., *Theocritus* (Cambridge, 1950), p. 41, describes the instrument as "a spoked wheel, or a disk, with two holes on either side of the centre." When a cord is passed through the two holes and alternately stretched and released, the disk will revolve back and forth. The instrument is illustrated on numerous Greek vases. On the wheel and its use in connection with love spells and other forms of binding, see Gow, p. 41, and Sarah Johnston, *Hekate Soteira: A Study of Hekate's Roles in the Chaldean Oracles and Related Literature* (Atlanta, 1990), pp. 90–110.

48. The names, of course, are the mysterious terms and words found in all spells and tablets of this period.

engraved with mysterious letters, remains of shipwrecked vessels[49] and also many limbs of dead and buried men.[50]

154. Apuleius, *Metamorphoses* 9.29. In a series of ribald tales involving betrayal and marital infidelity, a baker learns of his wife's adulterous affairs. Her efforts to save herself reveal a number of typical settings where *defixiones* were likely to be deployed: first, as love spells between married couples where one partner feared losing the affections of the other; and second, failing that, as a way to bring suffering and death on the lost spouse. Once again, as often in literary texts, both the client and the specialist are women.

> The baker divorced his wife and made her leave the house. . . . But she sought out a old woman who was believed to be able to do whatever she pleased by means of curse tablets and spells.[51] . . . The baker's wife promised her a large sum and urged her to do one of two things: either to soften her husband's heart so that they might be reconciled or, failing that, to invoke some spirit or infernal deity to put a violent end to his days. This powerful woman, able to control the gods, first tried the milder forms of her evil art and sought to influence the offended feelings of the husband.[52] . . . But when the results turned out differently from her expectations, indignant at the gods . . . she began to attack the very life of the poor man and to stimulate the spirit of a woman who had met a violent death to carry out his destruction.[53]

155. Artemidorus, *On Dreams* 1.77; a compendium of dream interpretations in the late second century C.E. This passage lists a series of dream images involving garlands (*stephanoi*) and gives the appropriate meaning for each. *Bibl.:* R. J. White, *The Interpretation of Dreams: Oneirocritica* (Park Ridge, N.J., 1975).

49. *PGM* VII, lines 465–66 (a recipe for preparing a love charm on a tin sheet), calls for a copper nail from a shipwrecked vessel.

50. The details correspond accurately with the material evidence: the common use of plants; metal strips with mysterious names, (the *defixiones* themselves); pieces of sunken ships, in this case the nails that commonly accompanied *defixiones;* parts of human bodies (the common association of *defixiones* with graves and tombs); and finally an elaborate set of ritual incantations whereby the *defixio* was consecrated before being deposited.

51. The terms are *devotiones* and *maleficia;* here, as elsewhere they probably refer to different aspects of a single action rather than to separate strategies.

52. The first effort was clearly a love spell, designed to rekindle affections in the husband. Thus we can safely assume that a number of love spells preserved on lead tablets were employed for intramarital reconciliation.

53. The second effort, following the failure of the first, took the form of deploying a *defixio* designed to harm or kill the target.

Garlands made of wool signify witchcraft and curse tablets because they are complicated/variegated.[54]

156. Harpocration, *Lexicon of the Ten Orators;* a lexicon of words and phrases, arranged in alphabetical order, and drawn from a wide array of orators. The date is uncertain; some have placed it in the late second century C.E.

To be cursed[55]: Deinarchos (uses this word) instead of "to be drugged" and "to be bound."[56]

157. Paulus, *Sentences* 5.23.15–18; an anthology of legal opinions, attributed to an important Roman jurist (ca. 210 C.E.). "Magical practices," however imprecisely defined and understood, had been outlawed in Rome from the time of the Twelve Tables (fifth century B.C.E.).[57] Under Sulla (81 B.C.E.), the *Lex Cornelia de Sicariis et Veneficiis* added *venena* (poisons/charms) to the list of prohibited offenses and made the possession, sale, gift, or production of poisons/charms, including those used for amatory purposes, subject to punishment; the punishments depended on the social standing of the offender. Both earlier and later prohibitions no doubt cover the use of curse tablets but this passage from Paulus is the only one to make specific mention of them.[58] The passage represents a significant expansion in the range of prohibited offenses. *Bibl.:* R. MacMullen, *Enemies of the Roman Order* (Cambridge, 1966), pp. 95–127; C. R. Phillips III, "*Nullum Crimen sine Lege:* Socioreligious Sanctions on Magic," in *Magika,* pp. 262–78.

54. The Greek term is *poikilos.* The connection may be that this term was often used of drugs; alternatively, it might lie in the fact that curse tablets sometimes used string to attach figurines to the metal tablets.

55. The verb is *katadedesthai,* to which both *katadesmos* and *katadô* which occur frequently in tablets, are related.

56. The two verbs are *pepharmakeusthai* and *dedesthai.*

57. In his *Apologia* (47), Apuleius remarks that "this magic of which you accuse me is, I am told, a crime in the eyes of the law and was forbidden in remote antiquity by the Twelve Tables because in some incredible manner crops had been charmed away from one field to another." The key was the prohibition of *mala carmina.* Some Roman authors took the phrase to be a prohibition against defaming persons through poetry—so Cicero, as cited by Augustine, *City of God* 12.9, and others; see Tupet, "Rites magiques," pp. 2592–2601. In support of Apuleius's reading stands Pliny, *Natural History* 28.10.

58. The Christian emperor Constantine strongly affirms the authority of Paulus's writings (*Theodosian Code* 1.4.2). To this he added, in line with earlier attempts to distinguish between "harmful" and "innocent" magic, that no punishment was due for remedies of illness or for spells to protect crops from natural disasters (ibid., 9.16.3).

Whoever performs or commissions unlawful nocturnal rites, in order to cast a spell, to curse or to bind[59] someone, will be crucified or thrown to the beasts. . . . It is the prevailing legal opinion that participants in the magical art should be subject to the extreme punishment, that is, thrown to the beasts or crucified. But the magicians themselves should be burned alive. It is not permitted for anyone to have in his possession books of the magical art. If they are found in anyone's possession, after his property has been expropriated and the books burned publicly, he is to be deported to an island or, if he is of the lower class, beheaded. Not only the practice of this art, but even knowledge of it, is prohibited.

158. Plotinus, *Ennead* 4.4.40. The Neoplatonic philosopher (205–269 C.E.) devotes a lengthy section of his essay "On Difficulties about the Soul" to the question of how spells (*goêteiai*) work. He assumes their efficacy but strives to empty them of all traditional significance by attributing their success to the natural workings of the cosmos. In addition, he denies their ability to affect the soul of the wise person. Like most intellectuals in the ancient world, Plotinus was well informed about the world of spells and curses. Some have wondered whether he was also a practitioner. *Bibl.:* P. Merlan, "Plotinus and Magic," *Isis* 44 (1953): 341–48; A. H. Armstrong, "Was Plotinus a Magician?" *Phronesis* 1 (1955–1956): 73–79; E. R. Dodds, *The Greeks and the Irrational* (Berkeley, 1963), pp. 283–311.

They also use figures with power in them and by placing themselves in the right position they are able to draw down powers upon themselves, since they are working in one and the same cosmos. But if you were to place one of them outside the cosmos, he would not be able to work his spells through curse tablets and love charms.[60]

159. Macrobius, *Saturnalia* 5.19.7; an aristocrat and intellectual of the early fourth century C.E. His *Saturnalia* takes the form of learned conversations by imaginary characters on every conceivable subject. At this point, the topic is Virgil's use of the Greek tragedian Sophocles. Macrobius goes on to indicate that Virgil was following a passage from a lost play of Sophocles, known as *The Root Cutters/Plant Gatherers (Rhizotomoi)*. *Bibl.: Macrobius: The Saturnalia*, trans. P. V. Davies (New York, 1969).

---

59. The verbs are *obcantare, defigere,* and *obligare*. Once again it is probably safe to assume that they cover a single action, namely, the use of curse tablets.

60. The terms are *katadesmos*, the normal word for curse tablets, and *epagôgê*, a technical term for love charms.

164. Cyril of Scythopolis, *Life of Saint Euthymius,* chap. 57; a monk and hagiographer, active in various monasteries in Palestine of the mid-sixth century C.E. The story narrated here is typical of hagiographic confrontations between the "higher" spiritual power of Christian monks and the "base" machinations of pagan wizards. The brother of a certain presbyter, Achthabios, fell seriously ill after an enemy persuaded a local wizard (*goês*) to kill him through a spell. The ill man, Romanos, prays to Euthymius for help. The saint appears in response to the pious prayer. *Bibl.: Kyrillos von Skythopolis,* ed. E. Schwartz (Leipzig, 1939).

> "I am Euthumius, summoned here in faith. There is no need to be afraid. Show me where your pain is." When he pointed to his stomach, the apparition (the saint) straightened out his fingers, cut open the spot as if with a sword and withdrew from his stomach a tin strip which had certain *charaktêres* on it and placed it on the table in front of him. He then wiped clean the spot with his hand and closed the incision.

165. Sophronius, *Account of the Miracles of Saints Cyrus and John* (*PG* 87.3, cols. 3541–48). Sophronius was patriarch of Jerusalem in the sixth century C.E. His life of two Alexandrian saints, Cyrus and John, narrates several episodes relating to spells and charms in which the saints show their superior power by overcoming curses invoked by pagan practitioners. *Bibl.: DT,* pp. cxxii–cxxiii; H. J. Magoulias, "The Lives of Byzantine Saints as Sources of Data for the History of Magic in the Sixth and Seventh Centuries A.D.: Sorcery, Relics and Icons," *Byzantion* 37 (1967): 236–38; Faraone, "Context," p. 9.

> *How Theophilos was bound hand and foot by magic.*
>
> Certain people wanted him (the devil) to do him (Theophilos) harm and enlisted him on their side. He was persuaded by their foul petitions and caused no slight harm to the poor man by binding him hand and foot and afflicting him with terrible pains. . . . (Local doctors fail to heal Theophilos and he appeals to the saints) . . . They appeared to him in a dream and gave him the following instructions: "Ask the *philoponoi*[74] to carry you and go to the sea early in the morning. There you will come upon one of the fishermen casting his net into the water. Agree on a fee with him to toss the net into the

passion" (ch. 41), brought about by her spurned suitor's visit to a local *magos*. Eventually, Irene prays for help to the martyr Anastasia and St. Basil. The story continues (ch. 48): "From the air there was let down a package . . . which contained a variety of magical devices (*periergeias*) wrapped in it: two idols made of lead, one resembling the suitor, the other the sick nun, embracing each other and bound together with hairs and threads, then some other contrivances of malignancy and inscribed on them the name of the author of the evil and the appellations of his servant demons."

74. An order of lay workers among whose responsibilities was care of the sick.

water for you. Whatever he catches will be the source of your cure." . . . After a short time he tossed the net and pulled out a very small box, secured not just with locks but muzzled with lead seals. . . . (There followed a dispute about ownership) . . . With much effort they opened the box before everyone's eyes and discovered a terrible and disturbing sight . . . a carved image in human form, made of bronze and resembling Theophilos, with four nails driven into its hands and feet, one nail for each limb. When the bystanders beheld this, they were astonished and did not know what to make of it. . . . One of them gave the command to pull the nails out, if it could be done. He took the statue and grabbed the nail stuck into the right hand and with much effort succeeded in drawing it out. Once it was out, Theophilos's right hand was immediately restored and he ceased suffering the great pain and the related condition of paralysis. And it became clear to all what abominable magic the charlatans had used against him in cooperation with those most evil demons, by throwing it (the box) into the deep waters so that it would not be recovered. . . . They hastened to remove the remaining nails. . . . As they removed them, the ill man was released from his bonds and suffering, until all of them were drawn out. Thus the sick man was relieved of the entire diabolical business. When they removed the nail on the left hand of the statue, the suffering man was able immediately to stretch it out. And when they pulled out the nails driven into its feet, the sick man was able to move with no pain at all.

166. Sophronius (same as no. 165). *Bibl.: PG* 87.3, col. 3625.

About Theodôros of Cyprus who was made lame by Magic.

Theodôros was a widely reputed physician of great skill. He became lame as a result of magic.[75] (He was unable to heal himself and the saints appeared to him in a dream, issuing the following command:) . . . "Send one of your servants to Lapithos and tell him to dig in front of your bedroom next to the doorway. There he will find the wicked instrument of the sorcerer.[76] Once it is uncovered, its maker will disappear immediately." Theodôros sent (him) to the place, as commanded, and he found the cause of the disability. Once it was brought to light, the sorcerer disappeared immediately, seized by death. . . . He was a Hebrew and as such was not above such suspicions.

167. Pseudo-Augustine, *Homily on Sacrilegious Practices* 5–6; a popular sermon, written in quite ordinary Latin. Geographically the text emanates from Germany and chronologically is from the eighth century C.E. The text is a mine of information for the survival of pre-Christian practices and for their continued appeal among Christians. *Bibl.:* C. P.

---

75. Although no details are given, the lameness must have been induced by a binding spell.
76. Again, no details about the precise nature of the instrument are given.

Caspari, *Eine Augustin fälschlich beigelegte Homilia de sacrilegiis* (Christiania, 1886).

> (*Chap. 5*) . . . (16) Whoever, during the time of the moon's increase, thinks that it is possible to avert (harm) through the use of inscribed lead tablets[77] . . . they are not Christians but pagans.

> (*Chap. 6*) . . . (19) Whoever produces writings of Solomon[78]; whoever ties around the neck of humans or dumb animals any characters, whether on papyrus, on parchment, or on metal tablets made from bronze, iron, lead, or any other material, such a person is not a Christian but a pagan.

168. Eustathius, an important ecclesiastical figure (he served as Metropolitan of Thessalonike) and scholar of the twelfth century C.E. Several lengthy commentaries on classical texts, including Homer's *Iliad* and *Odyssey,* have survived. In his commentary on *Odyssey* 19.455–58, where the sons of Autolycus skillfully bind (*dêsan*) Odysseus's wound and stop the flow of blood with a charm (*epaiodê*), Eustathius makes the following comment, in a remarkably offhanded manner, almost as if such things were common knowledge.

> For (the use of) binding spells[79] requires skill.

77. The Latin reads *per laminas plumbeas scriptas.*
78. Solomon was renowned in antiquity as a great man of power, who possessed knowledge of and control over the *daimones.* The apocryphal *Testament of Solomon,* originally a Jewish text though now preserved only in Christian versions, was among the most popular books on the use of spells in the world of late antiquity.
79. The Greek term is *katadesmos.*

Glossary of Uncommon Words

ABLANATHANALBA: a common palindrome on *defixiones* and amulets and in formularies (for example, *PGM* IV, line 3030), often occuring with AKRAM-MACHARI (see s.v.). Behind it may lie a Hebrew phrase, something like "father (*ab*), come to us (*lanath*)," although this must remain uncertain. See the discussion in D. M. Robinson, "A Magical Text from Beroea in Macedonia," in *Classical and Medieval Studies in Honor of Edward Kennard Rand,* ed. L. W. Jones (New York, 1938), pp. 250–51, and Martinez, pp. 108–110.

ABRASAX/ABRAXAS: one of the most common of all *voces mysticae,* used to name and address a deity with solar connections. On amulets, the deity is frequently represented with snakes as legs, an armored torso, and a cock's head ("the anguipede"). The numerical equivalent of its letters was said to equal 365, the ordinary number of days in the year. See the discussion in *GMP,* p. 331.

ADÔNAI: another common *vox mystica* found on amulets and *defixiones* and in formularies. It is used to designate and invoke a deity or cosmic spirit in numerous spells; its ultimate origins lie in the Hebrew Bible, where it serves as one of the most common divine names. Once it passed beyond Jewish circles and entered into the international glossary of *magoi,* it was used in various inflected forms, for example, Adonaêl and Adonaios.

AKROUROBORE: a variant of OUROBOROS (see s.v.).

AKRAM(M)ACHAMARI: another frequently occurring *vox mystica,* often used with ABALANATHANALBA (see s.v.) and SESENGENBARPHARANGES (see s.v.); according to Scholem, *Jewish Gnosticism* (pp. 94–100), it derives from an Aramaic expression, "uproot the magic spells." As such, its origins are probably Jewish.

ANOCH/ANOK: a term derived from Coptic *anok,* meaning "I (am)." Like many other terms of foreign origins, it seems to have become a deity in its own right on amulets and *defixiones.*

BAINCHÔÔÔCH: an Egyptian invocation ("spirit of darkness") which, like many other foreign phrases, appears to have become a deity or cosmic spirit on amulets and *defixiones.* See the discussion in H. C. Youtie and Campbell Bonner, "Two Curse Tablets from Beisan," *TAPA* 68 (1937): 57, and *GMP,* p. 333.

BARBATHIAÔ: a variant of the common terms ARBATHIAÔ and ABRAÔTH, related by a metathesis of the first two consonants. Various explanations have been offered for its origins. R. Wünsch, "Deisidaimoniaka," *ARW* 12 (1909), p. 31, sees it as a contraction of *hebraikos iao* ("Iao of the Hebrews"). Ganschinietz, "Iao," *RE* 9 (1916), col. 703, connects it with the mythological figure of Iabraoth in gnostic texts, for example, *Pistis Sophia* and *The Book of Jeu* (cf. the index, under Jabraoth, in C. Schmidt and W. Till, *Koptisch-Gnostische Schriften* [Berlin, 1959], p. 408); Cormack (see p. 137) derives it from a combination of Hebrew "*arbat iao,*" that is, "four" and "Iao," referring to the tetragrammaton or four-letter name of god in the Hebrew Bible (*yhwh*) whose Greek form was Iao; and Blau, *Zauberwesen,* pp. 102–5, regards it as an abbreviation of Sabaoth. For more recent discussion, see W. Fauth, "Arbath Iao," *Oriens Christianus* 67 (1983): 65–103, and Martinez, pp. 41f. and 76f.

BAROUCH: another term of Hebrew origin, "blessed"; used to address the god of the Hebrew Bible and later in Jewish prayers. See the discussion in Martinez, p. 77.

BARBARATHAM (sometimes BARBARITHA) CHELOUMBRA: part of a lengthy formula, regularly including BAROUCH and SESEGENBARPHARANGES; it appears in *PGM* III, lines 109–10.

BOLCHOSETH: a name or invocation of the Egyptian deity Seth, found on amulets, *defixiones,* and in formularies; it appears frequently with other terms, for example, IÔ and IÔERBÊTH. As for the origins of the term, Moraux, *Défixion* pp. 34–39, has proposed the following: BOL represents an Egyptian spelling of the generic Semitic word for god or lord, *ba'al;* SETH is obviously the name of the Egyptian diety, here named *ba'al,* as also in other Egyptian texts; and CHO derives from an Egyptian word meaning to hit or strike. Thus the general meaning would be "Ba'al, who strikes, (that is) Seth."

BÔRPHORBABARBOR . . . : this lengthy *vox mystica,* with its many variants, is known as the BORPHOR- series; it appears on *defixiones* throughout the Mediterranean region as well as on amulets and in formularies; it is generally associated with Hekate (see s.v.) and Selene (the Moon), as their secret name or as an invocation; see the full discussion in Jordan, "Agora," pp. 240–41. As for its origins, Morton Smith has suggested (Jordan, p. 214) that "perhaps they were (meant) to imitate the barking of the dogs thought to accompany the goddess: cf. Theocritus, 2.35."

CHUCH BAZACHUCH BACHUCH: these "words" or variants of them recur in texts throughout the Mediterranean world; cf. Youtie and Bonner, "Curse Tablets," p. 57. In a bilingual lead tablet from Carthage in North Africa, BACHACHUCH is followed by the invocation, "who is a great spirit (*demon*) in Egypt" (DTA 250A, lines 1–2). Some have sought to derive the root word from the Hebrew *kochav,* "star." In general, this series of *voces mysticae* illustrates a linguistic principle according to which a set of variations on a basic pattern is created by altering a limited number of vowels and consonants in successive "words."

DAMNAMENEUS: a common term, possibly derived from *damnazein* ("to tame"). For discussion, see Faraone and Kotansky (no. 133).

DAMNO DAMNA LUKODAMNA: a standard set of word variations, for example, *PGM* III, lines 434–35 read, "ARTEMI DAMNO DAMNO LUKAINA." The basis of the set is derived from one of the widely circulated *ephesia grammata,* i.e., *damnameneus* (see glossary): *lukodamna* = wolf-tamer; *lukaina* = she-wolf. The words are commonly associated with Hekate and Artemis.

Demeter: a popular Greek goddess, associated with her daughter Persephone/ Proserpina (also called Korê) and with Hades/Pluto, the god of the underworld.

Dikê: personification of justice and revenge; became a popular Greek goddess.

Dôdekakistê: a rare term of uncertain meaning. *Dôdeka* is the number ten, and *kistos* means basket; perhaps "the one (female) who encompasses the twelve (gods?)." See under HUESSEMIGADÔN.

Ephesia grammata: a term used in two quite different senses. The broad use, now less in vogue, designates all nonstandard forms of discourse in spells and charms; this use has now generally been replaced by *voces magicae* or, as in this book, *voces mysticae.* The narrow use refers to a set of terms said to have been inscribed on the great statue of Artemis at Ephesus—*askion, kataskion, lix, tetrax, damnameneus,* and *aision* (variant *aisia.* See the discussion in R. Kotansky, "Incantations and Prayers for Salvation on Inscribed Greek Amulets," in *Magika,* p. 111).

ERESCHIGAL: a female deity of Babylonian origin; the name sometimes appears together with NEBOUTOUSOUALÊTH.

EULAMÔN: this common term is used in a variety of ways. Various theories have been proposed regarding its origin and meaning. R. Wünsch (*Sethianische Verfluchungstafeln aus Rom* [Leipzig, 1898], p. 83) considers a possible connection with the Egytian god Ammon; he also argues that it stems from a command in Greek, "Destroy the body (of my enemy)!" in which the Greek letters have been reversed so as to read from last to first; that is, *sôma lue* becomes *eulamôs.* Others propose a derivation from a West Semitic adjective meaning "eternal." Still others would derive it from an Assyrian term, *ullamu,* again meaning "eternal." See the discussion in Bonner and Youtie, "Curse Tablets," pp. 62–63. Various other interpretations of this common term have been offered; cf. Youtie and Bonner, p. 62 and Preisendanz (1972), pp. 17–18.

Hekate: an important Greek goddess of the underworld and one of the most prominent figures in charms and spells; see Sarah Johnston, *Hekate Soteira: A Study of Hekate's Roles in the Chaldean Oracles and Related Literature* (Atlanta, 1990).

Hermes: a Greek god, associated both with the underworld and with a number of female deities, especially Demeter and Hekate. He was long associated with the sphere of charms and spells. He was further identified with the Egyptian god Thoth and the Roman god Mercury.

HUESSEMIGADÔN: the first word of a common formula that appears on several *defixiones* and in numerous recipes of *PGM* (e.g., II, lines 32–34 and V, lines 424–25). See Cormack, "A *Tabella Defixionis* in the Museum of the University of Reading, England," *HTR* 44 (1951): 31–33, and Martinez, pp. 39f.

IAÔ: derived originally from the famous tetragrammaton (four letter), the holy and unpronounceable name of god in the Hebrew Bible, YHWH. It came to be used widely on amulets and *defixiones;* it appears as well as a cosmic power in Gnostic texts. See the discussion in Fauth, "Arbath Jao," pp. 65–75, and Martinez, pp. 79f.

Korê: see under Demeter.

LAILAM: some connection with the Greek word *lailaps* ("storm, hurricane") seems likely, although others have proposed a derivation from Hebrew, *le-olam,* "forever"; whether our *vox mystica* was originally derived from it, or merely associated with it secondarily, is another matter.

MARMARAÔTH: this name occurs also in *PGM* VII, lines 488 and 598, 608. One variant is MARMARACHTHA. *Voces mysticae* built on the basic form of *marmar,* with numerous suffixes, are quite common; cf. *PGM* XXXV, line 2 and VII, line 572. Its origins probably lie in the Aramaic phrase, "Lord of Lords." See K. Preisendanz, "Marmaroath," *RE* 14 (1930), col. 1881, and Martinez, pp. 81f.

MASKELLI MASKELLÔ: one of the most common formulas containing *voces mysticae*. In its full form it reads as follows: MASKELLI MASKELLÔ PHNOUKENTABAÔ OREOBAZAGRA RÊXICHTHÔN HIPPOCHTHÔN PURIPÊGANUX (cf. *PGM* VII, line 302). It includes a number of recognizable Greek elements: *rêxichthôn,* "bursting out from the earth"; *hippochthôn,* horse plus earth; *puripêganux,* "lord of the fire fount." See the discussion in *GMP,* p. 336.

MELIOUCHOS: in various spells this term is used of a mythological figure associated with a number of deities, including Serapis, Zeus, Helios, Hekate, Ereschigal, and Mithras. The name may derive from the Greek word for honey, *meli,* or perhaps (so Harrauer) from an Egyptian phrase. For discussions see *GMP,* p. 336; C. Harrauer, *Meliouchos. Studien zur Entwicklung religiöser Vorstellungen in griechischen synkretistischen Zaubertexten* (Vienna, 1987); and the cautionary note (against hasty word derivations) of H. J. Thissen, "Etymologeien," *ZPE* 73 (1988): 304–5.

OSORONNOPHRIS: a common name for Osiris. Its meaning in Egyptian is "Osiris the beautiful."

Osiris: a popular deity of Egyptian myth and cult; the husband of Isis and the ruler of the underworld.

OREOBAZAGRA: see under MASKELLI MASKELLO.

OUROBOROS (AKROUROBORE is a common variant): a serpent represented biting its own tail; see the discussion in *GMP,* p. 337.

Persephone: see under Demeter.

PURIPÊGANUX: see under MASKELLI MASKELLO.

Sabaoth: one of the many terms used to describe the god of the Hebrew Bible. Its original meaning is "the heavenly hosts." In the later *defixiones* and amulets, it comes to designate an independent deity.

SEMESEILAM: a familar term in formularies (e.g., *PGM* IV, line 1805; V, lines 351 and 365; and VII, lines 645–50). The fact that this *vox mystica* is clearly associated with the sun makes it likely that it was originally derived from a Hebrew expression meaning either "sun of the world" or "eternal sun." G. Scholem, *Jewish Gnosticism*, p. 134, proposes a derivation from Aramaic, *shemi shelam*, "my name is peace."

SESENGENBARPHARANGÊS: a common *vox mystica* with solar connections. It is used frequently together with the palindrome ABLANATHANABLA. Scholem, *Jewish Gnosticism*, pp. 96–100, argues for Jewish origins. Josephus, *Bellum Judaicum* 7.179–80, speaks of a potent fig-tree located in a ravine (*tês pharangos*) near a place called Baaras. The roots of the tree produced powerful drugs. Thus our *vox* may have originated as a reference to "the Baaras-ravine." See the discussion in *GMP*, p. 339, and Martinez, pp. 78f.

Seth: the Egyptian chthonic god of wisdom and spells; the enemy of Osiris; associated with the desert and the color red; often represented with the head of a dog or jackal. See H. te Velde, *Seth: the God of Confusion* (Leiden, 1967).

THÔTH: with many variant spellings; an Egyptian god associated with the moon, the invention of writing, and the gods Seth and Osiris.

Index